The Genealogical Science

CHICAGO STUDIES IN PRACTICES OF MEANING

Published in collaboration with the Chicago Center for Contemporary Theory
http://ccct.uchicago.edu

Also in the series

Questioning Secularism: Islam, Sovereignty, and the Rule of Law in Modern Egypt
by Hussein Ali Agrama

Neoliberal Frontiers: An Ethnography of Sovereignty in West Africa
by Brenda Chalfin

Ethnicity, Inc.
by John L. Comaroff and Jean Comaroff

Inclusion: The Politics of Difference in Medical Research
by Steven Epstein

Political Epistemics: The Secret Police, the Opposition, and the End of East German Socialism
by Andreas Glaeser

Producing India: From Colonial Economy to National Space
by Manu Goswami

The Politics of Dialogic Imagination: Power and Popular Culture in Early Modern Japan
by Katsuya Hirano

The Moral Neoliberal: Welfare and Citizenship in Italy
by Andrea Muehlebach

American Value: Migrants, Money, and Meaning in El Salvador and the United States
by David Pedersen

Additional series titles follow index

The Genealogical Science

The Search for Jewish Origins and the Politics of Epistemology

NADIA ABU EL-HAJ

The University of Chicago Press
Chicago and London

The University of Chicago Press, Chicago 60637
The University of Chicago Press, Ltd., London

Paperback edition 2014
Printed and bound by CPI Group (UK) Ltd, Croydon, CR0 4YY

21 20 19 18 17 16 15 14   2 3 4 5 6

ISBN-13: 978-0-226-20140-5 (cloth)
ISBN-13: 978-0-226-15470-1 (paper)
ISBN-13: 978-0-226-20142-9 (e-book)
10.7208/chicago/9780226201429.001.0001

Library of Congress Cataloging-in-Publication Data

Abu El-Haj, Nadia.
The genealogical science: the search for Jewish origins and the politics of Epistemology / Nadia Abu El-Haj.
p. cm. — (Chicago studies in practices of meaning)
ISBN-13: 978-0-226-20140-5 (cloth: alkaline paper)
ISBN-10: 0-226-20140-6 (cloth: alkaline paper) 1. Jews—Identity. 2. Cohanim. 3. Genetics. 4. Y chromosome. I. Title. II. Series: Chicago studies in practices of meaning.
DS143.A15 2012
599.98′924 –dc23

2012033126

∞ This paper meets the requirements of ANSI/NISO Z39.48-1992 (Permanence of Paper).

CONTENTS

ACKNOWLEDGMENTS

This book has been a long time in the making and I have accumulated many debts. Research for this project would not have been possible without the generous support I received from the National Endowment for the Humanities, the Wenner-Gren Foundation, the Lichstern Fund of the Department of Anthropology at the University of Chicago, and Barnard College. I would also like to thank the Institute for Advanced Study at Princeton, where I spent a year as a fellow when I was just beginning the book. Most especially, I am grateful to Joan W. Scott and the late Clifford Geertz for being such generous hosts and interlocutors.

Many individuals generously agreed to share their insights, experiences, and expertise with me over the course of my research. In particular, I would like to express my appreciation for the willingness of research scientists to meet with me and explain their work. I would especially like to thank Jonathan Marks, who helped me very early on when I was conceptualizing the book, and Theodore Schurr, who, in the course of two conversations, clarified more than he can possibly know. A special thanks goes to Deborah Bolnick who read and commented upon several of the book's chapters. Her expertise in this complex scientific field has been invaluable. I only hope I can return the favor some day.

I would like to thank my former colleagues at the University of Chicago and my current colleagues at Barnard and Columbia for creating an enlivening intellectual atmosphere. I have also benefited enormously from the work of the very talented research assistants I have had over the years: Mario Yedidia, Irma Khoja, Sumayya Khalique, Amanda Himmeltoss, and Maya Wind, whose assistance gathering materials was essential. I am especially grateful to a few (former) Ph.D. students who dug up important materials, conducted a few interviews, and did the occasional stints of fieldwork on my behalf, and who read the manuscript, shared their insights, and helped me get it into production: Adam Bund, Clare Casey, Joshua Kaplan, Yifat Gutman, Maya Mikdashi, Don Navon, Harel Shapira, and Brian Horne. And then there are the colleagues and friends who have gone well beyond the call of duty: Rashid Khalidi has supported me throughout—from the time I was a graduate student to our being colleagues at not one but two institutions. Jean and John Comaroff have been as helpful to my thinking through this second book project as they were with my first: they have read many a proposal and chapter and they have been true friends. I am also immensely appreciative of Brinkley Messick and Nan Rothschild who, from the time of my arrival at Barnard, made sure I felt very much at home.

I thank Lila Abu Lughod, Bashir Abu Manneh, Samira Haj, Angie Heo, Lesley Sharp, Zachary Lockman, Claudio Lomnitz, and Gary Wilder for their very helpful comments on specific chapters. Then there are those who I somehow convinced to read the whole thing: thanks to Mark Mazower, whose keen eye as a historian made me think about the broad temporal shifts I was trying to map; to Brian Larkin who has an ability to articulate the arcs of a conceptual argument in a way that few others can; and to Gil Anidjar who convinced me to be less of a wimp (among other things). Special thanks go to Lisa Wedeen. She read this book so many times that the only word that comes to mind is exploitation: her seeming limitless willingness to read again and again is a true testament to her generosity, both intellectual and personal. Lisa has become as much a part of my family as of my intellectual and professional life. I cherish our friendship deeply. And then there is Beth Povinelli, who has been my colleague and friend for well over a decade. Beth too has read so many versions of so many chapters—and spent so many hours amusing my child with one kooky project or another—that I have no idea how to begin to reciprocate. Hopefully she will come up with a plan. Finally, I would like to thank my editor T. David Brent for taking on this book project, Priya Nelson for helping to shepherd it through the production process, and Richard Allen for a remarkably careful job in copyediting my manuscript.

No acknowledgments would be complete without mentioning my family, without whom my world would make no sense. I feel lucky to belong to not one but two trans-continental tribes. Relatives from the northeast corridor to Geneva, London, Beirut, and Jerusalem have lived through this book for far more time than they thought humanly possible. Thea and Tabatha are always there for me, for better and for worse, and god knows we have had a lot of the latter during the time I wrote this book. My husband, Amer, never quite believes me when I say I'm stuck and he inevitably turns out to be right. Always a good thing to be laughed at a little bit when one is taking one's intellectual angst too seriously. My daughter, Aya, is an extraordinary person and a source of boundless joy. For the life of her she can't quite figure out how I could be "finishing" a book for two out of the six years of her existence. My parents—I miss them more than I could ever have imagined. But there are moments when my daughter does or says something and I am reminded of the ways in which they live on. It is to the memory of my parents and to my daughter and her future that I dedicate this book.

INTRODUCTION

In early 2008, Shlomo Sand, a professor of history at Tel Aviv University, published a book that called into question the basic premises of Jewish national history. As argued by Maya Sela in the prestigious Israeli daily *HaAretz*, Sand's *The Invention of the Jewish People* "deals with questions that remain taboo in Israeli society, among them the ancestral origins of the Jewish people and the genetic lineage shared with modern day Israelis" (Sela 2009). In the preface to the English edition, Sand notes that "representatives of the 'authorized' body of historians fell on the book with academic fury, and excitable bloggers depicted me as the enemy of the people" (Sand 2009, ix). Nevertheless, the book remained on Israel's bestseller list for nineteen weeks. Its French edition sold twenty-five thousand copies in the first six months and won a prestigious prize (Sela 2009).

In *The Invention of the Jewish People*, Sand relies on original textual sources and secondary analyses of the textual and archaeological records in order to argue that there is no evidence that a collectivity called "The Jewish People" was exiled from ancient Palestine or that descendants of that collectivity lived for generations in the diaspora and then returned to the Land of Israel and founded the modern Jewish state. The story of an "exiled-people race" is the creation of modern Jewish nationalism, Sand argues, that reads back into the historical record a vision of peoplehood first articulated in the late nineteenth century.

The broad strokes of his historical argument are widely accepted in archaeological and biblical studies circles today. Nevertheless, Sand's account was immediately challenged by colleagues in departments of the History of the Jewish People, a field of study in Israeli universities squarely situated in the humanities and one that, Sand argues, remains a staunch defender of that nationalist narrative. More recently, however, criticisms of

Sand's conclusions have come from a different disciplinary perspective, one backed by the epistemological authority of the natural sciences. In a *New York Times* report on a recent genetic study of Jewish communities that concluded there is a substantial degree of "genetic similarity" among contemporary Jews, Nicholas Wade wrote that these results "refute the suggestion made last year by the historian Shlomo Sand . . . that Jews have no common origin but are a miscellany of people in Europe and Central Asia who converted to Judaism at various points" (2010). In a similar vein, a reader of the *Financial Times* wrote in response to a review of Sand's book by the late Tony Judt:

> The only reliable evidence as to who and what the Jews 'really' are is not any of the 'soft' stuff adduced by Shlomo Sand et al. but the hard facts of genetics. To simplify but not distort, the DNA record (at least on the male side) shows that Jewish communities everywhere, east and west, Ashkenazi and Sephardi, are more closely related to one another than to the non-Jewish populations they live among.
>
> Hence, ethnicity being a sense of common identity rooted in common descent as well as common history and common culture, including religion and language, there is indeed a common Jewish ethnic identity and a Jewish people that is scattered across the globe. (Silverman 2009)

The letter writer was referring to studies launched in the mid-1990s by scientists who focused on the Y-chromosome and mitochondrial DNA (the former tracks the paternal line, and the latter the maternal) in order to identify genetic markers that might shed light on the origins of "the Jewish Diaspora," the Jewish priestly line (the Cohanim), and specific communities who believe they are descendants of ancient Jews. On the basis of these two genetic systems and, more recently, of genome-wide genetic markers, prominent researchers at prestigious universities, medical schools, and research centers in England, the United States, and Israel have reconstructed the phylogenies of contemporary Jewry and Jewish subpopulations, determined or speculated about the geographic origins of the maternal and paternal lines of today's various Jewish communities, assessed the degrees of kinship among them, and tested whether or not groups of self-declared "lost tribes" fall within a Jewish genetic map. My project in this book, in contrast to Sand's, is not to assess whether or not particular accounts of Jewish origins and history are true. It is to analyze this increasingly pervasive and powerful field of scientific research and social practice.

Phylogenetic studies of population-specific origins constitute a proliferating field of mainstream scientific research. There are studies of the phylogenies of African Americans, Native Americans, Brahmins, Lebanese, the British, the Swiss—and the list goes on and on.[1] Moreover, as evidenced by the widely publicized documentaries produced by PBS in cooperation with Henry Louis Gates Jr, the Alphonse Fletcher University Professor and the Director of the W.E.B. Du Bois Institute for African and African American Research at Harvard University, the search for genetic markers of "one's" past extends to practices of knowing the individual self. *African American Lives* told the stories of the search for the African ethnicities of famous African American individuals, including Oprah Winfrey, and it provided an account of how scientists identify a person's ethnic origins on the basis of the genetic markers he or she carries within. In a more recent PBS documentary, *Faces of America*, Gates traced the genetic and genealogical origins of "12 renowned Americans," including Steven Colbert, Eva Longoria (of the CBS show *Desperate Housewives*), and Yo Yo Ma.[2] In the here and now, genomic knowledge is being harnessed in efforts to reconstruct histories, both individual and collective.

Anthropological genetics is a historical science. The discipline's operating assumption—that an organism contains evidence of its origins—is not new. That premise has guided evolutionary biology since its beginnings. What are new are the techniques, machines, and epistemic objects that anthropological geneticists use. New technologies have enabled researchers to "read" the nucleotide sequences of DNA (the order of DNA's four chemical components)[3] quickly and cost-effectively (see Wells 2006). By deciphering nucleotide sequences and comparing sample sets, anthropological geneticists reconstruct human origins and migration routes out of Africa, study the genetic diversity of the human species, and map the genealogies of particular populations. This book focuses on scientific efforts to identify population-specific origins and to trace the phylogenies of culturally and politically meaningful human groups. In order to highlight such efforts as a distinct subset of projects within the broader field of what I call "anthropological genetics," I refer to research on the origins and phylogenies of a specific population as "genetic history."[4]

This book is a study of the scientific making and social lives of genetic markers, genetic histories, and genetic genealogical selves. There is now an extensive social scientific literature on the impact of post-genomic medicine and biotechnology, fields that predict our individual risks for disease and that are devising techniques for creating life and averting or treating illness.[5] But as Stephan Palmié has noted, there is far less critical consideration of genomic practices that "have begun to reconfigure the past" (2007,

207). This book is such an exploration.[6] I analyze the scientific work of generating "genomic pasts" (207), together with commercial and activist projects and the practices of self-fashioning, both individual and collective, emerging on its terrain.

My analysis of the construction of genomic pasts focuses upon one set of genetic historical projects: research on Jewish origins, which has proved to be a particularly illuminating perspective through which to analyze the broader scientific and social phenomena at hand. But neither genetic history generally nor the specific forms of Jewish collectivity and politics being fashioned on its terrain can be understood without a look back in time. Therefore, I provide an account of the scientific disciplines and the social and political practices to which genetic history is an heir and in relation to which we might best understand its specificity, both scientific and social.

I examine three distinct moments in science and politics: race science, circa 1900, that relied on cranial measurements and phenotypic differences; population genetics, circa 1950, based primarily on blood group data; and genetic history, starting in the 1990s, which examined genetic differences at the level of the nucleotide, focusing on mitochondrial DNA and the Y-chromosome. More specifically, first, I analyze studies undertaken by European and American Jewish scientists at the turn of the twentieth century on the racial composition of the Jews, which fashioned a particular and ultimately enduring understanding of Jewish peoplehood and which articulated fundamental hurdles to be overcome for a new Hebrew nation and individual to be born. Second, I analyze the work of Israeli population genetics in the 1950s and 1960s in the newly founded nation-state, and I read this work as expressing a desire—indeed, a *need*—to find "a content" for the a priori nationalist belief in the fact of Jewish peoplehood: what biological evidence was there that the apparently motley collection of Jewish immigrants from various regions of the world really have and share a common origin in ancient Palestine? Third, and most centrally, I explore genetic history projects to identify the origins of contemporary Jewry that have been carried out, by and large, by self-identified Jewish scientists in the U.S., the U.K., and Israel, paying attention to the nexus of epistemological commitments, epistemic objects, the commerce in ancestry testing, and the making of genetic-historical selves. What cultural understandings of the (Jewish) self and what range of political projects—distinctly American, decidedly diasporic, and committedly Zionist—is genetic historical inquiry making possible and how?

In making historical comparisons, I do not stage a simple linear succession of eras. Instead, I trace the enduring effects of scientific practices, social imaginations, institutional contexts, and political worlds past. And at the same time I elucidate the shifts and novel configurations of biology, history, and selves emerging today. In order to do so, I highlight the substance of and meanings ascribed to "biological difference" in each of these scientific paradigms and social and political configurations, paying particular attention to the forms of evidence upon which scientific, cultural, ethical, and political beliefs and practices are built.

There are compelling arguments for why genetic historical studies of Jewish origins are an especially productive angle through which to ask broad questions about the new phylogenetic turn. As a category, "the Jews" has been essential to the work of the biological sciences since the birth of race science in the late nineteenth century. Much of racial thought was built on the terrain of the Jewish body, as Sander Gilman has shown, and few other populations suffered as direly the consequences of racial science and politics as European Jews (Gilman 1991, 1985; Lifton 1986; Mosse 1978; Proctor 1988). Moreover, Jewish populations remained key to the study of both human genetic diversity and disease incidence in mid-twentieth-century population genetics, a fact that remains true in the work of geneticists today: believed to be an unusually endogamous group, "the Jewish population" has long been considered an ideal epistemic object in efforts to understand not just the genetic etiologies of specific illnesses but more general processes of short-term human evolution and the nature of human biological diversity (see Wailoo and Pemberton 2006; Mourant, Kopec, and Domaniewska-Sobczak 1978).

Even more essential to the analyses and arguments in this book, studies of Jewish biological difference *undertaken as self-studies* have been ongoing since prominent Jewish scholars took up the mantle of race science at the turn of the twentieth century (chapter 2). That work of biological self-exploration and self-definition persisted, albeit in altered scientific terms and in different social and political circumstances, in the mid-twentieth century in the newly founded Israeli state. Moreover, with the technological developments, the scientific and social reach, and the epistemological authority of genomics in the contemporary world, since the mid-1990s such efforts have been given a renewed, in fact a reinvigorated, lease on life.

This long history of biological projects of Jewish self-fashioning is not unique. There were other communities that sought to study themselves as a race, as evidenced, for example, by the work of W.E.B. Du Bois and the Atlanta Conference (see Baker 1998, Du Bois 1897). Nevertheless, in contrast to other populations equally central to the late nineteenth- and early twentieth-century project of race science (say, "the Indian," or "the

Negro"), Jewish scholars were in a position—intellectually, politically, and economically—to study themselves in a far more sustained manner than were Europe's other racial Others.[7] There is an extensive archive of studies of Jewish racial difference carried out by scholars who saw themselves as intellectual and political leaders of the Jewish world. Moreover, after 1948, a state apparatus existed within which the scientific and medical institutions needed for biological self-studies were developed. In bringing a historical perspective to bear on my analysis of a contemporary science, I am not comparing anti-Semitic non-Jewish race scientists for whom Jews were an object of scientific research and disdain with genetic history studies that are born of projects and practices of self-definition and often self-celebration. In each of the configurations that I examine in this book—race science carried out in Western and Central Europe and the U.S. at the turn of the twentieth century, Israeli population genetics in the 1950s and 1960s, and contemporary work in genetic history—Jewish scientists have been understood by themselves and by others to be studying themselves.

This book operates at two overlapping levels of analysis. First, I give a reading of three distinct moments in modern Jewish culture and politics as seen through the lens of quests for the biological difference and the origins of "the Jews." Second, while a study of the ways in which the biological sciences have given form to and fashioned particular visions of a collective Jewish self, this book is simultaneously an exploration of and commentary upon the varied disciplinary, social, and political genealogies and effects of genetic history as a more general scientific and social phenomenon (see especially chapters 1 and 6). A particular science makes a particular set of worldly practices possible. *The Genealogical Science* explores the simultaneously shifting and overlapping worldly practices made possible on the basis of different kinds of biological evidence over the past century or more. I look back in time in order to better understand the ways in which contemporary practices of genetic history are perpetuating and/or reorganizing specific understandings of biology, culture, history, and the self. I want to understand the range of cultural and political projects this scientific field animates—and *might* animate. And I want to give an account of why it is that genetic historical projects and practices, scientific as well as commercial, are so socially felicitous today.

The Genealogical Science is perhaps best described as a history of science that merges with an anthropological account of a novel scientific and social formation. What makes this study anthropological is not the use of the discipline's arguably distinct method, ethnography. Nor is it a specific concern with tracking how various publics appropriate and reconfigure the meaning of scientific work. Instead, inspired by a tradition in the history, philosophy, and sociology of science, I pay careful attention to scientific epistemologies, past and present. But I read scientific epistemology via an anthropological sensibility trained to understand not just the epistemological, social, and political conditions of possibility of scientific work. I am interested in giving an account of the forms of life specific biological disciplines make possible or not. Moreover, in writing about genetic history I do not make a strong distinction between scientific and social practices. In using the terms "anthropological genetics," on the one hand, and "genetic history" on the other, I refer as much to scientific fields as to the cultural, political, and commercial practices to which they are giving form.

This book is thus an account of the entanglement of scientific work in larger scientific, social, and political frameworks. It is also an account of the circulation of scientific knowledge and objects (genetic markers of "identity") and their appropriation and interpretation by a variety of institutions and actors. I analyze the power of one scientific field—commercial as much as academic—and I seek to understand the ways in which it is forging distinct understandings of the (Jewish) self, both collective and individual. I build my arguments, first, by producing close readings of scientific studies through which I elucidate the specific epistemic objects, epistemological assumptions, and ethics that guide contemporary research on the origins of the Jews. I then compare those contemporary practices to the work of past Jewish scientists who sought to ascertain evidence of Jewish biological unity. Second, I analyze commercial structures (genetic ancestry-testing companies), social networks, and political projects within which genetic historical knowledge is appropriated by various publics—consumers and activists—who want to know something about "themselves," as individuals, as members of a family, and as members of a specifically Jewish community or world.

In contrast to much of the work in the anthropology and cultural studies of science today, I insist on the importance of exploring scientific epistemology *in detail*, even though my object of research is not a specific laboratory or research project (for detailed work in genomics laboratories, see M'Charek 2005; Fullwiley 2008a, 2008b). I do so because different forms of evidence make different kinds of cultural and political imaginations and practices possible. If we are to understand the power of genetic history in the social world, we need to understand the specific epistemological assumptions, concepts, and "working objects" (Daston and Galison 2007, 19) with which the discipline operates, and we need to compare these with those of race science and the population genetics of old. Publics—be they patients, consumers, or political activists—do not adopt scientific or medical knowledge wholesale, as many scholars have demonstrated (see Goodman, Heath, and Lindee 2003; Rapp 1999; Martin 1994; Dumit 2004). Neither, however, are the meanings of scientific work simply shaped by the "the social relations, cultural values, and political discourse within which it is deployed" (Rose 2007, 178; see also Nash 2006 upon whose work Rose is drawing). Social relations, cultural values, and political discourse do shape the meanings of ancestry testing, for example, as I will show in chapter 4, but the meanings and consequences of ancestry tests also and quite crucially derive from the science itself: from its forms of evidence and argument, and from the epistemological procedures and working objects on the basis of which the presumed facts of ancestry are revealed. While never divorced from social and political worlds, scientific practices nevertheless have a genealogy and a "life of their own" (Hacking 1983, 150).

There is a "determination" to science, in other words. In this book, I analyze the epistemological assumptions and evidentiary grounds of race science, population genetics, and genetic history in order to illuminate each discipline's distinct conceptions of human collectivities and agency, of the relationship between culture and nature, and of history and the ways in which "the past" is understood to be a constitutive element of the self. Genetic history enables certain epistemic objects and facts to come into being. In so doing it delineates horizons of possibility within which practices of self-fashioning and political activism take place, even as such practices simultaneously rework the meanings of scientific knowledge, objects, and work.

The Molecular Archive

Human population genetics entered a new era in the late 1980s when mitochondrial DNA (mtDNA), a circular strand of cytoplasmic DNA that is passed down the maternal line, was first deciphered.[8] As Rebecca Cann and her co-authors point out, mtDNA is especially conducive to understanding "genetic evolution *within* the human species" (Cann, Stoneking, and Wilson 1987, 131, emphasis added; see also Sommer 2008). Anthropological genetics took off with this new form of evidence. Mitochondrial DNA was understood to be uniquely able to specify human origins and migrations. Scientists maintained they could use the mitochondrial genome, a marker of lineal, maternal descent, in order to calculate how recently or distantly two populations had diverged from one another and to build "clean" (i.e., legible) phylogenetic trees. By the late 1990s a second genetic system had been added to the discipline's toolkit: many argued that the Y-chromosome, passed from fathers to sons, was even more helpful than mitochondrial DNA for studying the human past.[9]

Anthropological genetic breakthroughs and research projects have depended upon the biotechnological inventions and technological innovations of their parent field, genomics (see Rabinow 1996b, M'Charek 2005.) By the late 1970s, the invention of recombinant DNA technologies made it possible to isolate, clone, and analyze DNA sequences.[10] Such technologies enabled researchers to produce new epistemic objects on the basis of whose identification, isolation, and interpretation research in a variety of fields of genomics would proceed. Intrahuman phylogenies became easier to map and to assess (Sommer 2008). Moreover, by the turn of the new millennium, new machines and techniques had been added to genetic history's toolkit, making it cheaper and faster to read genome sequences and to generate large pools of data on a scale unavailable before (Wells 2006, 2–3).

Giving an account of the discipline of anthropological genetics, however, requires that we go back further in time to a much earlier history of developments in the biological sciences. Anthropological genetics is indebted not just to technological innovations driven by the needs of genomics. It is also beholden to conceptual developments in molecular biology that date to the mid-twentieth century. The late 1950s witnessed a shift in scientific understandings of living systems. An understanding of life as "organization," that is, an understanding of living systems as governed by "the interaction of the parts that gave meaning to the whole" (François Jacob quoted in Kay 2000, 40), was dislodged. In its place emerged the commitment that life is best characterized as an informational system. The work of molecular biology was conceived in the language of reading and, later, of editing the gene and genome, which were understood as informational texts. In what Lily Kay argues is "one of the most important and dramatic episodes in modern science," information—the genetic *code*—came to be understood as the locus of life (2000, xv).

This "molecular vision of life" (Kay 1993) took root in evolutionary biology in the 1960s. The architects of what was initially named "molecular anthropology" fought to establish a new evidentiary ground: no longer should evolution be studied primarily on the basis of morphological traits and fossil records. Evolution would be understood best by examining molecular evidence. Molecular anthropology would explore "primate phylogeny and human evolution through the genetic information contained in proteins and polynucleotides" (Sommer 2008, 480). The architects of molecular anthropology presented genetic information as a far more reliable and objective source of historical information than were fossil records or morphological traits, the evidentiary basis of evolutionary biology at the time (see chapters 1 and 6; see also Sommer 2008, Diaz 2007).

Over the next four decades, the meaning of finding information "in our genes" would shift—from an analysis of proteins and polynucleotides to a reading of DNA and its nucleotide sequences. But the basic principle has endured: DNA contains the most reliable information regarding our pasts, as a species and as distinct populations. In today's parlance, DNA is "a sort of molecular parchment on which an account of our species has been written" (Olson 2002, 5).

In writing about molecular anthropology and its presumption that there is "history in the gene," Marianne Sommer coined the term the "anthropological *gene* and *genome*," those "epistemological objects in search of answers to anthropological questions" (2008, 474). The anthropological gene and genome do not reveal facts about the nature of health and illness. Moreover, they are not objects that demand to be "edited" or altered in a post-genomic promise of gene therapy as a cure for future disease. The anthropological gene and genome reveal facts about "our" past as a species and as presumably discernible population groups. More broadly, in terms of its most basic epistemological assumptions about the meaning and function of genetic evidence, the anthropological gene and genome are quite distant from reigning understandings of DNA in biomedical fields (Sommer 2008).

By the turn of the millennium, the Human Genome Project had produced a sequence of the human genome. But in the aftermath of its success, geneticists have become increasingly convinced that knowing the sequence does not get us very far. Scientists began inquiring into how genes are actually expressed. What are the processes by which the "genetic code" generates proteins? Rather than focus on the genome and its genetic code, scientists are increasingly turning to analyze the cell apparatus as a whole. In the paradigm shift from mapping and sequencing the genome (genomics) to the study of development, function, and expression (post-genomics), the gene as "the central dogma," as the informational code from which biological expression and function proceed, has been challenged. Knowing the code cannot give an adequate account of life. It cannot give an adequate account of the development, growth, health and illness, and death of a living being.

In light of these recent scientific developments it would be easy to characterize the anthropological gene and genome as a "retrograde notion" (Sommer 2008, 475). After all, for anthropological geneticists the genome is still *the* "code"—*the* source of historical information about the past. The anthropological genome "resembles the classical molecular gene in that it is ontologically prioritized and in the sense that its bare sequence appears as the carrier of complex information" (475–76). Moreover, as I discuss in chapter 1, in anthropological genetics it is not *genes* per se that supply most historical information. Researchers focus on those parts of genetic systems believed to have little or no biological function, the noncoding regions of the Y-chromosome and the control region of mtDNA. According to the field's epistemological commitments, it is precisely because those regions have no known or very little biological function that they are reliable sources of historical information.[11] To quote Spencer Wells, this "functionally useless" part of the human genome—so called junk DNA—is "anything but junk to those of us who use the genome as a historical document. This is our text, and it provides us with the story of our ancestors" (2006, 15). In post-genomic efforts to understand the processes of life, however, an understanding of the genome's noncoding regions as "junk" is no longer intact. The past decade has witnessed a growing consensus that noncoding regions are essential to the biological processes of creating and sustaining life.

The anthropological genome, however, is not a retrograde notion in need of catching up to the more cutting-edge fields of biomedical post-genomics.[12] Although born of conceptual and technological developments in molecular biology and genomics, anthropological genetics is not a "derivative discourse" (Chatterjee 1986). It does not simply replicate either the epistemological commitments or the research goals of the post-genomic sciences that aim to understand and to intervene in the dynamics of disease and of life. Instead, anthropological genetics is heir to race science, evolutionary biology, and population genetics. It is a discipline organized around its own object of knowledge: DNA as "a historical document." If, as Timothy Lenoir (1997) has argued, establishing a discipline requires carving out a distinct epistemic object around which one can assert and develop one's unique expertise, the anthropological gene and genome are those objects for anthropological genetics.

Anthropological genetics is not just a biological science. Like race science and population genetics before it, anthropological genetics is also a historical science. The concept of information took a turn as it traveled in the 1960s from molecular biology to molecular anthropology. The concept came to refer to historical information (Diaz 2007). The anthropological gene and genome are molecular archives. Anthropological geneticists build narratives about origins, descent, and relatedness; they ask how members of a given group interacted with members of other groups over time; they investigate how the size and makeup of a given group has changed over the course of its history; and they do so on molecular terrain. Therefore, if we are to understand its scientific work and its social implications, we need to pay attention to the historical work that anthropological genetics does. As I argue at the end of the book, if we are to better understand the legacy of race science for contemporary work in genetic history (that subfield of anthropological genetics that focuses on "recent" and population-distinct genealogies), we need to focus on its historical and not just its biological commitments. We need to think about the distinct ways in which race science, on the one hand, and genetic history, on the other—as it gains ever more scientific credibility and commercial, cultural, and political reach in our time—articulate and biologize history (see chapter 6).

Practitioners of genetic history operate with particular assumptions regarding human collectivities, biology, culture, and agency on the basis of which historical narratives are built. Only after unpacking those assumptions, and comparing them to the assumptions that drove race science nearly a century ago, can we begin to articulate the meanings, the politics, and the possible effects of the contemporary practices of genetic history. More specifically, only then can we begin to understand what the social and political meanings and consequences of finding the truth of human collectivities in genomic facts—in this latest version of biological data—might turn out to be(come).

The Politics of Epistemology and the Question of Race

In scholarly debates about the political implications of anthropological genetics, as is true of debates about the genomic and post-genomic sciences more generally, the question of race and the legacy of race science loom large. In 2000, upon announcing that the human genome had been sequenced, the directors of the private and public ventures stood together with Bill Clinton and declared the end of race as a valid scientific category.[13] "Race," as argued in an editorial in the *New England Journal of Medicine*, "has become passé" (Schwartz 2001, 1393). The methodology behind the Human Genome Project (HGP) presumed human biological commonality. It was to be "*the* reference" sequence. It was to generate the "genetic terms in which *all individuals* would be expressed" (M'Charek 2005, 6, emphasis added). The Human Genome Project was to generate a genetic map of the human race.

Not everyone accepted this declaration of race as an invalid scientific category, however. Some geneticists, many of them self-declared minorities seeking to redress imbalances in health in U.S. society, "saw such statements not only as incorrect but also as irresponsible." For political and not just scientific reasons these scientists were determined to prove that "race is indeed bio-genetic" (Fullwiley 2008a, 687). During the first decade of the new millennium, the pages of major medical and scientific journals were full of debates about the biological status of race and its usefulness in biomedical research and practice (Collins 2004; Cooper, Kaufman, and Ward 2003; Burchard and Ziv 2003; Phimister 2003; see also Abu El-Haj 2007b). The most sustained arguments about the category of race were over the Human Genome Diversity Project, which was the first large-scale campaign to map population-based human genetic diversity.

In the summer of 1991, evolutionary biologists and population geneticists proposed an alternative to the Human Genome Project. Pointing out that the human genome being mapped by the HGP consisted of DNA taken from individuals who were for the most part of European ancestry, critics argued that it was imperative to "explore the full range of genome diversity within the human family" (M'Charek 2005, 6). The Human Genome Diversity Project was launched. Understood by many of its architects as an antiracist project—one interested in human diversity, one designed to undermine the Eurocentricity of the HGP (Reardon 2005, 11)—the Diversity Project soon came under attack.

In its design, the Diversity Project aimed to use the technologies and knowledge developed by the Human Genome Project and to redirect them towards the field of population genetics (M'Charek 2005, 7). Its architects wanted to create a genetic map of various human migrations out of Africa and to understand the migration routes of human communities and their genes across the globe. Scientists would "assign different populations to different loci on the [genetic] map" (7). In order to do so Diversity Project organizers set out to collect data from "isolated" populations with presumably pristine ("un-mixed") gene pools that were understood "to represent specific moments in the history of human migration" (9). On the basis of the genes of genetic isolates, researchers intended to build a better understanding of the history and genetic diversity of "the human species" (13; Reardon 2005).

As is well documented by Jenny Reardon, this project met a maelstrom of opposition. Nicknamed the Vampire Project, indigenous groups and scientific critics accused researchers of "propagating racism and colonialism by exploiting the genes of indigenous peoples" (Reardon 2005, 3). The Diversity Project was ultimately derailed by such criticism, although the genetic archive it produced endures and its goal of documenting human genetic diversity is ongoing in large and small, public and private, medical and nonmedical venues. The difficult questions the Diversity Project raised—about how to characterize human genetic diversity, about who has the authority to study whom, about the potential political and social consequences of such efforts, and about the ethical protocols for such studies—have not been laid to rest (see Reardon 2005).

The research projects that I focus on in this book are a subset of such studies of human origins and human genetic diversity. As a project in anthropological genetics, the goal of the Diversity Project was to understand the origins and diversity of the human species. By way of contrast, the goal of the studies of Jewish origins is to reconstruct the origins and phylogenies of contemporary Jewish communities. Mapping the genetic coherence and/or diversity of culturally and politically meaningful human groups—and not the species—is the aim of what I call genetic history.

Research projects in genetic history ask a distinct set of questions. What insights might genetic evidence lend to what we think we know or what we don't know about the origins of particular socially recognized groups? What historical insights or novel historical evidence might genetics provide? In the case of Jewish origins, researchers ask: Is the story of a Jewish population originating in ancient Palestine, from which it was exiled, plausible? Might the story of the Cohanim—the Jewish priestly line that, according to biblical and oral tradition, originated with the biblical figure of Aaron and is passed down from fathers to sons—be true? Are the various communities of the Jewish diaspora phylogenetically related to one another? Do they descend from a common ancient stock? How much have Jewish communities mixed with their "host" communities over the centuries? In the case of African Americans, by way of contrast, the central question is: What information might genetic evidence provide as to the specific *ethnic* origins of African Americans who, given the history of slavery, lost any specific knowledge of their particular cultural/ethnic African origins?[14] As a researcher in a laboratory working on African American origins once told me, whereas Jewish communities "know" where they came from, African Americans do not. We are trying to find out who "our Thomas Jefferson is," he said, referring to a recently published genetic study that had examined whether or not the male descendants of Sally Hemings, one of

Jefferson's slaves, were indeed fathered by Jefferson himself (Foster et al. 1998).

Moreover, in contrast to the anthropological genetic work of the now defunct Diversity Project and the ongoing National Genographic Project, the studies of Jewish genetic origins I examine in this book are best described as studies of the self. Most (albeit not all) of the main researchers are self-identified Jews, and in writing and interviews these scientists repeatedly highlight the importance of that identification in accounts of how and why they got involved in work on Jewish origins (see chapter 3). And the Jewish genetic origins studies are far from unique in that regard. Whether the African Burial Ground Project in Manhattan or the founding of genetic ancestry-testing companies that focus on African ethnic origins, these too are research projects and commercial enterprises undertaken by African Americans who frame their work as contributing to a knowledge of the self (see also Nash 2006).[15]

As I argue in chapter 3, self-studies are an increasingly common trend in a variety of genomic and post-genomic fields of research, especially in the U.S. This is a development that needs to be situated within the context of the rise of identity politics and the challenges to claims of objectivity that such movements have posed. Epistemological-qua-political critiques are not limited to the humanities and social sciences, influenced as they are by post-structuralist theory in its various instantiations. Questions regarding who is in the best position to produce knowledge about what and about whom, how one's "identity" structures the kinds of issues one is drawn to, the questions one may ask, and the intellectual and political commitments one may develop are also evident in specific domains of natural scientific work.

Biological self-studies, however, did not begin in the age of identity politics. As is well documented and widely known, "the Jews" emerged as a key object of study for physicians and anthropologists who, in the late nineteenth and early twentieth centuries, examined disease and pathology (and sometimes "health") within the parameters of the scientific study of race.[16] "The Jewish Body" became the site of much anxiety regarding the state of European civilization. Indeed, it became the site of much anxiety regarding the health and vitality of specific European nation-states (Gilman 1991). What is perhaps less well known is that it was not only Christian-European physicians and social scientists that studied the racial character of "the Jews." Jewish physicians and social scientists engaged in a sustained study of what they understood to be a distinct Jewish race, and they did so out of political commitments to improving the status and health of Jewish communities in the diaspora and/or to realizing the nationalist cause. By the turn of the twentieth century, Jewish physicians and social scientists were already engaged in what they understood as biological studies of the Jewish *self*. They argued that Jews are a race and they sought empirical evidence in support of that conviction. Moreover, most accepted the so-called fact of Jewish pathology and degeneration while working to provide explanations that were alternatives to prevailing ones (see chapter 2). At the beginning of the twentieth century, as Mitchell Hart writes, Arthur Ruppin, "arguably the leading social scientific authority on the Jews at the time, identified 'race' as the central theme of a newly created Jewish social science" (2005, 49).

As various scholars have argued, the turn to the language and science of race was for many a quest to articulate a secular conception of Jewishness. For Jews in Western Europe, where religious belief and practice were on the decline and where in the aftermath of the Emancipation it was not possible to speak of Jewish national difference, the language of race emerged as a way to assert a Jewish difference they believed in and desired to maintain (see Efron 1994; Endelman 2004; Hart 2000). Similarly, for key figures in the U.S. Jewish scholarly and political elite and, more generally, for many members of the U.S. Jewish public in the late nineteenth century, the language of race was a way of marking Jewish difference in the face of an anxiety about a community increasingly secular and assimilating into a Christian society. That embrace of racial self-definition declined in the early twentieth century, however, as the political and scientific leadership of the Jewish community began to worry that the arrival of their Eastern European brethren on U.S. shores threatened to place "The Jew" alongside "The Negro" in the U.S. racial hierarchy (Goldstein 2006).

As other Jewish Studies scholars have elaborated, arguing that Jews are a race was one response to the politics of assimilation that characterized the intellectual tradition of an earlier generation of European Jewish scholars of the *Haskalah* or Jewish Enlightenment who had embraced integration into European societies and argued that Judaism, like Christianity, is just a religion, "merely," in the words of their later Jewish critics, a matter of belief. By the late nineteenth century, a new generation of Jewish scholars understood the consequences of the Emancipation and of the intellectual and political commitments of the Jewish Enlightenment as posing a serious threat to what they often referred to as the survival of the Jews. There was increased anxiety—one paralleled in late nineteenth-century U.S. Jewish discourse—that with Jews integrating into European and U.S. societies and becoming increasingly secular, Jewish existence was at stake. The rise of a Jewish political project framed within the terms of central European nationalisms was central to articulating that anxiety. And in responding to the perceived impossibility and dangers of Jewish assimilation, many scientists and political activists embraced the language of race. As explicated most thoroughly and eloquently by Mitchell Hart, Jewish race science was a project of "self-criticism, reform, and regeneration, and for many Zionists, though not all, it was crucial for the reconstitution of Jewish national identity" (2005, 52; Hart 2000). Jewish physicians and social scientists rejected race science's assumption that "biological *determination* precluded Jewish transformation" (2005, 53, emphasis added). But as scientific men of their time, many embraced the concept of race (see chapter 2).

In the aftermath of Nazism and the Holocaust, racial self-definition became anathema to most Jewish scholars. More broadly, race was reconceptualized, and it receded as *the* object of inquiry in the biological sciences, even though, contrary to a long-accepted historiography of the transition from race science to population thinking, the scientific concept of race never actually disappeared (see chapter 1; also Reardon 2005). Biological studies of Jewish communities continued in the post-World War II era, but they were not often explicitly framed within the language of *racial* difference. "The Jews," now a presumably endogamous "population," were an important category of analysis for population geneticists who sought to understand human genetic diversity in the second half of the twentieth century. So too did the biology of "the Jews" persist as an object of a Jewish and now decisively *national* self-inquiry.

With the establishment of Israel in 1948, Israeli Jews emerged as a sustained object of genetic research for physicians and biologists in the state. Part and parcel of the ongoing project of nation-building, as I argue in chapter 2, Israeli-Jewish scientists sought to map Jewish genetic diversity and to find evidence of biological unity—of a shared origin—among the Jewish immigrant communities who had migrated to the new state: the Ashkenazim, the Sephardim (Jews of Spanish descent), and the various communities of *Edot ha-Mizrach* as they were named ("oriental Jews" in English translations). As I argue through a reading of scientific studies of "the genetics of the Jews" published in the 1950s and 1960s, while Zionism presumed the existence of the Jewish people, the founding of the Jewish state put that ideological commitment to the test. What is evident in the work in Israeli population genetics is a desire to identify biological evidence for the presumption of a common Jewish peoplehood whose truth was hard to "see," especially in the face of the arrival of oriental Jews whose presumably visible civilizational and phenotypic differences from the Ashkenazi elite strained the nationalist ideology upon which the state was founded. Testament to the legacy of racial thought in giving form to a Zionist vision of Jewish peoplehood by the mid-twentieth century, Israeli population researchers never doubted that biological facts of a shared origin did indeed exist, even as finding those facts remained forever elusive. On the one hand, I read this work in Israeli population genetics as expressing a prevailing anxiety among Israel's Ashkenazi elite about whether or not *these* Jews—oriental Jews, that is—could *really be* national kin. On the other hand, I simultaneously identify a second anxiety apparent in the search for genetic evidence of shared Near Eastern origins, one that stood at the very heart of Zionism: For a state in which the Ashkenazi Jew was the normative citizen, the search for their biological origins in the ancient Near East was crucial. What evidence was there that European Jews were actually *of* the Middle East?

In my previous book, *Facts on the Ground*, I argued that while a large body of postcolonial scholarship in anthropology and history had fruitfully explored the relationship between power and knowledge in constituting forms of rule, it was worthwhile breaking apart the general category "knowledge" and asking *which* disciplines emerged as particularly salient in *which* colonial contexts and why. In working on the biological sciences, I have come to understand something else: different disciplines provide different perspectives on the same political and cultural struggles. Looking at the history of Zionism through the lens of work in the biological sciences brings into focus a story long sidelined in histories of the Jewish state: Jewish thinkers and Zionist activists invested in race science as they forged an understanding of the Jewish people and fought to found the Jewish state. By the mid-twentieth century, a biological self-definition—even if not seamlessly a racial one, at least not as race was imagined at the turn of the twentieth century—had become common-sensical for many Jewish nationalists, and, in significant ways, it framed membership and shaped the contours of national belonging in the Jewish state.

More specifically, focusing on the biological sciences—as distinct from archaeology, as I did in *Facts on the Ground*—brings the Israeli *population(s)* into focus. In so doing, it draws attention to an irresolvable tension in the newly founded Israeli state, as I mentioned above: a commitment to the fact of a single Jewish nation now ingathered in Israel that sat side by side a profound and prevailing anxiety and doubt on the part of Israel's Ashkenazi elite regarding whether or not their "oriental" brethren could really be brethren at all. Insofar as Israeli population genetics in the 1950s and 1960s articulated that tension, it produced the Otherness of "oriental Jews," as we would expect given what we have learned over the past several decades from critical reconsiderations of Zionism from the standpoint of its Mizrahi "victims" (Shohat 1988; see also Chetrit 2010; Eyal 2006; Shenhav 2006; Shohat 1989). Nevertheless, Israeli population genetics simultaneously and repeatedly *summoned into being* the fact that Jews of oriental descent were indeed fellow Jews. In other words, every time population geneticists included studies of "oriental Jews" in their studies of "the genetics of Jewish populations," they not only fashioned oriental Jewish Otherness just as important and powerful, they iterated the fact that Jews of *Edut ha-Mizrach* belonged in the Jewish state and to the Jewish nation (even if they had to be "reformed"). That national summoning solidified the boundary between Jew and Arab in the Israeli imagination, and it did so in biological terms.[17] The possibility of kinship across the difference between Jew and Arab was quite simply unimaginable, in matters scientific and not only those of state.

Genetic historical studies of Jewish origins are the heirs to these earlier scientific efforts to substantiate the self in biological terms. As I demonstrate in chapter 4, the initial studies were born of a curiosity about the self, and such studies were carried out within an increasingly normative presumption in the study of human genetic diversity, especially in the U.S., that research scientists and medical doctors study "themselves" (see Epstein 2007; Fullwiley 2008a; Nelson 2009). In addition, the anxiety of assimilation and the fear for the "survival" of "the Jews" is today as prevalent amongst American Jewry (the largest Jewish community not living in the State of Israel) as it was for Central and Western European Jews at the turn of the twentieth century. With the decline of religious observance and the rise of intermarriage, the question of what makes Jews *Jews* now or a generation hence is an ongoing concern. The repeated refrain that genetic evidence has shown that the story of the diaspora is *indeed* true—and that it is true precisely because of a fidelity to religious traditions and endogamous kinship practices, as I demonstrate (chapters 1 and 3)—needs to be read in the context of this contemporary anxiety.

But there is more than an anxiety about the loss of Jewish existence at stake in this project of genetic self-exploration. The growth of genetic history as a scientific and social practice is part of a broader phenomenon of multicultural politics in what Elizabeth Povinelli terms the "liberal Diaspora" (Povinelli 2002) and in the attendant commodification of ethnicity found in places from Scotland, to South Africa, to Native American reservations—what Jean and John Comaroff have named "Ethnicity, Inc."

(2009). There may be far more companies offering tests of genetic ancestry in the U.S. than elsewhere, but this form of commerce is growing in Europe as well (Nash 2006; Sommer 2010). More specific to the U.S. and to my study of Jewish origins, the scientific and public appeal of genetic history and genetic ancestry testing can be understood within the rise, since the 1960s, of "white ethnic" politics (Jacobson 2006, 18). Genetic history's social felicity makes sense in relation to the desire of many American Jews—alongside many other white ethnics in the era of U.S. multiculturalism—to reclaim an ethnic heritage and to fashion a distinctive claim to identity and difference in the public sphere (ibid., 21; Brown 1995; Goldstein 2006).

The persistence of categories—repeated and resurgent attempts to ascertain the biological distinctiveness of the same groups—looms large in scholarly analyses of the politics of mapping and of buying and selling population-specific genetic diversity today. Stephan Palmié, for example, makes the point that reigning categories of racial classification form the basis of genomic technologies for producing the past, even as genomic "knowledge so produced will simultaneously feed back into" the U.S. racial system (2007, 207). Analyzing African American ancestry testing as a scientific pursuit and as a commercial product (i.e., ancestry kits that offer to identify one's African ethnic origins) thus serves to reinforce the U.S. racial system. Not only does it re-biologize what are fundamentally social and political facts. It replicates kinship logics of "structural amnesia" that have long grounded the U.S. racial economy, thereby specifying who will count as white or black (ibid.). Alan H. Goodman names this process "genetic racialization," that is, the process of producing races as "genetically meaningful entities" (2007, 227), a phenomenon that a variety of scholars believe to be growing ever more powerful today (Duster 2003, 1998; Fullwiley 2008a; Stevens 2002).

Social constructs or ideological commitments do not simply or directly shape the workings of science, however. Social forms and scientific categories inform one another. Analyzing the scientific production of "human genetic diversity" as an epistemic concept requires that we adopt an approach that examines the "proximate links" between various fields of practice (Daston and Galison 2007, 38). Race has never been wholly owned by the sciences. Nor was race simply relegated to the social domain following the paradigm shift from race science to population genetics. It has traveled across domains of practice—scientific and social—and has been entangled with shifting notions of personhood and citizenship. Race is an epistemic object that has been translated, made, and remade by the sciences, by states, by publics, and especially of late, by private companies and consumers.

Categories—the practices of classification—are essential to producing accounts of human genetic diversity, whether that diversity is talked about in the language of race, of population, or of "ancestry," as it is often referred to today. The fact that the same categories are examined repeatedly tells us something important about the racial economy of (this) science and of (this) society. As Duana Fullwiley demonstrates in her rich ethnographic account of the use of race in two medical genetics laboratories, "everyday broad American categories of racial groups" are crucial to how race is read "in the DNA." In producing admixture maps, which is a technology that assesses the percentages of different "ancestries" in a given population, "certain raced U.S. 'populations' ('Black'/African, 'White'/European, and 'Red'/Native American) and DNA markers with certain statistical frequencies in those populations are each posited as first principles to infer truths about the other" (2008a, 697–98; see also M'Charek 2005). What we get in the now-patented technology called Ancestry Information Markers is a map of human genetic diversity as seen through the lens of the U.S. racial order (Fullwiley 2008a).

Nevertheless, producing accounts and generating evidence of human genetic diversity have always entailed far more than just classifying groups. Therefore, if we are to more fully understand the ways in which the epistemic concept "human genetic diversity" has been reframed and renegotiated over time, we cannot focus on classificatory logics at the expense of all else. For race scientists in the late nineteenth and early twentieth centuries, the work of classifying was part of a project whose larger goal was to map a set of biological differences among social groups. Those biological differences were understood to generate differences in cultural achievements and intellectual capacities. Hierarchies of culture were generated by nature, by hierarchically ordered biological differences.

Scholars of race and science, of course, know that race science was committed to specifying biological distinctions that were read as identifying substantive civilizational and cognitive differences between human groups (Banton 1978; Barkan 1992; Gould 1981; Haraway 1991; Stepan 1982). But knowing it is not enough. Highlighting the causal assumptions that drove race science and not just its classificatory work will allow us to rethink the similarities and differences between race science, on the one hand, and population genetics and genetic history, on the other. It will allow us to explore more fully contemporary scientific quests to map human genetic diversity and the cultural and political grounds and effects of such work. If, in our analyses, we shift the focus away from scientific categories and the practices of classification, what else might we come to see?

In this book I focus on genetic history projects that began in the mid-1990s and that analyzed the Y-chromosome and mitochondrial DNA.[18] As I mentioned above, researchers focused on the noncoding and control regions of these genetic systems. Their forms of evidence differed from that of race science both in the shift from the phenotypic—the visible—to the genotypic (what Stephan Palmié calls "invisible essences")[19] and, more importantly, in a shift from understanding biological characteristics to have *causal properties* with *biological and social consequences* to understanding such characteristics as having little or no signficicance. As I elaborate in chapter 1, whereas for race science, biological difference was taken to *cause* cultural and cognitive differences, in anthropological genetics that causal logic is absent. One could argue it is inverted: noncoding genetic markers—particular mutations on the Y-chromosome and mitochondrial DNA—are read as having no *biological* meaning. They are read as "mere" indexes of ancestry and origins. They are read as signs of "history," history understood as referring to the origins, migration patterns, and descent lines (or kinship practices) of human groups. And it is precisely because they have no biological function that these markers are argued to be significant forms of evidence for efforts to reconstruct "origins" or "the past." Because the descendants of Aaron, the Cohen lineage, have remained true to the principle of passing priestly status from fathers to sons, for example, they carry Y-chromosome markers that distinguish them from other (Jewish) subpopulations. The repeated act of remaining faithful to that religious principle is what makes "the Cohanim" a scientifically legible group.

There are a variety of ways in which the logic of both "culture" and human agency, and with respect to Jewish origins, the language of "choice," are written into biological data. But biological data are always read as a sign of something other than biology itself. For example, based upon research in genetic history, ancestry-testing companies encourage individual African Americans to "discover" the "ethnic group" from which their ancestors were severed by the history of the Middle Passage and slavery. African "ethnic groups" are identified as the "ancestral" populations, the word "ethnic" signaling the originary and enduring cultural-linguistic groups from which individual African Americans descend. Those "original" populations *continue to exist,* and it is from them that contemporary African American individuals descend (see Fullwiley 2008a; see also Nelson 2008). In the case of Jewish origins the genetic historical imagination is somewhat different: given the history of living for centuries in "the diaspora," "the Jews" have endured only because of the decisions that human actors have made—the repeated choices made by one's ancestors who, against all odds, remained faithful to particular traditions of kinship and descent. In these genetic and commercial practices, culture simultaneously indexes something human actors and collectivities have made and sustained (ethnic groups, kinship structures), something other than biology, that is, and something we inherit and carry around within. That ambivalence—or incoherence—stands at the heart of this new scientific and social field and provides the grounds on which it is becoming ever more socially felicitous.

Noncoding regions are the perfect working objects for making and marking biological difference today because they do two things at the same time: they differentiate groups and, simultaneously, make no difference at all. This is a difference one can embrace because it does not seem to compel one to be defined by a biology, by a "biological difference" and potentially by a biological "inferiority," on the basis of which racial groups were excluded from political rights, subjected to slavery, eugenics, and genocide in eras past. In looking at junk DNA (and the control regions of the mitochondrion)—those genetic lines that endure only because they have been passed down from fathers to their sons and from mothers to their children—one sees "culture." One identifies the presumably "original populations," that is, the more pristine and by definition more endogamous (less admixed, in the language of anthropological genetics) cultures in the Old World from which contemporary populations of the "West" descend (see Fullwiley 2008a). And in the interpretation and narration of Jewish genetic history, one sees the choices ancestors have made over and over again: the choice to be endogamous, the choice to respect the religious edict to pass priestly status from fathers to sons, and as argued vis-à-vis mitochondrial evidence, the choice of women to be faithful to their (Jewish) men. It is those *cultures*—rather than a determining biological trait—that constitute our most authentic selves.

The commitment to culture and human agency built into genetic history's most fundamental working object—presumably biologically inconsequential genetic markers—is reiterated by the commerce in genetic ancestry testing. As this scientific epistemology merges with the rationality of the market, with the genetic ancestry test that we can purchase or not, we can "choose" to embrace this molecular difference as saying something about who we really are precisely because apparently we do not risk being caught in the determining logic of race. As a scientist, a consumer, an activist, one is not held to one's biology, even though it does tell one something very important, perhaps even the *most* important thing, about who one is in a genes-are-us world (Nelkin and Lindee 1995).

To be clear: I am not arguing that the anthropological historical subject is someone who just makes herself up. I am not arguing that subjects who self-test actually *choose* who they want to be. I am arguing that an ideology of human agency—as evidenced in "history," in "culture," and as often articulated in a language of "choice"—is built into the epistemic and commercial architectures of genetic history. And I am arguing that fact does a particular kind of political work: we can seem to have our cake and eat it too. We can find a truth in our bodies, and yet we can celebrate our own cultures as outcomes of who our ancestors were and/or of what they chose over and over again. Collectively and individually, we find our truest selves in genetic evidence of descent from a human collectivity that is defined not by some shared biological trait but by a shared and meaningful "culture." The burden upon us may turn out to be that we must embrace that culture to make sure that it continues to endure.

I am not suggesting that we should not examine the categories by which race scientists, population genetics, and anthropological genetics divide up the human world. It *is* important to track the production and use of particular categories of persons and to consider the reasons for and the significance of those enduring scientific quests for signs of group-based biological difference. In writing about and analyzing categories of biological classification, however, it is worth keeping in mind that even if the category is the same, its membership is not necessarily stable over time. As I illustrate in this book, the known Jewish world has shifted: which Jewish communities are considered normative or marginal and which are considered at all has changed over the past century, and those shifts in understandings of the boundaries of the Jewish world have been tied to the specific historical and political circumstances within which scientific knowledge has been produced.

In analyzing genetic history, however, we need to know more than *which* human collectivities this science—or any other (post)-genomic science—makes or reinforces and why. We also need to ask, what is it that *makes* a population *a population?* What makes a population a scientifically legible and meaningful human collectivity? And we need to consider how the answers to that question are the same or different for the various biological disciplines that have studied human collectivities over the past century or more. If not, we will not be able to understand the political and cultural imaginations and projects that have been—or *can be*—built on their terrains. And we will not be able to give an adequate account, as I try to do in this book, of why it is that individuals who identify as members of social groups who suffered dearly the violence of race science and eugenics in such recent memory seem so willing to embrace and even to promote a biological *self*-definition today.

Despite some scholars claims to the contrary, the anthropological genetic production of human genetic diversity—of groups—is not best described as a process of "geneticization" (Lipmann 1991). It is not best described as the reduction of social phenomena to genetic causes. The anthropological genome adjudicates historical and cultural authenticity and truth. But that genome, as I argued above, is understood to have no *biologically* causal properties. Genetic history's "determinism" emerges as a *post-facto* one. My true "*cultural*" self is legible in my genome. Upon learning who I have always already been—say, a Christian who discovers I have "Jewish markers"—I now desire (or *should desire*) to learn more about *my* culture, and in the case of Jewish origins, about *my* religion. I desire to become or to come closer to being that which I truly am. Choice and essence are re-sutured in an *a posteriori* determining logic of choices made in light of knowledge of who I have always already been.

Genetic history is not a level playing field, however, and the relationship between choice and determinism shifts in light of the individual, the group, and the social and political projects by and in which its knowledge and authority are deployed. The consequences of scientific studies and ancestry testing are not the same for all. Only certain kinds of groups are legible to the practices of genetic history: groups with molecularly identifiable origins. Other kinds of collectivities—for example, the Mizrahi Jew as an oppositional Jewish identity born in recent decades in the Israeli state—are not recognizable within its epistemological armature (see chapter 3). In addition, as I demonstrate in chapter 5, for those not recognized as members of the known Jewish world, falling within the Jewish genetic map can adjudicate, even if only at the level of background knowledge, whether or not one is likely to be recognized as a lost tribe and as such, for specific communities and individuals, whether or not one is likely to be recognized as "returning Jews." Finding "Jewish genetic markers" does not make communities who believe they have ancient Jewish ancestry recognized Jews. It can, however, make their beliefs plausible.[20] It can make possible the "choice" of Jewishness. It can make it possible for individuals to "return," following a proper Judaic education, to the Jewish fold. And while conversion is recognized in Judaism, the background knowledge that a convert is really a "returning" Jew is significant for members of non-European communities, especially vis-à-vis the Israeli state, its intra-Jewish racial politics, and its Law of Return. In turn, when those (potential) returnees emigrate to the State of Israel, join the messianic-nationalist movement, and live in settlements, a liberal antiracist self-fashioning that drives many U.S. activists to champion the rights of lost tribes comes full circle in support of a project of colonial settlement. The range of political practices that may be animated by genetic history is only beginning to be seen.

Genetic Historical Selves

In light of significant epistemological and technical shifts and the political-economic reorganization of research that have taken place in molecular biology since the late 1970s, scholars have argued that genetics is becoming increasingly constitutive of our understandings of life and our constructions of the self. According to Ian Hacking, "the intersection of medical, social, personal, and profit making interests ensures that the avalanche of genetic information about individuals and population has only begun" (2006, 89).

Genetic history is situated squarely within this avalanche of genetic information. But different kinds of genetic information can generate different effects and fashion sundry social imaginations. Sufficient attention to those distinctions—to the proliferation of various and often dissimilar fields of genomic and post-genomic research and their attendant practices of the self—is often absent in a scholarly literature that has taken up the mantle of analyzing the new genetic sciences and that has written about its consequences, broadly speaking, in the language of "biological citizenship" (Petryna 2002). In order to more fully understand the wide range of effects and practices emerging from this avalanche of genetic information, however, we need to pay a lot more attention to what had emerged, by the mid-1980s, as a fundamental commitment in the history and sociology of science: a commitment to the *disunity* of the sciences (Galison and Stump 1996). In this instance, we may need to pay a lot more attention to the disunity of what could be cast as a single scientific and social field.

Various scholars have argued of late that, as a consequence of developments in biomedicine in "the advanced liberal nations of the West—Europe, Australia, and the United States"—we are witnessing the emergence of a "new regime of the self" (Rose 2007, 154). We live in societies increasingly saturated by "novel practices of biological choice." Individuals are encouraged to act "prudently" in order to actively shape their individual lives, and quite centrally, in order to avoid disease risk and foster health (ibid.). "Biological citizenship," as this phenomenon has been named, operates in a "political and ethical field" in which "life has become a strategic enterprise"(Novas and Rose 2000, 487).

Genetic history is certainly part of this larger scientific and social field. One cannot understand either its emergence or its social power without recognizing its dependence on both epistemic and technological developments in what was once called "the new genetics" and on the increasing epistemological, cultural, and political authority of those broader fields. Nevertheless, genetic history's goals and techniques diverge from those of the mainstay of biomedical (post)-genomics in significant ways. Most broadly, as a discipline its goal is not to *intervene* in the human genome. Its politics are not a "*vital* politics." That is, genetic history is not "concerned with our growing capacities to control, manage, engineer, reshape, and modulate the very vital capacities of human beings as living creatures . . . a politics of *life itself*" (Rose 2007, 3, emphasis added). Instead, genetic history—and anthropological genetics more generally—is perhaps best described as a politics of "history itself." It is concerned with preserving "genetic markers," with collecting them before "isolated populations" disappear and take with them a whole treasure trove of knowledge about the human species and the human past. This is salvage genetics. This is a project to conserve a history believed to survive within each individual human body.

Genetic historical projects are proliferating in an epistemic culture (Knorr-Cetina 1999)—scientific, social, political, and commercial—in which genetic information is understood to tell us something profound and fundamental about ourselves, our health, our identities, our pasts, our futures. As a field of scientific practice, it constructs "groups" and the potential for cultural and political membership through a molecular optic believed to render phylogenies visible. In turn, on the basis of techniques and data initially generated by scientific research, genetic genealogical-testing companies offer individuals genetic ancestry tests on the basis of which new understandings of the self are being made. The public circulation and appropriation of the results of scientific studies and the commerce in genetic ancestry testing are generating new kinds of biological knowledge on the basis of which we can now know the "biosocial" groups to which we and others belong (Rabinow 1996; Hacking 2005).

But simply to refer to "biology" or to the *bio*social when analyzing contemporary anthropological research projects on human genetic diversity would obscure the fundamentally historical claims of both genetic history and the more encompassing field of anthropological genetics. To name genetic historical projects of self-fashioning "*biological* citizenship" would elide the fact that these are efforts to identify a *history* within. Therefore, while I begin the book by exploring what biology is in the epistemic architecture of genetic history (chapter 1), in the final chapter I focus on history and I ask, what is history and how is it made? I explore the translation of evolutionary theory as it is adopted and adapted to examine the "history" of specific groups, and I explicate what it is that history is made to be. While those who write genetic histories, scientists and journalists alike, often represent the project as a temporally deeper version of genealogical practices, there are substantive shifts in the kinds of questions that are asked, the evidence available, and the answers to be had. Moreover, narratives do not emerge from the molecular archive itself. I illustrate the problems involved in translating the mathematical notion of "a population" into a history of "the Jews" and in dating genetic evidence. I demonstrate the ways in which the genome is read through already existing historical narratives, often in a circular logic of discovery, interpretation, and proof. And I argue that it is not just that genetic historical data is "undetermined." It may be impossible to generate any detailed and meaningful historical *narratives* on its basis alone. Using the genome to generate narratives about the histories of particular cultural, political, and/or religious collectivities, I argue, is a category mistake. To return to where I began this introduction: the genome cannot prove Shlomo Sand wrong. Sand's account is not based on statistical distributions, and it does not presume a national story from a point of origin to the present that *could be* verified, if not easily falsified, on the basis of evidence of shared biological descent. Sand builds his historical narrative on the basis of an overall reading of extant political practices during the Assyrian and Roman empires, the historical accounts of the population of Palestine after the presumed exile(s) and migrations of the Jews, and historical accounts of Jewish communities that grew up around the world. And quite centrally, Sand's account relies on conceptions of polity and forms of consciousness that have shifted over the centuries, thus rendering the modern notions of Jewish peoplehood or ethnicity the wrong heuristics through which to read the historical record of the *longue durée*. And, to be clear, if the genome does not prove Sand wrong, neither can it prove him right. It is the wrong kind of evidence and the wrong style of reasoning (Hacking 2002) for the task at hand.

Moreover, rather than one instance of a novel and an increasingly widespread modality of personhood and citizenship that has been named biological citizenship, anthropological genetics is better understood as a rearticulation of the longstanding modern commitment that history is "the very medium in which beings develop and acquire characteristics" (Han-Pile 2005, 589; see Foucault 1970). Through a discussion of the emergence of the importance of origins in nineteenth-century thought, I demonstrate a continuity between contemporary practices of genetic history and those of various sciences—cell theory, history, psychoanalysis—a century or more ago. Not only are we (still) defined by our origins, in the rhetoric of anthropological genetics those origins never go away. The task of this science, as was the task of various sciences before it, is to render origins legible.

It is in their shared commitment to the significance of biological origins—that at some point we were born of the same ancestors and that that origin makes us, even if in complicated ways, who we are—that race science and genetic history may be most similar. As individuals and as groups our real selves are defined at our moments of birth. Therefore, if we are to better articulate the continuities and differences between race science, population genetics, and genetic history, we need to highlight not only the myriad evidentiary practices and styles of reasoning that have driven each discipline. In addition, we need to draw attention to each discipline's specific conception of history. Replicating the logic of racial and much national thought of the late nineteenth and early twentieth centuries, in the practices of genetic history vertical belonging trumps the possibility of horizontal membership. Descent from a common ancestor becomes the grid of intelligibility within which collectivities and individuals are—and are recognized for—what and who they really are.

But that may be where the similarity in the historical commitments of race science and of genetic history comes to an end. As many scholars of race have argued, racial thought entailed a theory of history. It was the destiny of certain races to succeed, to be civilized, to colonize, and to rule and of others to fail, to remain primitive, to be colonized, ruled over, even owned (Gould 1981; Duster 2003; Anderson 2006; Banton 1978; Goldberg 1993; Mills 1998). "Racist" discourse (as he names it) was, in Michel Foucault's words, a "historico-*biological*" discourse (Foucault 2003, 60, emphasis added), that is, a discourse in which history was understood to be driven by the facts of biology. By the second half of the nineteenth century, racial thought had cast history within the terms of "the theory of evolutionism and the struggle for survival" (60). Moreover, survival took on a biomedical cast: the survival of the race was spoken of not just in a language of purity but in a language of health.[21] In Jewish racial thought at the turn of the twentieth century, for example, the fear of Jewish *degeneration* "in the diaspora" emerged as an anxiety driving and helping to frame the Zionist movement and its vision of Jewish national settlement in Palestine, including the value of "Hebrew Labor," that is, of why it is that Jewish settlers needed to work on the land (chapter 2).

By way of contrast, genetic history is perhaps better named a biologico-*historical* discourse. History is evident in biological data, but it is not driven by the facts of biology per se. Therefore, to talk of "survival" in genetic historical terms is to index *cultural* survival. In the case of narrating Jewish history, it is to talk about the fact that despite a history of having lived for generations in the diaspora and often under very difficult conditions, "the Jews" have sustained themselves as a distinct cultural and religious group. Within the terms of a genetic historical imagination, if we are to be true to ourselves we must be true to those origins and that history. And within the terms of a biologico-historical discourse that embraces and performs the liberal commitment to human agency (seen to inhere in "culture") and to the self-made individual and simultaneously reveals biological-historical truths that the genome records and preserves, we must *choose* to be true to those origins if we are to be—and if we are to remain generations hence—who we have always already been.

Emergent Forms?

In writing of the challenges facing anthropology today, Paul Rabinow has argued for an "experimental mode of inquiry" in which one "confronts a problem whose answer is not known in advance rather than already having answers and then seeking a problem" (1999, 174). Rabinow's concern with experimental forms is directly tied to the substantive "events" (180) and practices he has been studying for the past two decades: the contemporary biosciences, in which technological capacities, scientific practices, and epistemological assumptions have been going through a rapid and continuous series of shifts. As do other scholars writing about the contemporary biosciences, Rabinow argues that accompanying those scientific developments are emerging forms of life that are as of yet still uncertain and unknowable. We do not know what the work of contemporary post-genomics will come to have meant. In part, we do not know—we cannot know—because we lack the philosophical and ethical language(s) with which to understand developments and novelties hitherto unseen (Rabinow 1999; see also Fischer 2003, 2009; Lakoff 2005; Rose 2007; Sunder Rajan 2006).[22]

As a science, anthropological genetics—and its subfield, genetic history—are certainly emergent. They partake of the larger genomic revolution. At the technical level, work in anthropological genetics is changing at a speed that does not generally characterize social life. And the social and political practices born of scientific research, including quite crucially the commercial market in genetic ancestry testing, are new and are proliferating at rapid speed. There may well be futures that emerge from those practices that we cannot imagine or understand today. More specifically, as I finish this book, new studies of Jewish origins are being published based upon a technique very different than the ones I discuss. Researchers have published the results of studies of Jewish origins on the basis of genome-wide surveys, which target certain kinds of mutations across the genome and that look at both coding and noncoding regions in order "to improve" our "understanding about the relatedness of contemporary Jewish groups" (Atzmon et al. 2010, 850).

As analytically useful as the concept of the emergent is, I am nevertheless given pause by all the talk about the potentially world-changing nature of (post-)genomics. I remain skeptical in the face of all the predictions and promises—the "hype" (Fortun 2008)—offered up by this world of "entrepreneurial science" (Shapin 2008). Moreover, even if it is true that as science changes so too does society (Latour 1987), it is also true that, as the sciences shift, there are fundamental scientific and social questions, cultural and political imaginations, and collective desires and forms of practice that endure (see Rabinow 1999). The search for biological evidence of the fact that "the Jews" are not "merely" a religious group is now over a century old. The desire on the part of scientists who study population origins to pry history out of biology has been going on for over fifty years. And those are but two of the most obvious and important social facts that persist. In addition, both the epistemic objects and substantive historical claims made by genetic historians at the turn of the millennium are out there in the world, circulating through domains of public discourse and political practice and through the commerce in ancestry testing. They now have a life of their own. They cannot simply be called back in if and when scientists decide that they were wrong.

As such, I have written a book about techniques and forms of evidence that were considered the best tools of the trade at the turn of the new millennium even as I recognize that novel technologies, shifting evidentiary assumptions, and new disciplinary configurations (evolutionary genomics, population genomics) continue to emerge. And I do so with the confidence that I am not sending into the world a series of arguments or interpretations that are already *passé*. The genetic historical subject has not been reconfigured by these latest techniques, certainly not in fundamental ways.

The problem of how to generate historical *narratives* out of biological data is not going to disappear. And even as the latest techniques and working objects available to the biological sciences are being harnessed once again to answer the same question, "What are the Jews?," the question of Jewish origins and relatedness is not going to be resolved once and for all.

CHAPTER ONE

The Descent of Men

> In an unusual marriage of science and religion, researchers have found biological evidence in support of an ancient belief: certain Jewish men, thought to be descendants of the first high priest, Aaron, the older brother of Moses, share distinctive genetic traits, suggesting that they may indeed be members of a single lineage that has endured for thousands of years.
>
> —(Grady 1997b)

In an article in the *New York Times*, speaking of the publication in *Nature* of the first Y-chromosome study of the Jewish priesthood (the Cohanim), Michael Hammer described the results as a "beautiful example of how father-to-son transmission of two things, one genetic and one cultural, gives you the same picture" (Grady 1997b). The author of the article, Denise Grady, wrote in a subsequent article in the *Times* that something (scientifically) new is happening here:

> Until now, the appeal and the fear of genetic testing have stemmed from its ability to predict the future. Will a baby be born with Down syndrome? Will a neurological disease or cancer strike a person down in the prime of life? While these tests have provided a window on the future and maybe even a way of avoiding tragedy, not everyone has wanted to know what they reveal. This month, though, researchers described a novel use of a genetic test—to look deep into the past—and people are clamoring to know more. (Grady 1997a)

In this chapter, I analyze research projects on the origins of contemporary Jews in order to illuminate more generally the forms of evidence that anthropological geneticists use to map group-based diversity, to construct population phylogenies, and to determine the "origins" of specific groups. More specifically, I examine Y-chromosome research, which, by the turn of the millennium, many anthropological geneticists considered one of the most reliable kinds of phylogenetic data. I draw upon my research on genetic studies of the origins and kinship of contemporary Jewish communities in order to sketch the broadest contours of the grammar—scientific, ethical, and political—of the science of anthropological genetics.

Critics have argued that anthropological genetics, or the study of population-based genetic diversity, is but the latest heir to race science, a scientific project begun in the nineteenth century that defined the human species in the language of biological kinds and delimited absolute categories of human difference (Duster 1998; Palmié 2007; Stevens 2002). What I argue in this chapter, however, is that despite apparent continuities in classificatory practices, anthropological genetics' relationship with race science is not so seamless. I highlight evidentiary logics other than practices of classification (the persistence of categories) and I ask, what exactly is being "discovered" through the identification of molecular differences? What are these biological markers of difference taken to be *signs of*? In developing my argument, I focus on the relationship between nature and culture presumed in population-specific genetic historical work.

Anthropological genetics divides humanity up into its "constituent parts"—its ethnic, linguistic, and racial groupings. In so doing, however, it does not produce the same understandings of human groups or of human difference as did race science. In what follows, I examine what makes a population legible and meaningful in anthropological genetics. I consider the political significance of the field's epistemological assumptions, evidentiary practices, working objects, and historical arguments. In so doing, I explore why it might be that genetic historical quests—as population-based research and as individual genetic-ancestry tests—are widely embraced today.

Priestly Descent

"According to biblical accounts, the Jewish priesthood was established about 3,300 years ago with the appointment of the first Israelite high priest. Designation of Jewish males to the male priesthood continues to this day, and is determined by strict patrilineal descent," explain Karl Skorecki and Michael Hammer in their first published paper on Y-chromosome research into Jewish origins (Skorecki et al. 1997, 32). If priestly descent (the Cohen lineage) has been passed from father to son originating with the biblical figure of Aaron, the paper reasoned, it should be possible to find evidence consistent with the biblical account through genetic analysis. A nephrologist at the Ramban Medical Center at Haifa's Technion, Skorecki contacted Michael Hammer at the University of Arizona because of Hammer's expertise in using the Y-chromosome to trace population origins.

As Karl Skorecki tells the story, the inspiration for this research was born of a whim: he was sitting in synagogue one day and a man—a Cohen from North Africa—stood up to perform his duties. Skorecki wondered what the two of them, both Cohanim but from vastly different regions of the world, might share. If the story of priestly descent were true, their common origins should be visible on the Y-chromosome. He started research into the Y-chromosome of Jewish priests as a hobby of sorts through which he would, ultimately, join forces with other researchers in an effort to expand Y-chromosome studies of Jewish history (see the PBS program, *The Lost Tribes of Israel* [2000]).

Skorecki and Hammer collected DNA samples from 188 Israeli, British, and North American Jewish men. They compared the Y-chromosomes of Jewish men who self-identified as Cohanim ($n = 68$) with men who self-identified as either Levites (a second priestly line)[1] or as "lay-Jews," who were named in the study, in accordance with biblical tradition, "Israelites." If the biblical and oral traditions of priestly origins and descent are historically accurate, "observable" differences should exist between the Y-chromosome haplotypes of "Jewish priests and their lay-counterparts" (Skorecki et al. 1997, 32).

The 1997 *Nature* paper announced just such an observable difference. Excluding Levites from the final analysis (Levites exhibited no pattern of patrilineal descent from a single ancestor), Hammer and Skorecki compared Cohanim and Israelites. On the basis of polymorphisms at two genetic loci,[2] Skorecki concluded that there is a difference in the Y-chromosome haplotypes of priests versus lay Jews, thus "confirm[ing] a distinct paternal genealogy for Jewish priests" (1997, 32). To further explore the results, Skorecki and Hammer, joined by colleagues at University College London who would emerge as central figures in studies of Jewish genetic history, designed a more expansive study, the results of which were published in a paper in *Nature* in 1998 (Thomas et al. 1998). An examination of the Y-chromosomes of 306 Jewish men and based upon a haplotype constructed out of twelve genetic loci,[3] this second study generated similar results. (A haplotype is a set of genetic markers on a given chromosome that are inherited together, or "linked.") "Despite extensive diversity among Israelites," the authors argue, "a single haplotype ([now named] *the*

Cohen Modal Haplotype) is strikingly frequent in both Ashkenazi and Sephardi Cohanim" (Thomas et al. 1998, 138, emphasis added). The Cohen modal haplotype, the most common haplotype found in Cohen men, is present in approximately 50 percent of Cohanim (0.449 of the Ashkenazi Cohen sample, and 0.561 of the Sephardi Cohen sample) (Thomas et al. 1998). "Given the relative isolation of Ashkenazic and Sephardic communities over the past 500 years, the presence of the same modal haplotype in the Cohanim of both communities strongly suggests a common origin" (ibid., 139).

Delineating descent is but one aspect of these projects of historical reconstruction. The question of time is at least as important: "To the extent that patrilineal inheritance has been followed since sometime around the Temple period (roughly 3,000–2,000 years before present), Y chromosomes of present day Cohanim . . . should derive from a common ancestral type no more recently than the Temple period" (Thomas et al. 1998, 138). In other words, the "coalescence time" (the time of origin) of the Cohen modal haplotype must date to *before* the "dispersion of the priesthood following the Temple's destruction" (138).[4]

Estimating coalescence time is a complex process. At a bare minimum, it depends on knowing the "normal" rate of mutations in the Y-chromosome, specifying what is referred to as the "molecular clock." In addition, it requires assuming the time of a generation—15, 20, 25, or 30 years.[5] In the 1998 *Nature* paper, the researchers determined that the Cohen modal haplotype originated approximately 106 generations ago. Multiplying that number by 25 (or 30) years, they concluded that the Cohen modal haplotype dates to 2,650 (3,180) years before present. The authors state that, "ignoring uncertainty in the mutation rate," there is a 95 percent confidence interval that the coalescence time of the Cohen modal haplotype is 2,100–3,250 years before present, "sometime during or shortly before the Temple period in Jewish history" (1998, 139).[6]

The conclusions of the 1998 paper went much farther than that, however. Might the Cohen modal haplotype indicate more than just priestly descent?

> The identification of haplotypes with restricted distributions may provide "signatures" of ancient connections *that have been partially obscured by subsequent mixing with other populations*. Gene flow from the Cohanim could account for the presence of the Cohen modal haplotype in both Ashkenazic and Sephardic Israelites, *or* it could be a signature of the ancient Hebrew population. The Cohen modal haplotype *may* therefore be useful for testing

hypotheses regarding the relationship between specific contemporary communities and the ancient Hebrew population. (Thomas et al. 1998, 139, emphasis added)[7]

While not definitive, the authors think that the Cohen modal haplotype—found in approximately 10 percent of Ashkenazi and Sephardic "Israelites"[8]—may well indicate ancient Hebrew and not just priestly origins. Having discovered a possible indicator of common Hebrew ancestry, the opportunity to use genetic information to study the patrilineal origins and descent of contemporary Jewish communities was opened up. Further examinations of Jewish origins would rely, at least initially, on the Cohen modal haplotype—determining its geographic origin, using it as a normative measure of ancient Hebrew descent. Researchers turned to the Cohen modal haplotype and to a search for other Y-chromosome types shared or "prevalent"[9] in contemporary Jewish populations in order to evaluate the historical relatedness of contemporary Jewish communities, the veracity of the history of the Jewish people as a history of diaspora born out of exile from ancient Palestine, and the claims of "potential" Jews, groups of Jews who believe they are descendants of ancient Israel.

In Search of Population Histories

The turn to DNA to pursue a "curiosity about origins" is not new, as two prominent researchers in anthropological genetics explain. DNA "has been passed down to us from our ancestors, accumulating mutations along the way." As such, "the DNA of modern humans are . . . different from each other, and these differences, or polymorphisms, provide a record of our relatedness and genetic history" (Jobling and Tyler Smith 1995, 445). And as records of our relatedness, two loci are believed by most, though not all, anthropological geneticists to provide the most useful data: mitochondrial DNA and the Y-chromosome.[10]

Mitochondrial DNA and the Y-chromosome are passed down unilineally. One inherits one's mtDNA from one's mother. Men inherit their Y-chromosome from their fathers. As explained by Jobling and Tyler-Smith, "Neither of these segments of DNA recombines at meiosis,[11] and this means that they each contain a particularly simple record of their past" (1995, 449).[12] As a result, the biological principles of descent are pried apart: one can track one's lineage via the maternal or the paternal line. The two lines remain fully independent of one another, and the "history" of each is distinct. By deciphering the sequence of nucleotides (the order of the chemical components of DNA),[13] anthropological geneticists reconstruct population histories by delineating lines of descent believed to be archived in the history of genetic polymorphisms (a variation or mutation in the sequence of nucleotides among individuals) as they are passed down from mothers to their children and from fathers to their sons.

Mitochondrial DNA was used in anthropological genetic studies long before the Y-chromosome (I discuss mtDNA-based phylogenetic research in chapter 3). The use of the Y-chromosome is actually relatively new. Jobling wrote in a paper published in 1994, "The human Y chromosome is poor in conventional DNA polymorphisms," hindering "studies of the paternal lineage" (1994, 107). For the Y-chromosome to be useful to phylogenetic analysis, sufficient genetic *diversity* must be present in human Y-chromosomes—and it was not until the late 1990s that a sufficient amount of diversity was found.

Anthropological genetics explores the history of human migrations and population-specific origins and relatedness by analyzing diversity at the molecular level. (By way of contrast, for example, population genetics in the 1950s and 1960s analyzed the phenotypic diversity evident in blood group distributions without any knowledge of the specific genotypes [see chapter 2]). Using genetic data to make inferences about population histories involves starting with the principle that, under certain circumstances and assumptions, "genetic similarity reflects common ancestry" (Relethford 2001, 68). But genetic *similarity* is derived from an analysis of polymorphisms, of genetic *differences*. As articulated by John Relethford, a paradox stands at the heart of genetics and evolutionary theory. Consider the case of mtDNA and the search for "Eve," the female (genetic) ancestor of all modern humans. If we all trace back to a single ancestor, must not our DNA—in this instance, our mtDNA—be identical? If so, how can we use mtDNA to untangle the different genetic relationships or degrees of relatedness among individuals—or among groups?

Descent from a common ancestor does not imply identity. Rather, it implies a presumably decipherable matrix of genealogical relationships "visible" in genetic polymorphisms. According to current scientific understandings of evolution, as organisms reproduce, molecular mutations are generated randomly: some are deleterious, some positive, and most are neutral. Researchers decipher (the chain of) those presumably step-wise (one mutation at a time) mutations in order to reconstruct genealogies, using genetic differences among pairs of individuals to evaluate their kinship: "the greater the length of time separating two individuals, the more mutations will accumulate, and the greater the genetic difference between them"

(Relethford 2001, 72). At its simplest, that is the underlying assumption of phylogenetic analysis. Genetic distance is a measure of the relationship of one population to another: the more genetically similar, the more recently two populations had a common ancestor, the more genetically dissimilar, the more remote the common ancestor.[14] By identifying specific polymorphisms on the basis of which distinct haplotypes (sets of linked polymorphisms on a given chromosome) are constructed, one individual or group of individuals is compared to another. With respect to studies of Jewish priestly origins, scientists ask: What is the relative frequency of the Cohen modal haplotype in Jewish priests versus lay Jews? What is the relative frequency of the Cohen modal haplotype in Jewish versus non-Jewish populations? What is the origin of the Cohen modal haplotype? In what other populations does one find either the same haplotype or a haplotype closely "related" to it (by being a few mutations removed)? And what might all this information tell us about the *geographic* origins of today's Jews?

Diversity has long played a central role in biological thought (Mayr 1982). From eighteenth- through twentieth-century biology, as Jenny Reardon has written, diversity has been a "key object" of speculation and research (2001, 361). "Is the diversity of the natural world meaningful? What is the appropriate unit of analysis" (361)? Moreover, "Where does diversity come from? Individuals or groups? If groups, how should these groups be defined, by whom and for what purposes" (358)? In the eighteenth century, Carl Linnaeus first codified and mapped biological diversity "by describing kinds of organisms in the terms of a strictly imposed formal system borrowed from classical logic" (Hey 2001, 7). He understood classes of organisms (genus, species) to be made of "distinct and unchangeable kinds of organisms that had been created by God" (8). Linnaeus's system for classifying organisms was typological in character, what Jody Hey has called a "well-codified version of Platonic and Aristotelian essentialism" (8).

In the late nineteenth century Darwin recast both the meaning and the significance of diversity. "The living world became a world in time, and both its occupants and its relational structure were refigured as products of its evolutionary history" (Keller 2000, 7). In contrast to Linnaeus, Darwin described "a continuum of variation—individual differences, to varieties, to species" (Hey 2001, 8). Darwin's evolutionary theory provided a mechanism through which the origin and transformation of species would be understood—natural selection acting upon individual variation. And in its emphasis on transformation and not just "origin," Darwin's theory of evolution paved the way for "a comparison of organisms not only in space but in time," which became the hallmark of modern genetics (Gudding 1996,

529). And as narrated retrospectively,[15] Darwin's shift of emphasis to individual variation laid the groundwork for biological and anthropological projects to come. The race concept was deconstructed on the basis of arguments that most genetic variation occurs at the level of individuals and not between so-called racial groups. Moreover, group-level differences came to be understood to be, by and large, *biologically* insignificant.

Nevertheless, questions about human history and a "population's" specific origins are explored by focusing on genetic diversity at the level of population groups—diversity observed on the genomes of individuals classified according to "ethnicity" or "geographic origins" (sometimes self-designated, as in studies like those of Jewish origins). Group-based genetic differences may only account for a tiny percentage of human genetic diversity, but they provide the most important kind of evidence for research projects that aim to reconstruct histories of human migrations and to delineate population-specific histories. And that, according to many geneticists today, should come as no surprise. Articulating an emerging scientific consensus on the significance of group-based genetic diversity, David B. Goldstein, a lead researcher in the Y-chromosome studies of Jewish origins and currently the director of Duke University's Center for Human Genome Variation, writes:

> For most of the past twenty-five years or so, the scientific community has followed [Richard] Lewontin in arguing that a 10 percent genetic difference among members of different groups amounts to a small amount of genetic variation. . . . The data I present here, however, showing clear genetic differences among the members of population after population, as well as hundreds of other published studies on the genetics of disease, drug response, and human variation, argue otherwise. Nor should we be surprised: there are ten million sites in the human genome that show differences among individuals. . . . If only a small fraction of these showed differences that correlated strongly with race or ethnicity (or, in less loaded terms, with geographic ancestry), that would still include hundreds of thousands of genetic differences. (Goldstein 2008, 9–10)

In addition, for studies of "our relatedness and genetic history" (Jobling and Tyler Smith 1995, 449) such as the Jewish Y-chromosome studies, the epistemic concept "human genetic diversity" means something quite different than it did to Darwin. Indeed, it indexes differences among groups rather than individuals—"races" of old, "populations" of late. Just as significant, the analytic significance of human genetic diversity is not Darwin's. Diversity is not scientifically important because it lends insight into understanding a living world *in time.* It is not important because it helps us to understand the *dynamics of* evolutionary change, although one can address issues of broader evolutionary interest from the perspective of specific historical studies. Instead, mapping genetic diversity is understood to provide a window into a history of the *longue durée.* In studying the origins and genealogies of contemporary population groups on the basis of DNA taken from living individuals, the quest for genetic diversity is actually a search for genetic signs that have *endured*:[16] haplotypes with restricted distributions, to quote Mark Thomas and his colleagues, may be "'signatures' of ancient connections [between specific populations] that have been partially obscured by subsequent mixing with other populations" (Thomas et al. 1998, 139).

A tension between stability and change stands at the very heart of modern biological theory. According to Evelyn Fox Keller, Darwin's concern with the mechanisms of transformation left a "fundamental mystery" in biological thought: "If change is the essence of life, how are we to account for the remarkable stability with which, in each generation, organisms develop and grow true to the type of their particular species, and with a certainty that endures over the lifetime of that species" (Keller 2000, 12)? How can we explain *stability*? Twentieth-century biology, Keller argues, focused on that task: it sought "to account for the persistence of individual traits through the genes" (13).

Anthropological genetics is beholden to the research efforts of geneticists who have sought to explain the mechanisms of stability and not just of change. As a field, it inhabits the tension between change and stability in human biology. For genetic history and relatedness to be traceable, variation must be detectable on the genome. At the same time, for phylogenetic trees to be "clean"—for origins to be *interpretable*—those mutations must be rare. They must occur *only once* over the historical time-spans in which researchers are interested. If not, there would be no way of determining whether two populations exhibiting a similar mutation are genealogically related or whether, as a result of the repeated occurrence of random genetic mutational events, they merely *appear to be.* This emerged as a problem for mtDNA phylogenetic analysis, for example.[17]

In the mid-1980s, researchers began to use mtDNA in the study of human origins and population histories precisely because it was a highly polymorphic site: given the levels of diversity evident in mtDNA, it was thought to be a good evidentiary source for tracking the origins of modern humans and the kinship and distance among contemporary population groups. Over time, however, mtDNA appeared to be *too* "variable," thereby leading population geneticists—including Luca Cavalli-Sforza, one of the father figures in the field—to turn to other evidentiary terrains (Stone and Lurquin 2005, 143). Mitochondrial DNA analysis was perhaps too complicated: with a mutation rate (a variability) far greater than was initially assumed, it was difficult to disentangle phylogeny from the phenomenon of "convergence"—"the apparently independent evolution of similar characteristics in different species, or in geographically isolated populations of the same species" (Woolfson 2004, 169).[18] Given the far less variable character of the Y-chromosome, by the late 1990s (human and population specific) paternity was believed by many researchers to be scientifically more knowable than is maternity.

Phylogenetic analysis requires "unique event polymorphisms" (UEPs): polymorphisms that are singular events, which can be used to root phylogenetic trees and to delineate their most fundamental branches—"deep splits in Y-chromosome genealogy," for example (Nebel et al. 2000, 630–31). More commonly occurring polymorphisms (mutations that occur more than once over relatively short time-spans, such as microsatellites on the Y-chromosome) can, against the background of UEPs, be used to delineate additional population divergences that reflect "more recent genealogical events" (ibid.). Against the background of UEPs, which are used to control for the phenomenon of convergence, in other words, microsatellites on the Y-chromosome (also known as short-tandem repeats or STRs) are used to sketch a more detailed history of the migrations and branching of specific human populations or subpopulations.

The nature of genomic variability is tied into a second fundamental issue of evidence in anthropological genetic work. Studying genetic history and relatedness relies, for the most part, on the *noncoding* regions of the human genome—that is, on genetic loci that do not encode proteins. And if one reason for the focus on noncoding regions is their increased variability in contrast to coding regions, there is a second reason I want to emphasize here: as genetic history's privileged working objects, noncoding regions are key to understanding how this discipline differs from—and how it generates its own distinction from—race science, the specter that hangs over its work and over all meta-discussions of contemporary research on population-based genetic diversity.

With respect to the Y-chromosome, genealogical questions have been assessed on the evidentiary terrain of what is colloquially referred to as "junk DNA." As Karl Skorecki explained in 2000 to an audience at the American Museum of Natural History, they are looking at "a set of markers on the Y-chromosome. . . . They are not in genes. . . . The Y-chromosome doesn't have many genes. . . . It is very useful in terms of the fact that it doesn't recombine, and has these *neutral markers* which don't really encode features or characteristics; however, it does serve as a tool in phylogeny," a "tool to uncover past histories."

Similarly, Vivian Moses, now Emeritus Director of the Centre for Genetic Anthropology at University College London, explained the importance of junk DNA at more length to an audience at the nineteenth annual convention of the Jewish Genealogical Society in 1999: "DNA is a coded message, written in [four] chemical letters. . . . These four letters are written in groups of three. . . . Buried in here is a message, a real message that you can really understand." Those "real messages," however, constitute only about 2 percent of DNA. The rest, "the parts without the message," is called "junk." Changes in "real message" DNA can have "deleterious consequences," and so are often evolutionarily selected against. They do not reliably survive from one generation to the next. "However, if you change a letter in the junk there is no consequence":

> The point about that change in the junk is that it is inherited by the progeny because . . . it doesn't matter whether you carry it or not. It's junk. . . . So . . . if somebody once upon a time, for some reason acquired such a change in their particular DNA, particularly if it is in the Y-chromosome of the male, then all the successive males descended from that chap will carry this particular change. . . . So, forever and a day, the person who carries that change will be marked. His progeny will be marked.

It is those enduring marks that anthropological genetic research projects find, categorize, and map. It is worth pointing out, however, that although "junk DNA" was long presumed to refer to genetic loci without significant biological function, that understanding of noncoding regions has changed radically over the past several years with the rise of post-genomic research. I discuss the shift in scientific consensus below in the postscript to this chapter. I now return briefly to scientific studies of Jewish descent in order to further explore the workings and implications of the epistemological commitments of this natural scientific field of historical research.

Threads to Antiquity[19]

Following upon the studies of priestly descent, various scholars launched additional Y-chromosome research projects into the historical and geographic origins of contemporary Jewry. If today's Jews are descendants of ancient Hebrew and Jewish communities who lived in and then fled ancient Palestine, the CMH and other Y-chromosome types shared by Jewish men must be closely related to other "Middle Eastern" Y-chromosome types, so the reasoning went. Contemporary Jewish populations, in other words, must be phylogenetically related to contemporary Arab populations.

Michael Hammer and colleagues published the first study comparing Jewish and "Middle Eastern non-Jewish" populations on the basis of Y-chromosome haplotypes. The paper begins with a historical account:

> Jewish religion and culture can be traced back to Semitic tribes that lived in the Middle East approximately 4,000 years ago. The Babylonian exile in 586 B.C. marked the beginning of major dispersals of Jewish populations from the Middle East and the development of various Jewish communities outside of present-day Israel. (Hammer et al. 2000, 6769)

Due to "numerous migrations of Jewish populations," a complex set of "genetic relationships" now exists among "the Jewish populations and their non-Jewish neighbors" (ibid.). As with previous genetic studies, Hammer's goal was to unravel "the numerous evolutionary factors"—common ancestry, genetic drift, natural selection, admixture—that have "come into play during the Diaspora": "Given the complex history of migration, can Jews be traced to a single Middle Eastern ancestry, or are present-day Jewish communities more closely related to non-Jewish populations from the same geographic area" (ibid.)? To re-word the question, are most contemporary Jews descendents of *converts to* Judaism or *descendents of a Hebrew population* that originated in ancient Palestine and founded the religion?

Genetic studies of Jewish populations thus far had not answered the question, Hammer explains. Some studies concluded that Jewish communities are more genetically similar to one another than to their host populations; others demonstrated "substantial non-Jewish admixture" (ibid.). Depending on the locus investigated, the degree (the mathematical calculation) of genetic similarity among Jews shifts: "This observation raises the possibility that variation associated with a given locus *has been influenced by natural selection*"(ibid., emphasis added). Recent genetic-genealogical studies, such as work on the non-recombining Y (NRY), have aimed to "circumvent some of the complications associated with selection" (ibid.). According to Hammer, in the current study "the DNA results . . . are less likely to be *biased* by selective effects" (ibid., 6773–74, emphasis added).

What precisely were the "complications" that Hammer and his colleagues were trying to circumvent? Why is selection understood as bias? In order to understand the specific evidentiary logic of genetic history, I turn to its similarities with and differences from race science. Specifically, how does each field understand the relationship between nature (or biology) and culture?

Racial Logics

Racial thought, as is well known, developed long before the rise of race science in the nineteenth century. Race emerged with modernity (in sixteenth-century Europe) when, as David Theo Goldberg has argued, "the forms exclusions could assume" changed radically from those that characterized the medieval Christian world (1993, 16). Race materialized as a fundamental technology of exclusion, which, like racism itself, is not a "singular and passing" concept or set of practices but a constantly transforming one (8).

Beginning in the late eighteenth century, naturalists—Linnaeus, Buffon, Blumenbach, to name a few key figures—generated classificatory schemes of human kinds, constitutive of an Enlightenment project that sought to find "order in nature" (Farber 2000; Banton 1978; Goldberg 1993; Stocking 1968, 1987). In the mid-to-late nineteenth century, race became a sustained object of scientific research: scholars conducted empirical studies that sought to quantify phenotypic differences that simultaneously characterized racial groups and were believed to generate the civilizational and cognitive distinctions between them. Researchers produced the truth of race through its measurability (see especially Gould 1981; also Banton 1978; Barkan 1992; Stepan 1982). As race science met evolutionary thinking, the classificatory scheme of racial "types" was maintained by specifying and by accumulating empirical evidence of those traits presumably not subject to environmental change (see Armelagos and Goodman 1998).

As is well documented, this scientific project was entangled with and formative of the practices of empire in the late nineteenth and early twentieth centuries, when both the discipline of race science and political projects of imperialism and settler-colonialism reached their zeniths. In addition, race science was central to the politics of various European and American nation-states as they grappled with the "problem" of immigration (of populations considered "non-White") and as they struggled to define and to demarcate the boundaries of inclusion and citizenship (Baker 1998; Haraway 1989; Goldberg 1993; Kevles 1985; Stocking 1968; Stoler 1995).

What is especially important for my argument is that in racial thought, as we all know, there were no clear distinctions between cultural and physical elements, between social and biological heredity (see Stocking 1968). Race theory was concerned with the "inherited capacity" of human groups (Barkan 1992, 187). The size of one's brain, the shape of one's head, for example, all those measurements for which anthropologists became famous, did far more than classify groups. They simultaneously signified and *explained* racial-cultural distinctions. During the first two decades of the twentieth century the focus of biological work turned inwards, but the significance of racial distinctions remained the same. Biologists (whose discipline had now become the center of this scientific quest) sought to isolate single genes, in Mendelian fashion, that coded for specific characteristics—height, color, intelligence (Barkan 1992, 5). As articulated in the science of eugenics, there was "a direct translation between superior genes and superior culture" (Marks 2002, 356). Nature and not nurture formed the foundation of race science as it sought to distinguish human groups and to place them along an evolutionary grid.

Analyzing the noncoding regions of the human genome, so-called junk DNA, does a different kind of work. It re-imagines the relationship between culture and biology and generates a different kind of history and self to be unraveled on the terrain of biological evidence.

The Meaning of Junk

For many researchers interested in reconstructing population-based histories—and for genetic ancestry-testing companies (see chapter 4)—there is an effort to create a firewall between questions of culture, ability, or behavior and questions of nature. They insist that their work is a matter of tracing *descent;* this is nothing more than a *mark,* and it has no bearing on the question of inherited characteristics. *This is not race science,* in other words. Genetic history—efforts to reconstruct population-based genetic histories—is a quest for "inheritance" stripped of the question of "capacity," an inheritance, moreover, no longer visible on the body itself. These are haplotypes that have no meaning—no function—in the biological domain. They are *neutral* markers. This is junk DNA.

Much has been written about the significance of the shift in biology from a science that conceived of organisms as organic and integrated systems to one that imagines life as a code. As Evelyn Fox Keller has argued, geneticists (often physicists who migrated into biology) adopted the cybernetic term "information" in order to analyze DNA and gene action, even if they used it in a more colloquial than technical sense. (In cybernetics "information" has a purely quantitative connotation. In molecular biology, however, "meaning" cannot be ignored if genetic information is to have anything to do with life [Keller 2000].) According to Donna Haraway, "The organism has been translated into a problem of genetic coding and read-out. . . . In a sense, organisms have ceased to exist as objects of knowledge, giving way to biotic components, i.e., special kinds of information-processing devices" (1991, 164).[20] The information metaphor, in other words, has radically transformed the meaning of life.

Nearly 98 percent of that genome, however, was long thought to be a code that does not code for any biological function at all. What happens to our understanding of that part of the code when one shifts research domains? In particular, what happens to the analytic significance of the genome's presumably noncoding regions in the context of anthropological genetic research on population-specific origins?

For researchers in genetic history to refer to these markers as either "neutral" or as "junk" is certainly a misnomer. But that misnomer all the difference makes: it helps to fashion the distinction between contemporary genetic historical inquiries and race science. Vis-à-vis the functioning of organisms, vis-à-vis the mechanisms of evolution, vis-à-vis phenotypic effects, these so-called neutral markers are read as having no genetic function. They carry no biological information that can be either read or rewritten through biotechnological interventions. Junk DNA is the residue not just of random but of biologically and selectively irrelevant evolutionary events. To borrow the terms in which François Jacob wrote about the dual nature of a computer program, neutral markers are carriers of "memory" (the traits of the parents, here purely *genotypic* traits) minus the capacity for "design" (the program's ability to control the formation of the organism and to determine its specific traits). As neutral markers, junk DNA cannot generate cultural, behavioral, or biologically consequential differences between human groups. This is not quite biology. It is not quite nature.

Nevertheless, researchers consider junk DNA deeply meaningful in a historical register. These haplotypes may not code for disease or behavior or phenotype, but they do carry information (they carry *meaning)* regarding ancestry, at least from the interpretive perspective of researchers seeking to establish points of geographic origin and specific lines of descent. According to Spencer Wells, director of National Geographic's Genographic Project, the largest ongoing project in anthropological genetics that seeks to collect and conserve a record of contemporary human genetic diversity and to plot the "history" of humanity, what geneticist's call "'junk-DNA' . . . is anything but junk to those of us who use the genome as a historical document. This is our text, and it provides us with the story of our ancestors" (2006, 15). And as explained by Brian Sykes, Professor of Genetics at Oxford University and the founder of Oxford Ancestors, a genetic ancestry-testing company:

> Mutations can be good, bad or indifferent. . . . Indifferent mutations, and they are in the majority, have no influence one way or the other on survival or success in breeding. They just get passed from one generation to the next, their fate entirely out of their hands. They risk elimination if they end up in someone who has no children or can do well if they find themselves in a large family. They might lead less dramatic lives than the mutations that bring success or devastation. But it is these, the silent passengers of evolution, that are its most articulate chroniclers. (Sykes 2006, 95–96)

The significance of these "indifferent" mutations, for my purposes, is that they configure a novel relationship between biology and culture. They refashion the biology-culture nexus that stood at the heart of race science. Consider the opening paragraph of Thomas and Skorecki's 1998 *Nature* article:

> According to Jewish tradition, following the Exodus from Egypt . . . male descendants of Aaron were selected to serve as Priests (Cohanim). *To the extent that patrilineal inheritance has been followed since sometime around the Temple period* (roughly 3,000–2,000 years before present), Y chromosomes of present-day Cohanim . . . should not only be distinguishable from those of other Jews, but—given the dispersion of the priesthood following the Temple's destruction—they should derive from a common ancestral type no more recently than the temple period. (1998, 138, emphasis added)

One can *infer behavior from DNA evidence*. If the oral tradition of passing the priesthood from father to son has been adhered to in practice, then, in the words of Vivian Moses, "from the Y chromosome point of view the descendants of Aaron form a progressively separate group through the ages." This "biology" indexes a particular set of cultural practices. The causal relationship is inverted: causal mechanisms move *from* culture *to* biology. In turn, one can assess the plausibility or "truth" of culture—of oral tradition, of religious and kinship practices—from biological data. In Karl Skorecki's words, "The [Cohanim] study suggests that a 3,000 year old oral tradition was correct or had a biological component" (quoted in Grady 1997a). Junk DNA is natural-cultural artifact that carries a genealogical message bearing witness to one's geographic origins and cultural past. It functions as evidence of what one might call cultural fidelity—of the fact that contemporary, self-designated Jews really do descend from a single ancient population, from a common history and long tradition of cultural distinction that is visible on the Y-chromosome only because their (male) ancestors remained true to their faith. Y-chromosome markers are "signatures" of ancient origins (Thomas 1998, 139). Such markers are not, by way of contrast, evidence of the "biological unity" of the Jews, a concept central to racial theories of Jewishness that dominated late nineteenth- and early twentieth-century thought.

Much has been written about race science and the Jews—mostly about non-Jewish European biological and anthropological scholars who investigated the biology of the Jews and found not just biological distinction but biological inferiority. What is perhaps less well known is the work of Jewish scientists on the biology of Jews. By the late nineteenth century, Jewish intellectuals and political figures (many but not all of them Zionists) embraced race science. Seeking to counter the politics and intellectual tradition of the *Wissenschaft des Judentums*, which they understood as too assimilationist, these scientists sought to redefine Jewishness "in biological and anthropological terms" (Efron 1994, 8). In so doing, figures such as Max Nordau and Nathan Salaman argued that "Jews were united by more than a common faith and a shared history; their identity was rooted in a fundamental physical, biological 'Jewishness,' which bound them—despite time and space—to one another and to their ancestral home" (Hart 2000, 193).

Embracing a Lamarckian perspective on the inheritance of racial traits—that is, the notion that the environment could create particular traits during a lifetime and that those traits would then be passed on to the next generation—these scientists believed in the "Jewish problem." But in their racial paradigm, Jewish "degeneration" was a consequence of historical and environmental circumstances and not of permanent racial traits: it was a consequence of years of living in exile and under difficult conditions. For the Zionists among them, only a return to Palestine would rejuvenate the Jewish race. In other words, for an influential strand of Zionist thinkers, the biology of the Jews was both a fact and in a state of decline. Only a return to Palestine (*Eretz Yisrael*) would redress that degeneration.

In contrast to the work of (Jewish) scientists a century or more ago, biology is not something additional in the contemporary logic of genetic history. It does not index a (Jewish) racial unity or identity *in excess of* a shared history and religion. Instead it is via "biological," noncoding genetic evidence that one can uncover history itself, that one can demonstrate that historical traditions may well be true.

The problem with selection, the "bias" introduced by coding regions of the human genome, to return to the Michael Hammer paper, has to do with the problem of biology, strictly understood. By focusing on junk DNA, scientists can control for the biological dynamic of natural selection whereby groups of self-identified Jewish men in different regions of the world become progressively more genetically different from one another and more similar to the populations with which they reside, not because of "admixture" (which, along the paternal axis, is what is being investigated), but because of selective advantage. Certain *genes* (as distinct from noncoding regions) are more biologically fit in certain environments and thus survive and proliferate.

The concept of biological fitness has been central to challenging the significance of the classic phenotypic traits of race science as saying something meaningful about the differences between human groups. Luca Cavalli-Sforza, a leader in the field of population genetics, has argued that racism is "a fallacy" because although genetics is "instrumental in shaping us," so too are "the cultural, social and physical environments in which we live" (2000, viii). Moreover, the genetic differences between populations or "so-called 'races'" are small. They are "*superficial*," "attributable mostly to responses to the different climates in which we live" (ibid.). Cavalli-Sforza takes skin pigmentation as his example and argues that although race scientists understood skin pigmentation to be important (because it was assumed to be a *permanent and thus a racial* trait), it is actually insignificant.[21] It is adaptive. Minus the influence of environment, the color of one's skin does not tell us anything truly important about the differences between population groups.[22]

Consider how the language of admixture and superficiality plays out in the following exchange between Karl Skorecki and a member of the audience during the session at the American Museum of Natural History in New York. A woman commented that she was very glad that "genetics has put to rest the question of race, which is clearly a social construct." She asked: "How can you claim that Jewish communities are really endogamous and don't intermarry given that when you go to different parts of the world, Jewish population groups resemble the population groups with which they reside?" Skorecki responded:

> That is an excellent question. . . . If there is common ancestry in the Jewish Diaspora from India to Eastern Europe to Northern Africa, what explains the phenotypic differences and the closeness in those appearances to the local, and not to Jewish communities? The answer? Recombination and admix-

> ture. This work that has enabled tracing back common ancestry has been based either on the Y-chromosome or on mitochondrial DNA. Those are two segments of the genome that don't recombine. . . . [They] don't swap genetic material with the other parental contribution. Therefore they maintain *a fidelity, a trace that goes back in history*, and therefore they are less affected by admixture. . . . But if you study admixture and its effects on phenotypes, a little bit of admixture goes a long way in a few generations. It doesn't take much admixture to have a change of phenotypic appearance and that also speaks to the issue of how superficial those phenotypic differences are. *Even though it looks striking to the eye, it doesn't reflect something very deep, and one can see through it by using non-recombining regions of the genome.* (emphasis added)

The goal of genetic historical research—of phylogenetic reconstruction—is precisely to "see through" all the genetic noise, the noise of recombination, of admixture, and of natural selection in order to uncover a truth about origins. Paralleling the logic of permanent racial traits, noncoding regions—precisely because they are not under selection—enable scientists to disentangle the origins and phylogeny of one population from that of another. (It is worth repeating that in contrast to presumably stable racial traits, noncoding haplotypes are understood to have no phenotypic significance, and especially in terms of cultural or cognitive effects.) In search of "ancestry," genetic markers that would necessarily render Jewish populations genetically *more similar to* "the local populations" with whom they have long lived are sidelined in favor of the search for an enduring trace believed to embody a history of the truly *longue durée*. And that historical *longue durée* is a distinctly internal matter. It is a genetic sign carried *within* the body, throughout time. Presumably, it remains discoverable regardless of what is added from without—via recombination, selection, and admixture.[23]

Jewish Origins

In order to resolve the question of Jewish (male) origins, Hammer and his colleagues (2000) compared the Y-chromosomes of Jewish and non-Jewish men, drawing on a sample of 1,371 men. More specifically, "the Jews," a population subdivided into a series of subpopulations (Ashkenazi, Roman, North African, Near Eastern, Kurdish, Yemenite, Ethiopian) were compared with (a) Middle Eastern non-Jews (Palestinians, Syrians, Lebanese, Druze, Saudi Arabians), and (b) Europeans, North Africans, Sub-Saharan Africans, and "Others" (Turks and the Lemba). The researchers wanted to measure "genetic distance"—the statistical rates of differences in haplotype frequencies—first, among Jewish subpopulations and, second, between Jewish and non-Jewish populations. Surveying eighteen biallelic polymorphisms in the above populations, researchers found thirteen represented in Jewish populations, eight at frequencies ≥5 percent. (The more commonly occurring polymorphisms are considered "diagnostic haplotypes"; see table in Hammer et al. 2000, 6773.)

There are two important findings for the purpose of establishing origin and kinship among today's Jews. First, Hammer and his colleagues conclude: "Diaspora Jews from Europe, Northwest Africa, and the Near East resemble each other more closely than they resemble their non-Jewish neighbors," despite what would be expected given their temporal and geographic separation.[24] "It is of particular interest," they write, "that the level of divergence among Jewish populations was low despite their high degree of geographic dispersion," the mean geographic distance among the six Jewish subpopulations being ~3,000 km (2000, 6771–72).[25] Second, they conclude that the Jewish population is closely related to non-Jewish Middle Eastern populations: the "Jewish cluster [minus Ethiopian Jews] was interspersed with the Palestinian and Syrian populations, whereas the other Middle Eastern non-Jewish populations (Saudi Arabian, Lebanese, and Druze) closely surrounded it" (6772). According to Hammer and his colleagues, "A Middle Eastern origin of the Jewish gene pool is generally assumed because of the detailed documentation of Jewish history and religion" (6773). On the basis of this study, "Several lines of evidence support the hypothesis that Diaspora Jews from Europe, Northwest Africa and the Near East resemble each other more closely than they resemble their non-Jewish neighbors" (6773). Common ancestry "is the major determinant of the genetic distance observed among Jewish communities"—and not genetic drift or gene flow (admixture), each of which would have increased genetic distances among Jewish populations. (In addition, admixture would have decreased the genetic distance between particular Jewish and non-Jewish populations [6774].) Moreover, given the genetic kinship of the Jewish and non-Jewish Middle Eastern samples, a "Middle Eastern origin is supported." From the perspective of the descent of men, the Jewish diaspora derives "originally" from a common ancient Middle Eastern gene pool.[26] The Y-chromosome evidence is "consistent with" the traditional story of Jewish diaspora. Relying on noncoding markers on the Y-chromosome, Hammer and his colleagues have pried history out of biological data and demonstrated that Jewish communities have maintained their distinctiveness throughout generations in the diaspora, as seen here in the practice of passing "Jewishness," and not just Cohen status, from fathers to sons.

The Politics of Epistemic Things

According to a long accepted historical narrative about the fall of race as a scientifically valid category, the shift from "race" to "population" occurred in the wake of World War II and the horrors perpetrated by the Nazi regime. Typological thinking was replaced by statistical thinking, the differences between human groups were framed as relative ("clinal"), and the scientific project shifted from classifying human groups to understanding the processes and dynamics of biological evolution of the "family of man" (Haraway 1989, 1997b; see also Banton 1977, Stepan 1982; for an earlier date for the transition, which ascribes it more to developments internal to the scientific field see Barkan 1992). Moreover, in contrast to race science's longstanding epistemic commitment to the belief that race-crossing led to degeneration, gene flow between populations became both biologically normative and the evidentiary terrain for studying the history of human origins, migrations, and kinship. Neither biologists nor physical anthropologists ordered human groups along an evolutionary grid. The race concept moved into the social domain, as this historical narrative tells it, persisting as politically powerful even though it was no longer authorized by the biological sciences.

Recent scholarship has challenged the sharp demarcation between race science and population thinking drawn in the historical account I sketched above. The race concept, some scholars have shown, was never abandoned. It was not even abandoned by the UNESCO Statements on Race, documents read as iconic of the scientific abandonment of race (see for example, Haraway 1989, 1997b). Jenny Reardon, for example, points out that the first UNESCO Statement on Race (drafted in 1951) accepted the reality of race as a *biological* category even as it denied that racial characteristics have anything to say about "meaningful" human differences ("mental characteristics" or "cultures and cultural achievements" (UNESCO 1952a, 100; Reardon 2001). By way of contrast, the second Statement (drafted in 1952) not only asserted the biological reality of race. It entertained the possibility that there may turn out to be meaningful human differences that occur at different statistical distributions in different races, even if we do not know of any yet (Reardon 2001; UNESCO 1952b). Other scholars have argued that the shift from typology to gene frequencies and probabilities did not necessarily free population genetics from the political and ethical limits of race science: there is no reason why racism requires typological distinctions between human kinds (Gannett 2001, 2004).

This recent rethinking of the relationship between population genetics and race science has been an important intervention into a literature that obscured continuities in scientific practices from the early twentieth century to today. Things *have* changed scientifically. Nevertheless as recent critical accounts illustrate, there is far less of a radical disjuncture in scientific epistemologies than was long presumed. Following World War II, race endured as an object of scientific study, even if its parameters and claims shifted, perhaps even in significant ways.

In what follows, I build on that re-reading of the transition from race science to population genetics in order to push its argument both further and in a somewhat different direction. No, the commitment to race as a scientifically valid category did not disappear. But instead of demonstrating the persistence of race as an epistemic thing, I trace the emergence of the neutral or "meaningless" marker, which is essential to understanding the science and the politics of population genetics in the postwar world. I present two distinct albeit overlapping origin stories of this new working object for students of human genetic diversity, one social and one scientific, and I explore the political and ethical work the neutral marker did. As I demonstrate, contemporary understandings of the apparent social neutrality of junk DNA echo an earlier discursive move. In short, there is now a long history of the search for neutral genetic markers—for signs of "evolution" or "history" *tout court* (even if those genetic markers turn out not to have been quite so neutral after all)—and that search has always been driven by epistemological commitments and political desires at one and the same time.

Scientific reevaluations of the race concept began far earlier than much of the historical literature on race science long recognized. It was not in the 1950s that scientific reevaluations started in earnest. By the 1930s the project of racial classification was collapsing under its own evidentiary morass, and the concomitant rise of molecular biology was beginning to provide a new framework for thinking about the diversity of human populations (see especially Barkan 1992; Kay 1993; Hutton 2005).

The 1950s was nevertheless marked by a significant shift for the biological sciences, but that shift was political far more than epistemological. In the wake of the Nazi genocide, biological studies of human populations were haunted by the violence carried out under the sign of race, and biological scientists were called upon to bear responsibility for that history. As argued in *The Race Concept*, which UNESCO published in 1952, "Since the beginning of the nineteenth century, the racial problem has been growing in importance. A bare 30 years ago, Europeans could still regard race prejudice as a phenomenon that only affected areas on the margin of civilization, or continents other than their own. They suffered a sudden and rude awakening" (UNESCO 1952a, 5). UNESCO took up the task "to study and collect scientific materials concerning questions of race; to give wide diffusion to the scientific information collected; to prepare an educational campaign based on this information" (6). And it did so "as the international institution best equipped to lead the campaign against race prejudice and to extirpate this most dangerous of doctrines" (5).

The UNESCO Statements on Race did not deny the reality of race. Instead, as Jenny Reardon demonstrated, the discursive logic of these documents was to assert the authority of *scientists* over race as an object of scientific scrutiny: scientists and not politicians or publics would be authorized to speak about race. And in saving the race concept for science, the drafters of the first Statement on Race sought to cordon off "hereditary particles (genes) or physical characters" (UNESCO 1952a, 99) from any socially significant qualities. Genes and physical characteristics can demarcate (although never in absolute terms) different populations, the document argued. Human groups do exhibit "certain physical distinctions," which groups have "by virtue of the isolating barriers which in the past kept them more or less separated" (98). But those physical characteristics say nothing about mental abilities or cultural achievements, about what were described as "socially significant traits." The second UNESCO Statement on Race was drafted in large part to respond to the claim that such traits are never socially significant. Its authors questioned whether or not scientists could be sure that they would never find any socially meaningful biological differences among races (Reardon 2005). But they did so with an excess of caution: "it is possible, though not proved" that "some types of innate capacity for intellectual and emotional responses are commoner in one group than in another"; even if true, individual variation within populations will remain paramount. Yes, genetics *may have* a role but so too does environment. It is always a dual process, one in which environment seems to be far more significant. Moreover, the second Statement insisted, the "ethical principles" of "equality of opportunity and equality in law in no way depend . . . upon the assertion that human beings are in fact equal in endowment" (1952b, 85). Replicating the ethical consciousness of the 1951 statement, the authors of the 1952 Statement on Race argued that, "having regard to the limitations of our present knowledge, all of us believed that the biological differences found amongst human racial groups can in no case justify the views of racial inequality which have been based on ignorance and prejudice, and that all of the differences which we know can well be disregarded for all ethical human purposes" (82).[27]

Ethical principles and political anxieties underwrote the claim that physical characteristics were distinct from meaningful—i.e., from so-called *socially significant*—human traits. But there also is a second story to be told here: the emergence of the biologically and evolutionarily "meaningless" marker as a working object in the biological sciences. That story is best understood from the perspective of competing scientific paradigms and professional or disciplinary disputes—in science studies terms, political struggles internal to the scientific fields (see Sommer 2008; Diaz 2007).

In the aftermath of the Darwinian revolution of the nineteenth century, a key project for twentieth-century evolutionary biology was the elaboration of phylogenies. Phylogenies were mapped on the basis of specific morphological traits, which were read in comparison with the traits of other species and the fossil record. By the early 1960s, however, evolutionary biologists found themselves facing a sustained scientific challenge. Studying evolution at the molecular level was now on the table. Over the next decade, advocates of a molecular approach would question the forms of evidence and the epistemological assumptions of the "modern evolutionary synthesis"—the "shift toward 'population thinking,' [that treated] . . . species as evolving lineages or groups of potentially interbreeding individuals" that characterized evolutionary biology at the time (Hagen 1999, 329).

The idea that one could study evolution through the "molecules of living beings" was not new in the 1960s. In the 1950s, serological techniques were widely used to study blood group distributions among populations in attempts to reconstruct their phylogenies (Diaz 2007, 652). Nevertheless, by the end of the 1950s there was a qualitative shift. Key scientific figures adopted the "most recent molecular techniques centered on the analysis of . . . protein and nucleic acids," and they did so in the name of founding a new field, "molecular evolution" (653). Powerful academics in evolutionary biology (Ernst Mayr, Theodosius Dobzhansky, and George. G. Simpson) defended their morphological approach against the molecular evolutionists.

This is not the place to narrate the history of that conflict or of the victory, by the mid-1970s, of many of the assumptions that the early advocates of molecular evolution proposed. Instead I want to recount but one aspect of that history of science: the emergence of "neutral evolution" as an epistemic thing.

The molecular approach to studying evolution presumed "a natural phylogeny preserved in DNA and proteins" (Sommer 2008, 477). It drew on a distinction between "evolutionary history in the raw" and "environmental 'distortion'" (478). Macromolecules, Emile Zuckerkandl and Linus Pauling insisted, were superior to morphological traits for the purposes of phylogenetic analysis because they allowed scientists to get closer to evolutionary history in the raw (478). In 1962, Zuckerkandl named this new approach to studying primate phylogeny and human evolution "molecular anthropology" (478). Moreover, in that same year Zuckerkandl and Pauling co-authored a paper that argued that it is not just organisms that evolve, so too do proteins. As such, one should be able to track the evolution of organisms by looking at the protein level: ancient kinship should be visible in the protein record, even if the groups of organisms have since diverged one from another (480). And in hypothesizing a rate of change specific to the molecular level, Zuckerkandl and Pauling proposed the possibility—the likelihood, as they saw it—that "there might be (nearly) indifferent substitutions that were selectively neutral" (488).

It was not until the late 1960s that the theory of neutral evolution was fully formulated. As articulated by Motoo Kimura in 1968, the neutral theory of evolution maintained that most changes at the molecular level are neutral, that is, they are neither selectively advantageous nor deleterious. Molecular change is for the most part a matter of random "genetic drift" (Dietrich 1994, 22). Evolution, as Kimura understood it, is a "mere physical statistical phenomenon at the molecular level." A "clock"—an "internal constant-rate mutational process"—regulates molecular evolution, which does not occur in an "irregular or a specifically adaptive way" (Diaz 2007, 662), as had long been assumed.

The hypothesis of neutral evolution flew in the face of the assumptions of the now "classical" evolutionary biology. According to Michael R. Dietrich, in 1964 when the first major conferences on molecular evolution were held, most evolutionary biologists believed that natural selection was the "dominant and most important mechanism of biological evolution" (1994, 21). They understood evolutionary biology to be a *historical* science, not a mathematical one. Evolution, as George Simpson, for example, argued was "contingent on environmental conditions that would never be repeated" (Diaz 2007, 661). By the 1970s, however, the neutral theory of evolution had gained wide scientific acceptance. And by the 1980s and 1990s, the theory that most evolutionary change at the molecular level is both random and "functionally insignificant" (52) emerged as a basic tenet of research programs in anthropological genetics.

The evidentiary terrain of neutral evolution was not just epistemologically significant. It is not just, as its advocates argued, that one could more easily "see" phylogenetic relationships through its lens. It was and it remains politically crucial. I am not arguing that the development of the theory of neutral evolution was politically motivated. It emerged from techniques, arguments, and professional battles within the biological sciences. It was the product of the molecular revolution that swept the biological sciences in the 1950s and 1960s. Nevertheless, the neutral theory of evolution forged an "ethical space" not previously available to studies of intrahuman phylogenies.

Highlighting the fact that this is a difference that makes no *functional* difference is, as I mentioned above, crucial to maintaining a distinction between genetic historical projects and race science. By insisting on the noncoding nature of the NRY (the non-recombining section of the Y-chromosome), scientists distinguish their field from race science, a discipline they do not wish to reinstantiate and yet the specter of which is continually raised by their work (see, for example, Reardon 2005; Gannett 2001).[28] In insisting on the functional neutrality of these markers, scientists are asserting the insignificance of *these* biological facts in a language reminiscent of that of the drafters of the 1951 UNESCO Statement on Race: there *are* biological differences between human population groups, but as noncoding markers these are not signs of *meaningful* (read, "social") human traits—intelligence, personality, culture (see Reardon 2005). Phylogenetic facts are thus cordoned off from the political implications and the ethical burdens of race science and racial difference.

Consider the following two comments from interviews with scholars involved in research on Jewish origins. One researcher told me he would answer the question of whether or not genetic-historical research is "objectively valuable" differently today (Fall 2004) than he would have a year and a half earlier. Why? "Before it was a mathematical thing—on junk lines. If [it is] not just segregating on junk lines but if [differences between populations] are biological features as well [as was becoming increasingly evident in human genomics, he argued], then it becomes a different kind of thing. So it is still useful in terms of telling us something about the origins of groups, but now it is more controversial, more damaging," that is, more damaging than finding differences segregating along junk lines. As articulated by a second researcher, once one starts analyzing coding areas of the genome that are linked to behavior, there is a potential problem. Take the MAOA gene, for example, "the locus for aggressive and violent behavior—or rather, locus of impulse control." (He explained that it is a Mendelian monogenic disorder carried on the Y-chromosome. Men who carry the mutation have no impulse control.) "What would happen if researchers find different frequencies of this haplotype in different parts of the world?" Coding regions carry far more dangers that do noncoding regions, he told me. They raise "ethical" questions.

Meaningful Human Collectivities

In his book *French DNA*, Paul Rabinow writes about the specific "life-forms and forms of life" that characterize the genomic age. In contemporary bioethics, Rabinow argues, "what is taken to be at issue is the fate of humanity, not only in the material sense (as some environmentalists see it) but precisely in a spiritual one." There is a palpable anxiety vis-à-vis the need to protect human dignity, the "symbol enshrined in the Universal Declaration of Human Rights as the bulwark against the justification of any future Auschwitz." According to Rabinow, "What is at issue today is neither docile bodies nor sinful souls (which is not to say that such states have disappeared). What is at issue are assemblages in the process of re-formation, ones that seek to bring together health and identity, wealth and sovereignty, knowledge and values" (1999, 12). In that matrix of new assemblages, Rabinow elaborates that "the biotechnology industry, the growing stock of genomic information, and the simple but versatile and potent manipulative tools" are key elements (13). But they are only one part of a larger set of transformations. We are also witnessing a transformation in conceptions of the self, collective and individual: we are witnessing the emergence of "biosociality"—a biologicization of identity that is "different from the older biological categories of the West (gender, age, race)." Biosocial identities are "understood as inherently manipulable and re-formable" (13), Rabinow argues. He then cautions (a caution often lost in subsequent invocations of the notion of "biosociality") that older biological categories of the West will nevertheless endure, even if not in identical terms (see Rabinow 1996).

The practices of genetic history that I have discussed in this chapter certainly reinforce or "buttress" (Brodwin 2002, 326) already existing social identities, groups long identified in the language of race. But they do more than that. They generate something additional. They can refashion categories of identity (although in the case of Jewish origins the category itself endures even if which groups are recognized as belonging or as central has shifted over time). And they certainly reconfigure fundamental conceptions of collective and individual selves.[29] Genetic historical quests generate not just substantive claims about group identity. They generate ontological claims about the nature of human subjects: about what at the level of "background" makes us who we are, about what it is that *can signify*—that *authenticates or makes true*—our human collectivities.[30] The question we must ask then is not just *which* human collectivities the practices of genetic history or any other post-genomic science makes or reinforces. We also need to explore *what it is* in this scientific imagination that makes a human collectivity and that makes it *meaningful and enduring*.

As I have argued in this chapter, within the epistemic grammar of anthropological genetics, a population emerges, acquires genetic "characteristics," and develops an identifiable trace through a succession of evolutionary events in tandem with the cultural practices human actors sustain. It is the repeated *decision* to adhere to kinship practices, to pass the Cohen lineage from fathers to sons, that scientists reveal through their analyses of neutral markers or junk DNA. If in the design of these studies what makes Jews a meaningful population at the outset is self-designation, what makes Jews a meaningful population at the end is the fact that they share particular Y-chromosome haplotypes, and those haplotypes are visible only because Jewish men chose over and over again to pass priestly status and/or Jewishness down the paternal line.[31] Those haplotypes are what substantiate a Jewish collectivity as a "population" in a genetic historical epistemology. And in substantiating a Jewish collectivity, those haplotypes become a way of knowing Jewish *history*.

If we are to analyze the social and political entanglements and the emerging meanings and effects of genetic historical work, we need to take seriously that this is a scientific practice that, while looking at "biological" data, is hoping to see "historical" signs. And we need to specify, within the field's epistemic grammar, what those historical signs are taken to be evidence of. To place anthropological genetics in a simple continuum with race science would be to fail to understand the specificity of this scientific field. Moreover, as I discuss at length in chapters 3 and 4, it would fail to explicate why it is that genetic history—whether scientific research on population-specific origins or commercial ventures that sell individual genetic ancestry tests—are so socially felicitous today.

Anthropological genetics presumes the existence of a bodily memory. The memory believed to be stored in a genomic archive is not just a record of biological processes and facts, however. Random genetic mutations are read as records of *human consciousness*, and as records of cultural collectivities and of fidelity to kinship practices and religious traditions. If human differences are written into our biology, it is the difference *of culture* (however reduced) that is sought and that is sometimes, apparently, found.

History is not understood to be a record of "the destiny of societies and peoples," as it was for race science (Balibar 1990, 287). Insofar as history consists of a record of the genetic origins and relatedness of identifiable "populations," it is made by the actions, the cultural traditions, expectations, and practices, and the religious choices of men and women (in the case of mtDNA) who have lived before us. And in sustaining cultural traditions and making particular choices, so too did those men and women end up "making-up" themselves as members of a meaningful and *now scientifically legible* human collectivity (Hacking 1986).

Postscript

By the late 1970s it had already become clear that only a small proportion of DNA in higher organisms actually coded for proteins, the standard definition of a gene. In 1977, scientists reported evidence that the coding genes of higher organisms varied from those of bacteria in a significant way: they were discontinuous; "most structural genes in higher organisms are broken up into stretches of nucleotides that code for amino acids and stretches that do not code for anything. The 'silent,' noncoding regions were called 'intervening sequences' or 'introns'" (Wright 1986, 326), regions that came to be referred to as "junk."

Over 95 percent of the human genome is made up of nucleotide sequences that do not encode proteins. And for a long time that junk was presumed to have no biological function. The reason why the "idea that junk was uninteresting," Adrian Woolfson explains, is an artifact of the history of genetics: having focused their research on "simple organisms" in the 1950s and 1960s, specifically on bacteria which are comprised of "wall to wall genes and contain little junk," geneticists had become used to focusing on the protein-encoding sequences. When noncoding sequences were discovered in humans and other higher organisms, it "was dismissed as detritus" (Woolfson 2004, 106). The extent of that disinterest in the junk was evident during initial debates regarding the design for Human Genome Project. Some researchers argued that it would be a waste of time and money to sequence the junk. That view did not prevail.

The completion of the human genome project in 2003 raised serious questions about the previously reigning assumption regarding junk DNA: the number of genes that code for proteins in humans does not differ radically from the number in simpler organisms. Geneticists initially assumed that the human genome would contain approximately 100,000 genes. As it turns out, the number is far less. By 2001 the estimate was 30,000, "only about twice as many as in a worm or fly" (Lander et al. 2001, 860), and the estimate continues to drop. Moreover, the number of proteins generated by coding regions is far greater than the number of identified genes.

As researchers in post-genomics turn to understand gene expression, the assumption that noncoding regions are "junk" is increasingly being proven wrong. For example, noncoding regions provide the mechanism for the regulation of protein production; they hold chromosomes "together," thus maintaining the genome's structural integrity; and they shuffle or "reprogram" genetic information in ways that lead to the evolution of organisms and species (see Woolfson 2004; also McDonald 2005). As Barry Barnes and John Dupré write, "whatever else, junk DNA functions as part of the physical material of the chromosomes; and this gives it a role in defining the 'space in which genes happen' and of locating other pieces of genomic DNA within that space and separating them from each other therein" (2008, 82). Moreover, noncoding regions are affected by natural selection. "Junk DNA" is biologically important stuff.

In light of emerging technological capacities and shifting understandings of the genome as a whole, newer techniques, such as genome-wide surveys, are being used in projects to reconstruct population origins. Genome-wide surveys target a much larger array of single-nucleotide polymorphisms (SNPs) across the genome, and they examine both coding and noncoding DNA. Recent studies of Jewish origins, for example, the Jewish HapMap project, have used this technique (Atzmon et al. 2010). Nevertheless, the epistemological struggle remains the same: how to separate out ancestry from selection, even if as post-genomic understandings of the genome become more and more complex, some coding regions may turn out to be indicative of the former and some noncoding regions indicative of the latter.

What effect the growing consensus about the biological significance of the genome's noncoding regions will have on the evidentiary assumptions and the scientific practices of anthropological genetic research is yet to be seen. As one researcher told me in an interview, junk DNA *is* subject to natural selection. The key question is, *how much?*

CHAPTER TWO

What Are the Jews?

Not long after the founding of the State of Israel, Israeli researchers began studying the state's Jewish population. They collected data on Jewish immigrants to Israel, along with some data on Arabs and Druze.[1] The oft-stated goal was to take this "unique opportunity" to study the different Jewish "communities" (*edot* in Hebrew) before their assimilation into Israeli society made such a scientific endeavor no longer possible. Alongside their Israeli colleagues, human geneticists from the United States and Europe argued that the recently founded Jewish state presented an opportunity that should not be missed. As asserted many times during a 1961 international conference in Jerusalem on "The Genetics of Migrant and Isolate Populations," Israel was a "perfect laboratory" for studies of human genetics: it was said that the state had good medical records on its population, despite the fact that the state was little more than a decade old. Israel's was a population whose history was, researchers argued, "well known." And it was a country of "well-defined *populations of different origins*, some of which still exist as isolates" (Goldschmidt 1963, 8, emphasis added). Specifically, the study of Israel's Jewish population(s), researchers believed, would contribute to understanding the general processes involved in short-term human evolution (see Falk 1998, 2006b). In this work of human population genetics, the Jews of Israel were treated simultaneously as a single population with a shared (ancient) origin and as a collection of subpopulations who had migrated to Israel from various points of (more recent) origin.

In the aftermath of World War II, leaders of the Euro-American discipline of human population genetics, many scholars have argued, took up the mantle of studying the "Family of Man" (Haraway 1989; Steichen 1955). In contrast to race science's focus on biological difference, as a long standard historical narrative told it, population geneticists asserted the biological unity of the human species. That presumed unity formed the epistemological and political grounds for studying human evolution and the dynamics of biological diversity. From the perspective of the international field of population genetics, Israel's Jewish citizenry was regarded as an ideal population in its effort to find diversity within the unity of the human species. Believed to be a largely endogamous community and set of subcommunities, Israel's Jewish population was presumed to share a relatively recent historical origin. Now gathered in a single place, their unity *and* diversity (evidence of short-term human evolution) could be studied at one and the same time. As indicated by the fact that the Rockefeller and Ford Foundations funded much of the work by Israeli researchers in the 1950s and 1960s, interest in their work clearly traveled well beyond the boundaries of the newly founded state.[2]

From the perspective of Israeli researchers, however, studying the genetics of Israel's ingathered Jewish population was also a project of state building. In a newly founded state that sought to "ingather" the "exiles" and merge them into a single polity, the project of population genetics was wedded to the urgency of the state's nationalist project. If generating statistical data about one's population has been a key biopolitical strategy of the modern state (Foucault 2007; Hacking 1990), the State of Israel was no exception: knowing the genetics of Israeli Jews was part and parcel of epidemiological studies and medical management, most especially vis-à-vis the massive influx of Jewish immigrants arriving on Israel's shores, many of whom were widely viewed as a risk to the health and vitality of Israel's existing Jewish citizenry (Shvarts et al. 2005). More specific to my interest here, generating knowledge regarding the genetics of Israel's Jewish population was simultaneously a practice wedded to the work of *imagining* the nation: what evidence is there that the Jews are a nation with a shared origin in ancient Palestine? Faced with communities of immigrants who from the perspective of the state's Ashkenazi political and scientific elite seemed so radically different, that question took on urgency in the early state period. For population geneticists who operated within an intellectual tradition within the Zionist movement that imagined peoplehood in terms of European racial and eugenic categories, on the basis of what kinds of *biological* evidence could Jewish unity be rendered legible and true?

What is apparent in the early work of Israeli researchers is a struggle to reconcile their belief in the biological unity-qua-shared historical origins of the Jews with the "fact" of phenotypic evidence to the contrary. The Jews were presumed to be "a people" descended from the Israelites who were exiled from ancient Palestine. That vision was crucial to the ideology of settler-nationhood—to an understanding of Jewish settlement in Palestine as a project of *return*—that formed the bedrock of the Israeli state. However, a speaker before the 1961 Jerusalem conference pointed out:

> The visitor to Israel is usually struck by the marked differences in the external appearance of the various Jewish groups. It is difficult to decide how many of these seemingly distinctive features may be ascribed to apparel and hairdress, to nutrition or sun exposure, and to traditional demeanor, which shapes figures as well as faces. However, the blood-group studies have revealed certain striking differences among the Jewish communities. These are undoubtedly genetic, but this audience need not be reminded that they may, nevertheless, reflect selective influences of environmental agents which have been at work for many generations. (Nelkin 1963, 18)

On what basis could one claim that the Jews form a single biological population, given that "external appearances" (and blood group data) seemed to indicate the existence of different groups? What was the biological evidence that Jews actually are a single population with a shared geographic origin in the ancient Middle East? Implicitly or explicitly, such questions dominated Israeli studies in population genetics in its early years. Various "physical characteristics" (blood group systems) and genetic mutations (disease-causing genes) were sought and used in order to substantiate the origins and biological-historical unity of the Jewish people, and it was against the apparent contradictory evidence of phenotypic difference (in which the biological and the cultural often merged analytically one into the other; see Hirsch 2009) that researchers framed their work.

Throughout the late nineteenth and the first half of the twentieth centuries, attempts to provide an answer to the question "What are the Jews?" were iterated over and over again on the terrain of the biological sciences. The question itself was not new: what made Jews *Jews* was a question that had been asked many times before, most prominently in the modern period by scholars of the Jewish Enlightenment. But that an answer would be provided in biological terms was novel indeed. In this chapter, I give an account of two projects of Jewish self-definition carried out on the evidentiary terrain of the biological sciences. First, relying primarily on a growing body of scholarship on the topic, I consider the work of prominent European and American Jewish scientists who, in the late nineteenth and early twentieth centuries, grappled with the question of the racial identity and the biological health of the Jews. Second, through a reading of scientific papers on the genetics of Israeli-Jewish populations produced by Israeli researchers in the 1950s and 1960s, I analyze the scientific effort to discern some evidence of a shared biological-qua-historical origin among Israeli Jews.[3] In so doing, I unpack the methodological and epistemological practices and assumptions of the biological sciences, and I analyze the social and political struggles of which those practices were a constitutive part.

During both the earlier and later periods, scientists sought to study the biology of "the Jews." Nevertheless, the persistence of that category only tells us so much. On the one hand, the search for biological evidence was ongoing. It was (and is) an enduring quest. On the other hand, what it was that actually made Jews *Jews* for race scientists and population geneticists did not simply remain the same. What were the criteria for the distinctiveness of "the Jews" operating in these different scientific practices and historical circumstances? What forms of evidence were turned to and what were the facts that needed to be overcome if the category of "the Jews"—as a race, and subsequently, as a population—was to emerge as a biological-historical fact? Moreover, what were the broader political struggles within which these scientific debates took place, which they reframed and in which they intervened? In examining the work of earlier Jewish and Israeli biological scientists, I generate a comparative historical perspective in relation to which, over the next four chapters, I read the contemporary projects of genetic historical self-fashioning that stand at the center of this book. Out of the convergence of Zionist politics and scientific work, what was established by the mid-twentieth century was a distinctive vision of Jewish identity that has gained renewed power and saliency today: that Jews are the biological descendants of an ancient people, and that evidence of that origin and shared descent, of that peoplehood, will be revealed by the biological sciences.

In what follows I explore the tension between a definition of who the Jewish population is or is supposed to be that structured research into Jewish population(s) and the actual practices of trying to establish an a priori *fact of* a shared Jewish biological-historical origin. From the beginning of Jewish racial science at the end of the nineteenth century through the work in Israeli population genetics in the 1950s and 1960s, (most) researchers assumed that the Jewish communities "of the diaspora" are biologically related. In the actual work of studying and analyzing Jewish populations, however, that a priori commitment to Jewish biological commonality was not so seamless: trying to answer the question of how it is—or if it is—that Jewish communities are biologically related involved and generated shifting understandings of the Jewish race/population. It generated various accounts of the evidentiary grounds for Jewish (biological) peoplehood. It produced different configurations of the Jewish world—who was included, who was marginal or dubious or absent or "problematic." Moreover, race scientists and population geneticists harbored different understandings of the meanings and political significance of biological distinction.

Amidst all those shifts and all that (discovered) diversity, however, the question itself—What are the Jews?—was asked again and again in biological terms. Its continual reiteration performed the fact of Jewish peoplehood as grounded and/or evidenced in biological distinction even as biological data that might resolve the question once and for all was never found. By the mid-twentieth century in the work of Israeli population geneticists and to a large extent that of their European and American colleagues, the question was no longer whether or not Jews *are indeed* a biological collective, a question that had been vigorously debated in the late nineteenth and early twentieth centuries. What remained open to deliberation was merely a matter of details: Which biological markers distinguished Jewish from non-Jewish populations? Which Jewish communities were outliers in the Jewish genetic map?

I want to be clear: in examining the connections between the biological sciences and political projects of Jewish self-definition I am not arguing that Zionism is—or for that matter, is not—racism. Race science and its stepchild, eugenics, were integral to the intellectual traditions and cultural milieu of scholars in Europe and the United States at the turn of the twentieth century. Jewish scholars partook in that context as much as did anyone else. Zionism itself was a diverse and evolving project, and one powerful stream within it became increasingly invested in the concept—if not always the science—of race. As Zionist activists sought to realize Jewish settlement in Palestine, many came to articulate a racial self-definition of the Jews. By the early decades of the twentieth century, the link between biology, national self, and soil in Jewish national thought became ever more influential and robust. The national question was increasingly framed within the language of biology, and powerful political and intellectual figures came to understand the Jewish Question in eugenic terms: Zionist activists aimed to cultivate a "new Hebrew" who would be radically different—biologically and not just culturally—from the "diaspora Jew." And they took the biological piece of that regeneration seriously: the Jewish *body* had to be rejuvenated; Jewish degeneration had to be countered, even as who it was who qualified as the most degenerate Jew would shift over time.

Moreover, once this racial self-conception migrated to Palestine and confronted a series of non-European Others—not just "Mizrahi Jews," the focus of the growing literature on Zionism, race science, and eugenics

(Hirsch 2009; Bloom 2007; Falk 2006a, 2006b; Shvarts et al., 2005), but Palestine's non-Jewish inhabitants—the power of a racial self-definition came to matter to the kind of state Jewish nationalists would produce.

The Jewish Question

As is well known, "the Jews" occupied a particular place in late nineteenth- and early twentieth-century race science. Race scientists wrote extensively about the biological properties of the Jews and their attendant social and political implications (Proctor 1988; Gilman 1991; Mosse 1978). Steeped in debates regarding the political status and rights of the Jews in late nineteenth-century European nation-states, race science emerged as the primary terrain of (scientific) anti-Semitism. And it was not just Christian Europeans who generated racial definitions of the Jews. So too did self-identified Jewish scientists who took up the mantle of studying "themselves." Drawing upon the reigning paradigms of racial science and its concepts of the normal and the pathological, health and disease, degeneration and regeneration, Jewish scholars generated their own discourse regarding the (racial) character of the Jews. Are Jews best understood as a "pure" or at least, a stable race? Are they a mixed race? Are Jews best categorized as a religion? A *Volk* or *Stamm*? In other words, this was a debate regarding what kind of a collectivity the Jews are. Could that question be answered in biological terms?

In constructing their own evaluations and understandings of "the Jewish Question," Jewish medical doctors and social scientists were responding to anti-Semitism. But in their responses, they did not challenge the racial paradigm's most basic terms: that human collectives are biologically constituted, that key biological properties are inherited from one's ancestors (even if subject to environmental pressures), and that those properties are politically, socially, and mentally significant. Moreover, most accepted not just the basic discursive terms of race science. Most Jewish scientists accepted its *evaluation of* "the Jews": that Jews were a degenerate and a degenerating race. What many of these Jewish "men of science" (Hirsch 2009, 592) did was reconfigure the relationship between nature and nurture in their assessment of the racial status of the Jews and provide alternative explanations to mainstream racial accounts of Jewish degeneracy.

It would be a mistake to understand Jewish scholarly and political debates about the biology of the Jews as being solely a response to anti-Semitism, however. It would "elide the degree to which" many Jews embraced the notion of a Jewish race and believed that "racial identity was not something imposed on them by anti-Semitic opponents, but resided in very real physical and mental attributes" (Hart 2007, 14). The late nineteenth century witnessed the rise of Jewish nationalism, which increasingly formulated the fact of Jewish distinctiveness—of Jewish peoplehood—in biological-cum-racial terms. While no longer credible, or even thinkable, to many of us today, a commitment to the reality of race was pervasive at the time. As Amos Morris-Reich points out, "the intense controversy in the social sciences . . . in the first half of the 20th century, did not deal with 'race' in general but . . . with the principle of 'racial determinism.' That is, it did not challenge the existence of differences, it challenged their significance" (Morris-Reich 2006).

In the late nineteenth and early twentieth centuries fundamental questions about the nature of Jewishness were being asked: *What* are Jews—a race, a religion, a *Volk*? I first outline that debate within Jewish social science. I do so by summarizing the arguments of a few key figures in those debates.[4] In my presentation I gloss over the distinctions between Central European, British, and American social scientists in their engagements with the Jewish Question. Despite the differences between the Jewish communities in each of these locales, the question "What are the Jews?" occupied them all. And directly and indirectly, they engaged one another scientifically and politically when making the case for or against the Jews as a distinct race. Jewish social scientists constituted an "interpretive community" (Hart 2000, 139), and they participated in a single "problem space." That is, they all participated in "a context of argument . . . of intervention . . . [and in] an ensemble of questions and answers around which a horizon of identifiable stakes (conceptual as well as ideological-political stakes)" hung (Scott 2004, 4).

Jewish Social Science and the Politics of Emancipation

By the early twentieth century, Maurice Fishberg was a key figure in this debate. An American physician and anthropologist, Fishberg was a proponent of the position that the Jews are not a race, and he produced the most extensive social scientific case arguing for assimilation as the solution to the Jewish Question. In sketching the contours of his argument, I approach Fishberg's work as an instance—if perhaps the most articulate one—of a more general scientific case against the racial character of the Jews and of the political stakes of the anti-racialist position.

In his book *The Jews: A Study of Race and Environment* (1911), Fishberg notes that Jews have occupied a particular place in anthropological discourse. In a field in which it has been "accepted . . . that there are no pure races among civilized peoples and nations," the Jews are an exception. They are "alleged to have succeeded in maintaining their race in a state of purity for the last four thousand years." Due to endogamous marriage practices the "Jews of to-day [are argued to] present a uniform physical type wherever they may be encountered" (1911, 21).

Fishberg wanted to disprove that proposition, and he set out two grounds on which to do so. First, he argued for the non-hereditary (i.e., environmentally influenced) character of certain presumably permanent racial characteristics (height, size and shape of face and nose, complexion, for example). Second, he demonstrated heterogeneity among Jewish communities in physical characteristics. Rather than explicating his environmental interpretation of so-called permanent racial traits, I will briefly engage his argument regarding the diversity of physical characteristics among the Jews. I do so with reference to the characteristic understood at the time to be the most stable marker of racial difference: head-form.

According to Fishberg, "The most noteworthy characteristic of the skulls of modern Jews is their great variability, according to the country from which they are derived. It may, in fact, be stated that there is no single type of head which is found among the Jews in all countries in which they live" (1911, 49). He identified three different Jewish groups: from Tunis, the Caucasus, and Lithuania, which "represent three different races," and, "when further inquiry reveals" that each of these Jewish head forms corresponds to that of the non-Jewish population with whom they reside, "the inference forces itself on us that the differences in head-form between these three groups of Jews have been acquired in the countries in which they live." The source of that diversity is "racial intermixture"—or "intermarriage" (511; see also 55–56). Taken together with other kinds of evidence—that "Jewry possessed statistically significant numbers of individuals with blonde or red hair, blue eyes, and black skin" (Hart 2000, 165)—Fishberg insisted that the evidence is unequivocal that Jews have mixed with other races throughout their history. There is absolutely no evidence that Jews are a race.[5]

This question of the racial character of the Jews was of far more than academic interest for Fishberg. "It is of vital interest to the Jews, as well as the people among whom they live, whether they really differ radically, whether they are of different race stock when compared to the *Homo Europoeus*, and whether their prepotency is so strong that they can never be assimilated by people of different origin" (1911, 505). (Note that "racial stock" and "origin" are two sides of the same coin in race discourse.) Fishberg's interest in showing that Jews are not a pure and distinct race was tied most immediately to the political project and social milieu of which he was a part. In the latter part of the nineteenth century, American physicians and social scientists weighed in on the debate regarding U.S. national identity, and they did so in the language of race science and medicine (Hart 2000, 150). Eastern European Jews were coming to U.S. shores in large numbers, and it was their bodies that were studied in attempts to specify the Jewish racial type. Defined by "American racialists" as Oriental or Asiatic (ibid.), the leadership of the existing Jewish community feared that their standing in the U.S. was threatened: "[Eastern European Jews'] perceived racial peculiarities threatened to put . . . [all Jews] . . . beyond the pale of whiteness" (Goldstein 1997, 54). In response to that changing reality, many American Jews abandoned what had been a widespread commitment by the late nineteenth century: that Jews *are* a race, an embrace of racial identity within the American Jewish community in response to the tide of rapid assimilation into Christian society (Goldstein 1997, 2006). By the early twentieth century and in the context of a changing political climate, many in the U.S. Jewish community distanced themselves from that prior embrace of Jewish racial difference *in order to protect their standing as white.* The leadership of the organized Jewish community enlisted Maurice Fishberg to explode "the notion that Jews formed a distinct racial group" (Goldstein 1997, 54).[6] Scientific authority was to be brought to bear on the political struggle in an effort to undermine the image of the Jews as nonwhite, in sum, as a lesser race.

In a paper on "Intermarriage between Jews and Christians," Fishberg addressed a prevailing anxiety in white-racialist circles in the U.S. that intermarriage between Christians and Jews threatened the degeneration of the nation. Although such marriages are on the rise, he argued,[7] no harm will come from the flow of "Jewish blood into the veins of other white people" because it does "not introduce any *new or alien racial elements*" (Fishberg 1923, 132, emphasis added). Jews are not Semites. They are white. They are Europeans. "It has been argued by nearly all competent anthropologists that the Jews are a composite ethnic unit; a racial blend in which nearly all the types of white humanity have entered" (132). Jews are not a distinct race. They *belong* among Europeans.

In his argument in defense of the assimilation of Jews into European and U.S. Christian society, Fishberg went even farther. As he saw it, intermarriage was *eugenic*: "Intermixture of the European ethnic elements" is of "immense advantage." Note, by way of contrast, "the homogenous populations of central Asia, and Africa, free from any racial admixture, but in a low stage of mental evolution" (1923, 132). Fishberg was not alone in advocating the eugenic possibilities of (specific) racial mixture: by the early twentieth century, many race scientists argued that endogamy was potentially deleterious, an assimilation of Darwin's understanding of descent and natural selection. A new distinction had emerged: that between mixing "distant" versus "related" races (Hirsch 2009, 598). Embracing that distinction, Fishberg presented what he considered ample evidence that the offspring of Jewish-Christian marriages are often exemplary persons. He generated a list of persons—politicians, astronomers, biologists—"of half-Jewish origin who have attained distinction and eminence" (Fishberg 1923, 132). Jewish-Christian intermarriage, a likelihood only increased with the immigration of Eastern European Jews, should not be regarded as a problem. It is eugenic. It is good population policy.

Fishberg's work was not addressed to public debates in the U.S. over Jewish immigration alone, however. He was also engaged in an argument with European Zionists and with "what he believed to be the dangerously mistaken racial definition of Jewry advanced by some Jewish nationalists, and the 'separatist' conclusion they drew from it" (Hart 2000, 159). That problem had traveled to U.S. shores with the arrival of Eastern European Jews, many of whom were Jewish nationalists.[8] Taking the example of "Western" (i.e., Western European) emancipated Jewry, Fishberg argued there was ample evidence that the Zionist racial definition of the Jews was wrong: emancipated Jews "prove that there is nothing *within the Jew* that keeps him back from assimilating with his neighbours of other creeds" (1911, 479, emphasis added). Mentioning, for example, the decline in religious observance and in the use of Hebrew among emancipated Jews, Fishberg wrote: "it is evident that they have practically discarded all their particularisms in exchange for the culture, civilization, habits, customs, and manners of the populations among whom they live" (480). It is "evident" that the "Jews of to-day cannot be considered a nation." The "only thing they have in common is their religion. [And] it is the consensus of opinion of all modern statesmen as well as ethnographers that *religion alone cannot be considered a basis of modern nationality*" (480, emphasis added).[9] Moving seamlessly back and forth between invocations of "ethnicity," "race," and "nationality," Fishberg concluded: "There is no more justification for speaking of ethnic unity among the modern Jews, or of a 'Jewish race,' than there is justification to speak of ethnic unity of the Christians, of Mohammedans, or of a Unitarian, Presbyterian, or Methodist race" (514). Jews are but a religious community.[10]

Not "Merely" a Religious Group[11]

The classification of Jews as "merely" a religious group was precisely what, by the late nineteenth century, many Jewish social scientists, and most consistently Zionist social scientists, sought to contest. As much scholarship has demonstrated, the emancipation of the Jews in the late eighteenth century brought with it the transformation of Jewish identity (Hess 2002; Herzberg 1968): the Jews were fashioned as a religious community, no different from Christians, and religion was conceived of as a realm of "choice," of "belief"—as "optional," in Talal Asad's words (1993, 49). Emancipation conferred particular rights on Prussian and French Jews. In turn, it demanded particular transformations. And those demands precipitated efforts within the Jewish community to reform Jewishness and to render Jews more easily assimilated into state and society. The Emancipation produced—demanded in fact—a conception of "Judaism as a mode of belief or as one faith among others," and scholars of the *Wissenschaft des Judentums* took up that demand. As Patchen Markell has argued, the Emancipation "transformed the meaning of Jewishness itself, working to convert it *into* a matter of religious consciousness" (2003, 136).[12]

The 1880s saw the first systematic attempts of Jewish men of science to collect statistical data on the Jews. As various scholars have argued, such social scientific efforts were "part of a reconceived Jewish scholarship [that] became an instrument with which to challenge the ideology of emancipation" (Hart 2000, 16). Most of the work in "Jewish statistics" stood in contrast to that of Maurice Fishberg. It represented a scientific and political break with the earlier tradition of modern Jewish scholarship, the *Wissenschaft des Judentums* and its attendant political and social visions. This community of Jewish scholars, who researched and wrote primarily but not exclusively in Central Europe, by and large rejected "the liberal, assimilationist goals" of that earlier tradition (ibid., 39). Jewishness could not be "'reduced' or limited to a religious identity and community" (39). In its place, these scholars fashioned the distinctiveness of the Jews—as a nation, a *Volk*, a race—most often if not exclusively in light of a nationalist cause.

Todd M. Endelman has argued that, in comparison with the Central European tradition, the social scientific and racial literature produced in Britain during the late nineteenth and early twentieth centuries focused far less on Jewish communities. It was more concerned with the colonized subjects of the British Empire and the urban poor at home (2004, 53). Nevertheless, "Anglo-Jewish scientists, however small their number" did take up the question of Jewish racial difference. They did so, according to Endelman, because talk of race, however imprecise its referent, circulated in the Jewish-British public; it was "too pervasive in cultural and political life to allow [Anglo-Jewish scientists] . . . to escape or ignore its influence" (53). Moreover, both anti-Semitic race science and the responses to it produced by Jewish scholars in Central Europe, Russia, and the United States traveled to Britain, including quite centrally Maurice Fishberg's work. Whether or not an extensive literature on race and the Jewish Question was being produced in Britain, in other words, it was circulating in a transnational field.

In Britain, the most prominent scholar to take up the question of Jewish heredity was Redcliffe Nathan Salaman, a medical doctor and a geneticist, who by the end of the First World War had become committed to the Zionist cause. Salaman had retired early from the medical profession because of an illness. Once recovered, he went in search of another scientific career. Living near Cambridge University, a center for the study of Mendelian genetics, he took up genetics (Endelman 2004, 59). Following upon a published study of the "heredity of the potato" (59), and long committed to the belief that Jews are indeed a race, by 1910 Salaman turned his attention to the Jewish Question.

Over the next two decades, Salaman used Mendelian genetics to weigh in on the question of the character of the Jewish collectivity. He was "active in the ranks of European Jewish scientists, who, in the debate on Jewish self-definition, argued that Jews were a race" (ibid., 59). Salaman published his first scientific paper on the topic in 1911, and in it he sought to ascertain the "racial position of the Jew" (1911, 274). How can one explain all the apparent diversity among the Jews *and yet still sustain* the belief that Jews are a race? Salaman sought to resolve that paradox on the terrain of historical, iconographic, and genetic evidence.

The "racial position of the Jew," Salaman wrote, has "engaged the attention of all modern ethnologists." "The problem is extremely difficult because, on the one hand, we have the oft asserted and by no means easily disproved statement of the Jews themselves that they are pure Semites, whilst observers . . . point out that the Jew of to-day has no uniform cranial characteristics, that he is decidedly brachycephalic whilst the typical Semite such as the Bedawyn is essentially dilochocephalic" (1911, 274). Salaman turned to history to resolve the problem. In racial thought, after all, the question of "racial character" was tethered to the question of origins: "Most authorities agree," Salaman argued, that "up until the time of the destruction" the Jews had "freely intermarried with the surrounding people" (274). That original mixture (which included Amorites, who Salaman claimed were non-Semites and thus might be "related to the Central European people") was the source of Jewish phenotypic diversity. For example, how can one explain the fact that there are blond Jews? Salaman ascribed it to racial mixture between Semites and Amorites prior to the Jewish dispersal from ancient Palestine. Having placed the cause of elements of Jewish diversity *in ancient Palestine*, Salaman insisted—and this was his key move—that following the "destruction" and the dispersion of the Jews from their ancient homeland, "intermixture has been absolutely minimal" (276). Racial "purity" proceeds from that point of departure

Writing in the early twentieth century, Salaman did not need to prove the Jewish race to be pure. That the Jews were a "mixed race" had been widely accepted in race science. The German anthropologist Felix Van Luschan, for example argued that "the ancient Hebrews consisted of at least three distinct races—the Aryan Amorites, the Semites, and the Middle Eastern Hittites" (Hirsch 2009, 596–97). But whereas Van Luschan insisted that mixing with "additional elements" continued throughout life in the diaspora (ibid.), Salaman took the opposite position. In the logic of race science, the argument for racial "origins" and "purity" necessarily proceeds from a given temporal and geographic point of departure. For Salaman that origin had to be life in Palestine *prior to* Jewish exile and dispersion.

Salaman's logic for resolving the Jewish Question in favor of the existence of a Jewish race was not novel. In 1891, Joseph Jacobs—a British-Jewish scientist who was the first to conduct social scientific studies of Jews in London (Endelman 2004, 78)—had taken a similar albeit not identical position: following their exile from Palestine, Jews have not intermarried or mixed with others, Jacobs wrote. In contrast to Salaman's subsequent work, however, Jacobs insisted there had never been any *racial* mixing. All the groups with whom Jews intermarried in the pre-Exilic period were Semites. Therefore, Jews had never engaged in racial intermixture. "The distinction between Jews and other Semites was religious, not racial," he insisted (Jacobs 1891, xviii).

In "Heredity and the Jews," Salaman offered historical and contemporary evidence to support his belief that Jews had "intermixed" very little following their exile from Palestine. And in seeking to prove that belief he returns us to the Cohanim: "At this point one might with advantage consider the relation which the existence of the *Kohanim* has to the question of Jewish type" (Salaman 1911, 279). After explaining the origins of the Cohanim and the traditional lines of priestly descent, Salaman argued that "it is most improbable that anyone could, and much less would, assume the title of *Kohen* without having a right by birth because it conveys neither social distinction nor advantage, whilst on the other hand, it brings in its train some undoubted disabilities" (279). Based upon that rather large leap of faith, Salaman concluded that thereby "we have a sect whose descent may be regarded as strictly Jewish" (279). For Salaman, the Cohanim function as the known Jewish type against which other Jews could be measured: "Now if we review the physiognomies of the various *Kohanim*, it will be found that they exhibit no type in any way distinct from that of other Jews." The very high "value" which can be ascribed to the "purity of descent" amongst Cohanim can also be ascribed "to their brethren among whom they live." The Jews, in other words, are "pure." They have not intermixed very much since their ancient exile (279). They are not *of Europe*. They are *of* (ancient) Palestine.

For Salaman the Mendelian geneticist, however, such a "historical" study was not enough to place his argument on sound scientific footing. He devised an "experiment." Having established that what distinguishes Jews is not cranial form (he admitted that Jews differ in their cranial measurements, and he ascribed that fact to their mixed origins in ancient Palestine) Salaman turned to "facial expression" (278). In this choice of physical characteristic, Salaman followed in the footsteps of Joseph Jacobs, who had, as requested of him by Francis Galton, made composite photographs of Jewish schoolchildren in London (Endelman 2004, 60).

Salaman gathered a number of Jewish participants to "observe" the facial types of the offspring of mixed Jewish-Christian marriages. He chose Jewish observers because he hypothesized they could more easily identify the Jewish type. Salaman wanted to disprove two conclusions made by critics of the notion of a Jewish race: first, that there are Jewish individuals "who are indistinguishable or at least practically indistinguishable from North Europeans," and second, that this phenotypic fact is evidence of "mixture with the surrounding people" (280). On the basis of his experiment, Salaman came to the opposite conclusion: Gentile and Jewish facial expressions are segregating Mendelian types. One inherits one or the other. One cannot inherit some "mixed" form. Moreover, among the offspring of 138 Jewish-Christian marriages, observers classified 328 as Gentile, 26 as Jewish, and 8 as Intermediate. After going into a set of reasons why the intermediate category is an unreliable set of observations (unknown provenance of parents, observers being a little confused as to what the Jewish facial type really is, and so forth), Salaman declared Fishberg among others wrong. Vis-à-vis the Jewish facial expression, he declared, the Gentile one is dominant and the Jewish recessive. Therefore, "If the Jew had freely intermixed with the European races as some authors think is the case, it is obvious that, the characteristic facial type being recessive, it would have been rapidly swamped" (288). And that is not the case: the Jewish expression remains identifiable to nearly all observers, scientists and lay-people alike.[13]

The Jews *are* a "type." Whatever racial mixture there was in the ancient past, the Jews are a race that has remained true to the type formed in ancient Palestine. The Jews are not a composite of the various races with whom they have resided through thousands of years of "exile." And it was not just historical and genetic evidence that Salaman brought to bear on this debate regarding the racial character of the Jews. Following what was a widespread scientific practice of the time, Salaman turned to iconographic evidence to further support his theories of Jewish post-exilic racial purity. Comparing the "expressions" of photographs of contemporary pure Jews and those of mixed parentage with depictions of "ancient" (thirteenth and ninth century B.C.E.) Jews on archaeological monuments, Salaman concluded that contemporary Jews resemble their ancient brethren (see also Jacob's use of iconography, 1891).[14]

Salaman's commitment to Jewish biological difference was shared by many a Zionist activist and scientist. While Salaman's full-blown commitment to Zionism came later (in the aftermath of the First World War; see Endelman 2004), for many Central European social scientists the belief in the reality of the Jewish race and the commitment to Jewish nationalism were cut from the same cloth. As Dafna Hirsch writes, "among Jewish physicians, anthropologists, and other 'men of science' in Central Europe, proponents of the idea that the Jews were a race were found mainly in the ranks of Zionists, as the idea implied a common biological nature of the otherwise geographically, linguistically, and culturally divided Jewish people, and offered scientific 'proof' of the ethno-nationalist myth of common descent" (2009, 592).

Race and nation were not isomorphic in early twentieth century biological and political thought. But they were deeply entangled, and efforts to align them were ongoing in scientific and political spheres alike (see Hutton 2005), especially in Central Europe, where many of the leaders of the Zionist movement were born and schooled. Nevertheless, this nationalist commitment to the racial character of the Jews did far more than scientifically substantiate the ethno-national myth of common descent. It framed and generated substantive questions for the Zionist project: How is it that a renewed Jewish nation would be built? What were the problems that had to be overcome? Studying Jews as a race involved specifying their *biological character*. And for many Jewish physicians and social scientists, that involved "recognizing" Jewish degeneracy, thus concurring with the anti-Semites of the day. But Jewish degeneracy, they argued, was a consequence of the "environment"—of Jewish displacement from their homeland and of the social and economic conditions in which they lived (the Eastern European shtetl for some; for others, the modern, alienating city in which most Western European Jews lived). Degeneracy, anti-Semitic racial theories notwithstanding, was not a consequence of the inherently and permanently inferior character of the Jewish race. The fact of Jewish degeneration would be stopped only by a change of environment.[15] For Zionists, that required a return to Palestine, the imagined locale of Jewish *regeneration*.[16]

The attempt by Jewish scholars to study and assess the state of contemporary Jewry was first institutionalized in Central Europe in 1903 with the establishment in Berlin of the Verein für Jüdische Statistik (1903). And the Verein was interested in producing "usable knowledge"—knowledge that would improve the state of the Jewish masses. As Amos Morris-Reich reminds us, in the late nineteenth century the social sciences were founded not "primarily, as a means of contemplating society, but in order to create . . . the rationalization of society and better handling of social tensions" (2006, 3). The vision of science as a means of *intervening in* and not just representing the world, to borrow Ian Hacking's distinction (1983), is evident in all these efforts in Jewish statistics in the early twentieth century, whether in the United States, Great Britain, or Central Europe.

Producing usable knowledge was not an exclusively Zionist goal, as evidenced by Maurice Fishberg and, for that matter, by the work of Anglo-Jewish scientists who sought to improve the (biological) state of Eastern European Jewish immigrants to London's East End. The work of Jewish statistics in Central Europe was, likewise, not exclusively a Zionist project. Over the first two decades of the twentieth century, both the Verein für Jüdische Statistik in Berlin and other similar institutions included both Zionist and non-Zionist Jewish scientists and perspectives. For that matter, some Jewish scholars who produced evidence for the fact of the racial character of the Jews were proponents of a diasporic nationalism: they were committed to demanding rights for Jews as a political minority in their countries of residence (Wiendling 2006, 134). Nevertheless, the birth of Jewish statistics in Central Europe was a constitutive moment for the Zionist movement. As Max Nordau declared before the Fifth Zionist Congress in Basel in December 1901, the success of Zionism required scientific facts:

> We must know more. We must know with greater precision about the national material with which we have to work. We need exact anthropologi-

> cal, biological, economic and intellectual statistics of the Jewish People. We must have quantitative answers to the following questions: How is the Jewish People physiologically constituted? What is the average size? What are his anatomical characteristics? What are the numbers of diseased Jews? . . . How many Jews are city-dwellers, how many rural? What sorts of occupations do Jews pursue? . . . All of this one must know if one wants to really know a people. As long as this remains unknown, whatever one seeks to do for the nation will be a fumbling in the dark; and whatever one says about this nation will be, at best, poetry, and at worst empty chatter. (Quoted in Hart 2000, 30)

The Verein für Jüdische Statistik was founded as a scientific organization devoted to collecting statistics on contemporary Jewry. And while relatively short-lived (by the mid-1920s its influence had waned as other Jewish institutions devoted to studying the Jews had been established throughout Central Europe; see Hart 2000), it had brought about a fundamental transformation. The first scientific institution both funded and administered by Jews, the Verein "succeeded in defining the social scientific study of Jewry as a *Jewish* scholarly endeavor" (Hart 2000, 72–73). Moreover, within the first few years of its existence, the Berlin Bureau for Jewish Statistics (established under the auspices of the Verein) organized "the first bank of data of Jewish statistics and demography, which became the basis for the sociological researches and the numerous diplomatic and juridical negotiations concerning the Jews in the twentieth century" (Bloom 2007, 338). The data bank was a "crucial step in the unification of the Jews as a modern nation represented by the Zionist movement" (338).

While committed to "objective" research and intentionally inclusive of non-Zionist Jewish scientists, the Verein provided the Zionist movement with crucial information. Its goal was to produce statistical knowledge that could be harnessed for projects of social and political intervention. For its non-Zionists participants, that meant improving Jewish life in the diaspora—however "improvement" itself was understood.[17] For Zionists, it meant understanding Jewish demography and biology as a step toward realizing the nationalist cause.

"Understanding" the demography and biology of the Jews as characterized by numerical decline and physical and mental degeneration, many Zionist scientists and political leaders embraced eugenics in formulating their vision of Jewish immigration to and settlement in Palestine. For Max Nordau, for example, "the 'typical' Jewish physique was degenerate." He believed that fact to be a consequence of "generations of persecution and the degenerating living conditions imposed on Jews," and he insisted that Jewish biology would change—would improve—when "life conditions" did (Falk 1998, 594). "A people cannot, in the long run, remain healthy and strong if it does not again and again, at least temporarily return to the rejuvenating soil" (quoted in Falk 1998, 594). A return to Palestine—and more specifically, a return to *laboring* the land—would produce Jewish regeneration (Falk 2006a).

There is an extensive secondary literature on the Zionist commitment to fashioning the "new Hebrew," who would be different from "the diasporic Jew," physically as much as culturally (e.g., Almog 2000; Zerubavel 1995). Unlike the diaspora Jew, who was weak and ill and who never engaged in any manual labor, the new Hebrew would be physically strong, a laborer and a soldier. But even though scholars have written extensively about *physical* regeneration, the literature has sidestepped the role of racial theory and eugenic thought in helping to frame that vision and desire. In standard histories of Jewish settlement in Palestine, the origins of "Hebrew labor"—the commitment that Jews engage in manual and agricultural labor themselves to build their own economy, society, and state—was long argued to be a consequence of Labor Zionism's socialist commitments, best exemplified in the form of the kibbutz. Gershon Shafir was one of the first scholars to challenge that claim: "Hebrew labor" emerged as a fundamental practice of Labor Zionism, he argued, because of competition *on the ground in Palestine* between Jewish labor and a much cheaper Arab labor pool (Shafir 1989). But as this discussion of Jewish race science indicates, there was also another source of inspiration for the commitment to Hebrew labor that likewise had nothing to do with an *a priori* dedication to socialism: the assessment by Jewish physicians and social scientists of "the Jews" as a degenerate race and their eugenic framework for imagining a "solution" to that problem, that is, a revived and reborn Hebrew nation in Palestine (on eugenics and Jewish settlement in Palestine, see Bloom 2007; Hirsch 2008, 2009; Falk 2006b).

Arthur Ruppin's scholarly and political work perhaps best exemplifies the convergence of race science, eugenics, and Jewish nationalism. A founding figure in the social scientific study of the Jews—his 1904 publication *The Jews of Today* was widely regarded at the time as a founding text in the emerging field of Jewish statistics—Ruppin was the first albeit quite short-lived director of the Bureau for Jewish Statistics in Berlin. Moreover, Ruppin has been referred to in the classic Zionist histories as the "Father of Jewish Settlement in Palestine" (Bloom 2007, 331). Having resigned from the Berlin Bureau in 1907, Ruppin conducted a preliminary study of the possibilities for Jewish colonization of Palestine on behalf of the Zionist Organization. He was subsequently appointed an official representative of the Zionist movement in Palestine and was the first director of the movement's Palestine office, founded in 1908 (ibid.). Over the next several decades, Ruppin emerged as one of the main architects of Jewish settlement in Palestine, exerting considerable influence over everything from land purchases to the forms of production and labor relations in Jewish settlements, and the control of Jewish immigration (Hirsch 2009, 604).

For Ruppin, the Jews were a race, and that fact grounded his commitment to Zionism: "I joined the Zionist movement under the slogan 'against political Zionism' [i.e., Herzl's idea of a charter] and for practical work in Palestine. I wanted to base the right of the Jews to come to Palestine not on some 'political' agreement and concession, but on their historical and racial connection to Palestine" (quoted in Falk 2006a, 140). In an effort to force a recognition of that historical and racial connection, Ruppin, like other Zionist social scientists, considered his work in Jewish statistics key: his scientific work would "offer empirical support to Jewish nationalist claims" (Hart 2000, 61). As Ruppin wrote in the revised version of *The Jews of Today* (1910), social science "shall be the theoretical foundation for my practical work, and hopefully it will also contribute to a general clarification of the Jewish Question and Zionism" (quoted in Hart 2000, 64).

As data that would "clarify" the Jewish Question and Zionism, Ruppin turned to Jewish statistics in order to understand the threat that the Jews of his time faced. The goal of Zionism was to save the Jews from "annihilation"—and it was the danger of rising rates of conversion and intermarriage far more than the threat of anti-Semitism to which Ruppin ascribed the risk, even in his writings that appeared as late as 1940, as Amos Morris-Reich points out (2006, 13). In fact, for Zionist social scientists like Ruppin, it was the Jewish race's "internal enemies" (Hart 2000, 77) who worried them most. To quote Ignaz Zollschan, an Austrian physician and racial anthropologist: "The Jewish people is now engaged in a struggle not only with the world around it, but also within itself." Assimilationists were bringing about "'the end of Judaism, the absorption and disappearance of the Jews among other peoples'" (quoted in ibid., 78). The Zionists, for their part, were fighting for Jewish survival.

The significance of Jewish statistics—in effect, the study of race and demography—was in part a matter of mere numbers for Zionist social scientists. Jewish statistics indicated a declining number of Jews due to the rising tide of assimilation. But it told of far deeper troubles still: Jewish statistics demonstrated the fact of Jewish degeneration. Scientific evidence made apparent all that was wrong and all that had to be done. Ruppin and his allies were committed to bringing about the "*biological* transformation of the Jews"—and, for Zionists, that transformation could only take place in Palestine (Bloom 2007, 340, emphasis added). As is now well documented by a growing body of scholarship in Hebrew and English, prominent Zionist figures envisioned their work in eugenic terms, and that had real world consequences for Jewish settlement in Palestine during the pre-state period. For example, as Ruppin argued in 1919, "it would be better if only healthy people with all their needs and powers would come to Palestine so new generations will arise in the country that are healthy and strong" (quoted in Falk 2006, 155). By the early 1920s, potential immigrants were routinely subjected to medical examinations at their points of departure and, when that failed, in Palestine (Bloom 2007, 2008; Falk 2006a, 2006b: Shvarts et al. 2005). And the discourse of health and hygiene—for example, on the part of Jewish nurses in Palestine who tried to inculcate hygienic habits primarily amongst Jews of the "Old Yishuv" (both Eastern Jewish communities and Ashkenazi Orthodox Jews long resident in Palestine)—was central to visions of how best to build a new, and a modern, Hebrew polity (Hirsch 2008). As argued by Etan Bloom in his work on Arthur Ruppin and the influence of German intellectual culture on the contours of Jewish settlement in Palestine, "the Jewish body . . . was perceived in Zionist-Palestine as the heart of a new Hebrew tradition" (2007, 349).

For those intellectuals and political figures who embraced a racial self-definition of the Jews, race science served a double function for Jewish nationalism in the late nineteenth and early twentieth centuries. First, it sought to provide evidence that Jews are not *merely* a religion, a scientific weapon in a struggle to render the nationalist understanding of what it meant to be a Jew definitive. Second, in specifying the biological distinctiveness of the Jews, it also identified a substantive "problem" that nationalist project would need to overcome: not just Jewish annihilation due to assimilation, but the "fact" of Jewish degeneracy. Thereby, Jewish statistics gave content and direction to a political struggle that desired to make Jewish peoplehood anew, inserting eugenic commitments into the struggle to create a Jewish state and a Hebrew polity.

Israeli Population Genetics: The Early Years of Statehood[18]

In the years between 1948 and 1951, Israel's Jewish population doubled.[19] And more than mere numbers were at stake: there was a dramatic change in the regions from which Jewish immigrants came, and thus, from the perspective of Israel's political and cultural elite, there was dramatic shift in the character of the state's Jewish citizenry. Of the 101,819 Jewish immigrants to Israel in 1948, 75.5 percent came from Europe and the U.S. By 1954, that was only true of 14 percent; the remainder had emigrated from either Asia or North Africa (Shvarts et al. 2005, 11).[20] The "absorption of immigrants" took on urgency in these early years. And especially vis-à-vis Israel's non-European immigrants, but also vis-à-vis Holocaust survivors, who Israel's first prime minister, David Ben Gurion, believed were so traumatized that they had been "reduced to . . . 'human dust'" (ibid., 17), the fear of biological unfitness was a prominent issue in public policy and political debate.

As Dafna Hirsch has argued, in the context of Palestine during the British Mandate, it was in medical discourse and practice that racial thought, and more specifically, the commitment to "Jewish biology" and a belief in the need for "physical regeneration," was most fully articulated (Hirsch 2009, 593). An emphasis on the medical selection of immigrants was one venue for managing the biology of the Yishuv. So too were medical efforts to teach proper hygiene to (particular) Jewish communities in Palestine (Hirsch 2008) and debates over "mixed marriages" and the consequences of such mixture (Hirsch 2009, 595).

In the context of Mandate Palestine, the debate over "mixed marriages" indexed a very different set of groups than it did in Europe and the U.S. A medical doctor, for example, wrote in the journal *Moznayim* in 1934: "the ingathering of the exiles in Eretz Israel permits mixed marriages not between Jews and non-Jews, but between Jews and Jews, between Jews of different ethnic and territorial origins. Does not that alone promise a great extent of eugenic salvation?" (quoted in Hirsch 2009, 595). That is, mixed marriages among the different Jewish communities offered the potential to "elevate" oriental Jews to the standards of the European. The "oriental" Jew, because of longstanding local Jewish communities in Palestine which European immigrants confronted upon arriving and because of the immigration of Yemeni Jews to the Yishuv between 1912 and 1918, was already a presence for Jewish medical doctors and racial theorists working and living in Palestine (see Hirsch 2008, 2009; Bloom 2007, 2008; Falk 2006a, 2006b; on Yemeni immigration and labor, see also Shafir 1989).

By way of contrast, while the distinction between "East" and "West" had as much resonance for Jewish scholars and activists in Europe during these same years, that polarity had a very different set of referents.[21] East and West—East European and West European—was a fundamental division and important distinction: from the perspective of Western Jews, and, quite specifically, from the perspective of those Western European Jewish scholars who engaged the Jewish Question through scientific practice, Eastern Jews, *Ostjuden*, were traditional, poor, and degenerate. They were utterly different. In the wake of their emancipation, German Jews projected Jewish "pathologies" onto their Eastern brethren, a distinction between native and foreign Jews having been institutionalized in the Prussian Edict of Emancipation (Markell 2003, 147–48; see also Aschheim 1982).[22] It was within that context that much of the scholarship—be that of Maurice Fishberg, of Joseph Jacobs, or of Nathan Salaman—was written. And in that work, it was in the image of the Western Jew (leaving aside the problematic nature of their assimilationist tendencies) that the Eastern Jew was to be remade: modern, middle class, emancipated, educated, even if for Zionist thinkers in the first decades of the twentieth century that commitment would become far more ambivalent.

But my point here is that the known Jewish world—perhaps more accurately, the relevant Jewish world—rarely reached beyond Europe itself, Ashkenazi and Sephardic, for Jewish scholars of race in Europe and North America. As evident in the writings of Nathan Salaman and Joseph Jacobs, for example, the central characters in their drama of Jewish racial unity and diversity were European Jews, Eastern and Western. There were, however, some European scholars for whom non-European Jews played a central analytic and political role. Samuel Weissenberg, for example, went off in search of Jews "either in, or within relatively close geographic proximity to Palestine" in order to find the "ur-Jew" or what Efron calls the "ur-typus" (1994, 88). And there was a longstanding tradition of the "myth of Sephardic supremacy," that is, that Sephardic Jews were more true to the ancient Israelite racial type than were Ashkenazi Jews (Schorsch 1989), although it is important to remember that the term "Sephardic" did not refer to what came to be known as "oriental Jews" in the Israeli state. The known Jewish world for Ashkenazi Jews who had migrated to Palestine was much broader (Bloom 2007, 2008; Hirsch 2008, Shenhav 2006).[23] And that reality became radically more real following the establishment of the Israeli state.

If eugenic visions and a concern with "medical management" influenced policies of the official Yishuv regarding Jewish immigration in the prestate period, similar issues resurfaced following the massive immigration of Middle Eastern and North African Jews. Immediately after the Second World War, Zionist agencies relinquished their criteria for medical selection of immigrants in order to allow European-Jewish refugees to immigrate to Palestine (Shvarts et al. 2005). But following Israel's establishment and the "huge wave of immigrants entering the young country," a strident debate ensued over whether or not to enforce medical criteria selecting which Jews could immigrate (ibid., 10).[24] Israel's political leadership by and large opted for "quantity" over "quality" in the state's early years. Nevertheless, that did not mean either that medical inspections or interventions ceased or that eugenic anxieties evaporated. Between 1949 and 1952, for example, of the 700,000 immigrants who passed through Sh'ar Ha'aliyah, a camp near Haifa, 400,000 were subjected to medical examinations (Davidovitch and Shvarts 2004, 156). More broadly, "the health system constituted a key component in immigrant absorption and the melting pot policy," in which "building a preventative medicine scheme and health education network going beyond immediate treatment of various diseases" were considered key. Hygiene education emerged as one focus of that project (Shvarts et al. 2005, 153), as did vaccination campaigns and the control over children's health, with state authorities removing children from their (Eastern) immigrant parents in order to provide what they considered proper medical treatment (ibid., 161–62) and sometimes, more generally, a presumably better life in the homes of Ashkenazi parents (Shenhav 2006; Weiss 2001).

If medical management and eugenic anxieties constituted one axis of the interest in the biology of Israel's Jewish citizenry, the other was an intense investment in and anxiety about Jewish peoplehood. For a state founded upon the "ingathering" of communities who shared little in the way of culture, language, and even religious practices, the work of building *the nation* and not just the state loomed large. It was in that context that blood group and epidemiological-genetic data were used in order to reconstruct Jewish history, determine "the common genetic characteristics of the Jews," and "establish links with the more 'esoteric' communities" (Falk 2006b, 157). And given the fact that medicine and public health were key components of managing immigrant "absorption," the reach of medical institutions and personnel extended deep into the initial "reception" camps in which immigrants were put and into the *maabarot* or transit camps intended to provide temporary housing to the state's new (Eastern) Jewish arrivals but which became longstanding places of residence. Blood was being collected for a variety of medical reasons. And it was apparently made available for population genetic research.

Blood Group Data and the Question of Jewish Peoplehood

By the 1950s, blood groups were the mainstay of population genetics. The first blood groups to be identified were the ABO blood groups (circa

1900),[25] and while the "practical importance" of blood group systems has been primarily about assessing compatibility for blood transfusions, 1919 witnessed the "first time blood-group analysis was applied to anthropology" (Mourant 1961). At the close of the First World War, a Professor Herzfeld and his wife, as reported by Mourant in a 1961 Huxley Memorial Lecture, were in Salonika, "a great cross-roads for the movement both of troops and of refugees," and they tested the blood of "large numbers of persons from many lands and most continents" (ibid., 155). What they found was that while all four blood groups were found in nearly all populations, the distributions of each varied significantly (155). Their work marked "the first application to anthropology of a totally new method, the study of gene distribution" (155).[26]

Blood groups were understood to be "physical characteristics" that "are stable throughout life and appear to have little if any effect on survival" (Mourant 1954, xx). As Mourant explained, particular blood groups had certain "advantages" for anthropological studies:

> [Blood groups] are fixed for life, at the moment of conception, by the genetical constitution of the individual. Also, unlike such features as the size of various parts of the body, they are unaffected by the subsequent history of the individual. . . . Moreover, while visible characteristics of the body, and especially the colour of the skin, have become associated in some quarters with racial prejudice, and allegations of inferiority and superiority, the blood groups have hitherto gathered no such unscientific accretions. (1961, 155)

Funded by a grant to the Royal Anthropological Institute in London to study the use of blood groups in "understanding human diversity," Mourant sought to explain the different blood group systems and their possible usefulness for this purpose. He compiled the available data on the distribution of blood groups and systems in order to present some general conclusions about their frequencies in different populations and different regions of the world. He published his book, *The Distribution of The Human Blood Groups*, in 1954. Together with his 1958 publication *The ABO Blood Groups*, that book became a canonical text and a standard reference for global distributions of the blood groups in human populations. For the field of Israeli population genetics, Mourant's work provided essential data in terms of which the blood group data on Israel's Jewish population(s) would be interpreted and read.

In the 1950s and 1960s, a team of Israeli researchers led by Yosef Gurevitch of the Hebrew University–Hadassah Medical School conducted a series of studies on the blood groups of Israeli populations. Gurevitch was medical doctor who immigrated to Palestine in 1921, joined the staff of the Rothschild Hadassah Hospital in Jerusalem in 1930 (as a bacteriologist), and in 1949 became a lecturer in the medical school. He subsequently directed the Department of Microbiology at Hadassah Hospital, and perhaps most important for our purposes, was also in charge of the blood bank (Kirsh 2007, 182).

As explained by a speaker before the 1961 Jerusalem conference, Professor Gurevitch, who had by then passed away, "felt that the gathering of exiles from all over the world offered a unique opportunity for anthropological studies" (Nelkin 1963, 18). Israeli researchers never questioned whether or not *Israel's* Jewish population was representative of worldwide Jewry (even though particular communities of Jews migrated in much larger numbers than did others), although that was a question raised by Mourant (1954).[27] In the context of Israeli statehood in its early years and the state and society's complex and critical relationship to the Jewish diaspora, it is no surprise that such questions were not raised: the *real* Jew was the new Hebrew, the Jew who lived in Israel. The diaspora was a temporary condition to be reversed and the diaspora Jew a problematic Jew to be remade.

Gurevitch's team collected blood group data from "3,500 individuals from ten communities," data they wanted to be sure to compile before "the communities merged" (Nelkin 1963, 18). This was salvage genetics. In practice, those ten communities were not of the same scale: whereas the Ashkenazim and the Sephardim were single categories, the remainder of the Jews—oriental Jews who were understood, at least formally, to belong to the "third" large division of Judaism (Margolis, Gurevitch, and Hermoni 1960a, 201)—were studied as a collection of distinct and disparate populations: as Jews from Yemen, Cochin, Baghdad, Kurdistan, Persia, Morocco, Tunisia, and Tripolitania (in Libya), a collection of categories that moved from city to country to regions without any consideration of the different classificatory regimes that were in play. Moreover, "the Jews"—a category that presumably incorporated all of these groups and which, by definition, was the population being examined—rarely operated *in practice*: who was compared to whom or what quandaries or problems were being explored is far more complex than the a priori goal of studying "Jewish communities" would suggest.

Increasingly over the past several decades, critics have drawn attention to the history of Ashkenazi dominance of the Israeli state. It has been argued that Mizrahi Jews, formerly often referred to as Sephardim,[28] were marginalized by the Ashkenazi establishment and subjected to massive discrimination and a civilizing mission which was intended to drive a wedge between Mizrahi immigrants and their Arab and "Eastern" cultures and to assimilate them to "the Israeli"—that is, the Ashkenazi Jewish—social and cultural order (Swirski 1989; Shohat 1989, 1988; Shenhav 2006; Eyal 2006; Chetrit 2010). The view of Mizrahi Jews as marginal to the Jewish state and to the Jewish "mainstream" is certainly evident in this work of Israeli population genetics: oriental Jewish communities were explored as Jewish communities less known, their histories less clear, their relations one to the other and to the Jewish "mainstream" a primary object of scientific inquiry.[29] I begin by reading the publications of Gurevitch and his team from this perspective on the Ashkenazi state and its view of the oriental Jewish Other.

Gurevitch and his colleagues published a series of papers from the mid-1950s through the early 1960s, the results of blood group studies of particular Jewish communities: the Jews of Cochin, of Iraq, of Persia, of Morocco and Tunisia, for example, and in 1961, of the Sephardim and of the Ashkenazim.[30] In analyzing these papers, I explore what the population being studied is or is made to be by virtue of the comparisons drawn and the silences evident in each publication.

As with all of the blood group studies published by Gurevitch and his team, the paper on "Blood Groups in Jews of Iraq" begins with a lesson in history. "Iraqi Jews have been considered to be the descendants of the Jews deported from Palestine to ancient Babylonia after the destruction of the first Temple" (Gurevitch and Margolis 1955, 257). Various regimes ruled over the area and various populations came and went:

> So many destructions, displacements, persecutions and massacres took place, that it is difficult to assume that the Jews remained there so isolated as to be considered an anthropologically pure community. Nevertheless, Iraqi Jews themselves have insisted on their being a non-assimilated Jewry, devoted to Jewish religion and tradition. Intermarriages have not been reported. (1955, 257)

The history of "Iraqi Jews" is somewhat ambiguous, according to this account. There is the claim that Iraqi Jews are not "pure" Jews, and yet that account sits side-by-side with a cultural narrative that the communities have not assimilated and have sustained endogamous marriage practices. As presented in the paper's opening paragraph, both of these sorts of claims—those based upon the history of Iraqi Jews written in encyclopedias of Judaism and those based upon the self-understandings and narratives of Iraqi

Jews—were granted equal evidentiary standing. But the reader soon learns that "the Jews of Iraq" are not a single population. There is not one Jewish community but two. "The Jews of Iraq" refers to first "Iraqi Jews," who are "more accurately" named "Jews of Baghdad," and, second, to Jews of Kurdistan or "Kurdistani Jews." According to Gurevitch and his team, Kurdistani Jews are the "pure" ones. It is they who more plausibly descend from ancient Israelites with little admixture. Once again combining information from Jewish encyclopedias with the immigrants' self-representations,[31] we learn that the Jews of Kurdistan, by and large, have been saved from persecution and massacres and from cultural influences. They have "to a great extent preserved the language, customs and religion of their ancestors" (ibid.). They consider themselves to have lived in Kurdistan from ancient times, being descendants of the ten lost Jewish tribes. Kurdish Jews "might be considered as a pure stock" (ibid.; see also Gurevitch, Hermoni, and Margolis 1953).

In the context of these apparent historical facts, the paper offers two sets of comparisons of the blood group data: first, the paper presents a comparison between Kurdish Jews and Europeans. It is worth emphasizing here that this is a comparison between Kurdish Jews and "Europeans," not European Jews. For example, with reference to the frequencies of their ABO, MN, and Rh frequencies among Kurdish Jews, "a very low O percentage and relatively high AB and B percentages were noted, as compared with results found in Europeans" (Gurevitch and Margolis 1955, 257). With regard to the MN groups, "the percentages were similar to those found in Europeans" (258).[32]

Second, the paper presents a comparison of the data on the Jews of Baghdad with the data on Kurdish Jews. In this comparison, there is no mention of Europeans or, for that matter, of any population residing outside of Iraq. Some figures are similar between Kurdish and Baghdad Jews, ABO frequencies, for example. Others exhibit "more of a difference" (MN system distributions).[33] It appears as if the results from studies of Baghdad Jews are being evaluated against the pure(r) stock, the Kurdistani Jews. This paper presents *two Jewish populations*—one that has a potential relationship to Europeans worth exploring, the other worth evaluating solely in terms of its sameness with or difference from the other *local* Jewish population.[34] Strikingly missing is any reference to any Jewish community outside the Iraqi fold, let alone to a category of "the Jews" writ large.[35]

Each paper on the blood group of one or another oriental Jewish community is structured in a similar way. Each opens with a history of the community. Some histories were more fully known (Moroccan and Tunisian

Jews, Tripolitanian Jews) than others (Cochin, Yemenite, Persian). Each paper presents comparisons with other populations. And in most cases those comparisons are limited to other oriental Jewish communities. Occasionally there are comparisons with either European populations or Sephardic Jewish populations.

What might an analysis of Gurevitch's published papers on Jewish-Israeli blood group data reveal about practical understandings of the Jewish world among Israel's scientific elite at this moment in time? And what might these papers reveal about the central questions and problems regarding Jewish origins that required resolution or proof?

As a population category, oriental Jews posed the biggest problem: What *is* their history? What is actually known about them? Are oriental Jewish populations related one to another? Is there any evidence of unity among members of this seemingly disparate and widely scattered category? As the scientific publications point out, oriental Jews do not share culture, language, or distinct religious rites. For Gurevitch and his colleagues the "oriental" is a residual category, invoked only at certain moments, such as in the paper on the Ashkenazim, which I discuss below, in a talk by D. Nelkin to the 1961 conference, or in comparisons with the blood system data of other groups, say of Persian Jews. In actual practice, however, data on "oriental Jews" is never presented as a single sample. None of the papers ever define or justify the classificatory logic of "oriental Jews." And in contrast to that of "Sephardim" and "Ashkenazim," the name "oriental" is almost never capitalized. There is no speculation on "the history of" *Oriental Jews*. Each community has a history of its own, however poorly known. What makes these communities "oriental" for scientific researchers—as was true for the Israeli state, its demographic imagination, and its treatment of these new immigrant communities—is quite simply that they are not in or of the West.

In contrast to the historical narratives that open the papers on the Sephardim and Ashkenazim, the researchers barely sound mildly self-assured regarding the history of each group (with the partial exceptions of the Moroccan, Tunisian, and the Tripolitanian Jews that I discuss separately below). As in the paper on the Jews of Iraq, each paper begins by explaining what is "known" about each oriental community, often "very little." Operating within the logic of an anthropology of the primitive or, in the language of population genetics, the logic of an "isolate," the papers (most often) proceed to speculate upon the anthropological status of each group—pure or not. Here is a second example, an account of the history of the Jews of Persia:

> Very little is known about the history of the Jews in Persia and about the social and cultural conditions of their life. It is definite, however, that a Jewish community was established there during the Babylonian exile. The community appears to have existed continuously since then. Its members sought to guard their cultural individuality, carefully guarded themselves against all foreign influences and observed all of their religious laws, despite many vicissitudes during which many of them were killed or converted to Mohammedanism. Nevertheless, it is difficult to assume that the Persian Jewish community isolated itself so successfully that it can be considered anthropologically pure. (Gurevitch, Hasson, and Margolis 1956, 135)

Or to take the example of what seems to have been considered the most obscure of the oriental Jewish communities, Gurevitch and his colleagues present the following account of Cochin Jewish history:

> Some historians believe that the first Jewish settlers in Cochin, south India, arrived about the 10th century B.C. with King Solomon's fleet. Others hold that Cochin Jews are descendants of the Ten Lost Tribes of Israel. It is also believed by some that they are descendants of the Jews exiled to Babylon by Nebuchadnezzar. The Jews of Cochin themselves maintain that soon after the destruction of the Second Temple (70 B.C. [*sic*]) ten thousand Jews escaped their Roman captors and landed on the Malabar Coast. There are indications that towards the end of the second century B.C. [*sic*] several thousand persecuted Jews from Yemen emigrated to India. (Gurevitch, Hasson, Margolis, and Poliakoff 1955a, 254)[36]

The historical accounts of Sephardic and Ashkenazi Jewish communities are nowhere near as speculative. Moreover, in the publications on the Sephardic and Ashkenazi blood group data, what it is that makes each "community" *a community* is clear. For example: "Sephardic Jews are the descendants of the Jews who were expelled from Spain in 1492 and from Portugal in 1497, and who settled in all parts of the Ottoman Empire" (Margolis, Gurevitch, and Hermoni 1960b, 197). The authors present the broad strokes of Sephardic history through the centuries: the lack of "definite proof" of Jewish existence in Spain in the early Christian era; the evidence of Jewish presence in Spain by 306 C.E; and their expulsion from Spain and Portugal in the fifteenth century and subsequent emigration to southern France, Holland, Italy, North Africa, and Asia Minor. "With so extensive [a] geographical and political distribution . . . it was only natural that the character of the cultural, economic and social life of the Sephardic

Jews developed great heterogeneity. Nevertheless there were certain common traits of language, religious customs and general demeanor" which they sustained "wherever they settled," preserving their separation from the "indigenous Jewish communities" (ibid.). There is absolutely no mention of their racial status: whether or not they are anthropologically pure. This is not a Jewish isolate. It is a Jewish community formed in history by language, customs, and general demeanor. The differences in the types of historical accounts and in the assessment of each population's "anthropological status" are not the only notable distinctions between the papers on oriental communities, on the one hand, and those on Ashkenazi and Sephardic Jews, on the other. In addition, in contrast to the analysis of blood group data of Sephardim and Ashkenazim, it is almost entirely among themselves that the different Jewish communities of the Orient are compared. I want to offer several possible and partial explanations for these differences in scientific practice.

First, the general lack of broader comparisons seems a logical consequence of the fact that the category is a residual one: how can one compare "oriental Jews" with Sephardim or Ashkenazim or for that matter with Europeans when the category doesn't actually exist in any substantive or practical sense? Moreover, it is not just that no such single population is evident in the manner in which Gurevitch and his team designed the research and collected the blood group data, i.e., on the basis of individuals classified according to their membership in Iraqi, Cochin, Persian Jewish communities, and so forth. No population of "oriental Jews" emerges as a consequence of their work either. In other words, *on the basis of the biological evidence,* Oriental Jews are not "a population." I return briefly to the paper on Cochin Jews.

After learning that we know virtually nothing for certain regarding this group's history (except for the origins of Cochin's "White" Jews),[37] Gurevitch and his colleagues present the data. As the authors note, they compared Cochin blood group data with "findings on Jews from Yemen, Kurdistan, Iraq and Tripolitania." The authors present various conclusions. Similar to Yemeni Jews, Cochin Jews have "unusually high" O frequencies. They have a very low A frequency, "the lowest figure found so far" in any of the studies. B frequencies are very high, similar to Tripolitanian Jews. As for the M and N frequencies, the former is high and the latter low, and both are "quite different from findings in other oriental Jewish communities." Data analysis proceeds in this manner, and no coherent picture of a discernible relatedness among Jews of these five oriental Jewish communities emerges. This outcome is evident in each paper on the blood group data of a specific oriental Jewish community: *communities* are compared; no single "oriental" community or population exists, either with reference to a shared history or with reference to shared blood group frequencies.

By way of contrast, one could assume that the blood group data on Ashkenazi and Sephardic Jews *confirmed* their existence as recognizable, distinct populations. No such confirmation could be derived from the data, however. Given the research designs of Gurevitch's studies of Ashkenazi and Sephardic Jews, one cannot assess—or "test"—the presumption that the Ashkenazim and Sephardim are meaningful biological groups. Gurevitch and his team treated each group as a population *at the outset*: they did not collect blood group data using a method that would test the hypothesis that Jews from Poland, Germany, Austria, Hungary, and so on are *a* population, that they carry sufficiently similar distributions of blood group markers to indicate a shared origin. The data on blood group frequencies among the Ashkenazim, as was true of the blood group data on the Sephardim, were collected and reported precisely as that—as *Ashkenazi blood group data*. In other words, the data was not divided up according to countries of origin (see Kirsh 2003, 643). The classificatory categories "Ashkenazi" and "Sephardic" were black-boxed in the very design of the studies (Latour 1987). Those categories were assumed a priori to exist. In turn, the work of collecting and analyzing the data on each "group" reiterated the biological truth that the Ashkenazim and the Sephardim are identifiable populations which can, in principle, be compared with other presumably equivalent populations—Jews of Cochin, of Yemen, of Baghdad, and so forth. And in making such comparisons, the researchers paid no regard either to the differences in the diversity of methods used to study each of these statistical compilations of individuals and their blood group data or in the scale or types of categories used.

Second, the fact that oriental Jews are compared, by and large, only among themselves can be seen to be a consequence of a persistent racial logic that structured science and politics alike in the mid-twentieth century, and more specifically among European Jews in the Israeli state: in practice it was hard to imagine that *these* Jews—oriental Jews—were actually the biological kin of their European counterparts. They "looked" different, after all. They were primitive. And the more "far flung" each community was the less likely any connection to European Jews, Sephardic or Ashkenazi, became. The studies of Kurdish, Baghdad, Persian, and Cochin Jews, for example, are never compared (at least not directly) with their European brethren. But the inability to imagine kinship between oriental Jews and European Jews is probably most striking in the study of the Tripolitanian community. That paper displays a radical disjuncture between what the authors tell us about the history of this Jewish community, on the one hand, and the Jewish groups they chose as relevant for the purposes of comparison, on the other.

"Jews have lived in Tripolitania," the paper begins, "since the 10th century." Originally founded with about eight hundred families, the Jewish community there grew when more Jews emigrated in the seventeenth century from Livomo, Italy. By 1910, the Jewish population in Tripoli had expanded to about fifteen thousand. But the Israeli team did not collect blood samples from these urban Jews. Instead, they collected the data from Jews of Gibbel-Gefren, one of two small villages in which "a small part of Tripolitanian Jews were scattered . . . living a primitive life in caves" (Gurevitch, Hasson, Margolis, and Poliakoff 1955b, 260). Those were the Tripolitanian Jews who had emigrated to the Jewish state.

Although there are no historical records regarding the geographic origins of the "original" Jewish families in Gibbel-Gefren, given what we have been told regarding the migration of Italian Jews to Tripoli in the seventeenth century, it seems plausible that at least some of the ancestors of these "primitive" Jews were European. Relying on purely geographical criteria, it should have been southern European Jews who were the most relevant comparative Jewish group. Nevertheless, the comparison with Italian Jews is never made. The "Tripolitanian" blood group data is compared with data from "Europeans" generally, but, as in the case of the Kurdistani Jews, not with European Jews. As far as the Jewish comparisons go, the data from the Jews of Gibbel-Gefren is analyzed only in relation to other oriental Jews: Cochin Jews, Yemenite Jews, Kurdish Jews. Gurevitch and his team note differences with the data from other oriental communities, with only one possible exception (a relatively high frequency of the B phenotype) common to both Tripolitanian and Cochin Jews. By way of contrast, despite the fact that various similarities emerged between the blood group frequencies of Tripolitanian Jews and those of Europeans, the next logical step was never taken: in this paper, Gurevitch and his colleagues never compare the data from Tripolitanian Jews with data from Sephardic Jews (the most likely comparison) or with Ashkenazi Jews. *By definition*, these Tripolitanian Jews are oriental Jews—on the basis of their (traditional) cultural and (Arabic-Hebrew) linguistic practices.[38] The conceptual and political leap to comparing their blood group data directly with that of their European Jewish "kin" was apparently impossible to make.

The treatment of Moroccan and Tunisian Jews, however, presents a somewhat different picture. As in many of the "oriental regions," there are actually two Jewish communities in Morocco and Tunisia. First, there are "the so-called town and village Jews of Morocco and the 'Tonensa' Jews of Tunisia . . . [who] are supposed to be the descendants of the first Jewish settlers," presumed to have arrived "before the destruction of the Second Temple" (Margolis, Gurevitch, and Hasson 1957, 65). Second, there are the descendants of Spanish and Portuguese Jewish exiles. Gurevitch and his team studied the former community. It was those Jews who migrated to Israel in significant numbers. Most important for my purposes, the data on this group of oriental Jews was analyzed with a much broader comparative brush.

Given that the Moroccan and Tunisian communities were connected to southern Europe through the history of Muslim rule in the Iberian Peninsula and the subsequent deportation of Jews (and Muslims) from a re-Christianized state, the researchers seem to have imagined a wider comparative framework through which to evaluate their blood group frequencies: the data was analyzed in relation both to that of other non-Jewish populations and to that of Sephardic Jews. To be clear: even though this blood group study was a study of "the first Jewish settlers" (ibid., 65), the Jewish community classified as the first Jewish settlers was a more complex population than the term suggests. According to the authors, the Jewish populations of Morocco and Tunisia expanded in the seventh century "by the arrival of Spanish refugees escaping from persecution of the Visigothic kings and under the Arab rule in Africa a new immigration of Andalusian Jews ensued" (ibid.), that is, well before the arrival of Jews expelled in 1492 from Spain and in 1497 from Portugal. Moreover, upon the arrival of those fifteenth-century Jewish exiles, "at first, owing to their small number . . . they mixed with the older settlers . . . [even if] towards the end of the sixteenth century they formed separate organizations" (ibid., 66). In other words, even Morocco and Tunisia's "first" Jewish inhabitants had a long history of mixing with communities who came from afar. Tunisian and Moroccan Jews were compared to one another and to "other peoples of the Mediterranean"—the populations of Spain and Catania (Italy), and to Sephardic Jews of the Balkan states. In addition, one very specific historical question was asked: are the thirty thousand members of a contemporary Jewish-Berber tribe descendants of ancient Jews who became Berber in terms of language and culture, or are they Berber converts to Judaism? (Based on the similarity of their blood group frequencies to Berbers, researchers concluded the latter explanation to be correct.)

In analyzing this group of "oriental Jews," a far wider range of presumably relevant comparative data was available, and far more specific historical questions were explored: Who is this Jewish Berber tribe? How integrated did Jewish exiles from Spain and Portugal become into the long-standing local community? While the comparisons drawn in the paper on Moroccan and Tunisian Jews did not invoke a single Jewish world as the relevant category of analysis, neither did this group of presumably oriental Jews remain an isolate: there is a world—both Jewish and non-Jewish—beyond Morocco and Tunisia of which these Jews were understood to be a part. And there was no mention of any other oriental Jewish communities at all. Despite definitions, the "oriental Jew" was not an operative category in this study.

What emerges from these studies is evidence of a collection of different and disparate Jewish communities. The data are anything but cohesive. No unifying picture can be drawn on their basis, and Gurevitch and his colleagues made no attempt to do so. As Nurit Kirsch (2007, 190) has argued, in the face of that disparate evidence, the researchers often avoided drawing any conclusions at all. In short, there is no *single* Jewish population *made* on the basis of the blood group data. There was not even a single Oriental Jewish population that emerged out of this practice of collecting blood samples and analyzing blood group data on recent Jewish immigrants to the Jewish state, although those a priori categories—the "Jews," "oriental Jews"—were nevertheless not abandoned. They were never recognized as having been undermined by the blood group data.

These papers on blood groups can be read as part of a project to produce a *content*—in this instance, a *biological content*—for a Jewish peoplehood presumed a priori to exist. And what is evident in the work of human genetics, as was evident in many other policies and practices for integrating oriental Jews into the Ashkenazi state (see Swirski 1989; Shenhav 2006; Eyal 2006), is a practical ambivalence and an uncertainty about the very presumption or ideology of Jewish peoplehood upon which Zionism was built. Are all Jews *really* kin? Was it possible *in practice* to imagine and to sustain such an expansive understanding of Jewish kinship and of the Jewish world?

But I also want to propose a second, more counterintuitive reading that lends insight into a different problem faced by the newly established Jewish state, given its belief in and commitment to a national ideology of "return." A major problem for Israeli human geneticists in the 1950s and 1960s was the historical origins of European—and, more specifically, of Ashkenazi—Jews. Consider the various logics and practices evident in the paper "Blood Groups in Ashkenazi Jews" (Margolis, Gurevitch, and Hermoni 1960a), the only paper in which the results from *all* the various Jewish communities are mentioned and compared with the blood group frequencies of the Ashkenazim.

As is true of all of Gurevitch's publications, the paper on the Ashkenazim begins with a historical account of who they are. In this paper, however, the authors present comparisons with the other Jewish groups up front. "The term Ashkenazim is used to denote one of the great divisions of Jewry in contradistinction to the Sephardim (Spanish Jews) and the Oriental Jews, from whom they differ in many respects" (1960a, 201). (This is the one context in which I have found the term Oriental Jews capitalized.) The authors then give an account of Ashkenazi history—persecution in France and Germany, beginning with the Crusades, and flight toward Northern and Eastern Europe. I want to focus on what comes next: "Anthropologically the Ashkenazim differ from the Sephardim and their oriental brethren: they have a larger proportion of blonds, have rounded faces and heads and are shorter especially in comparison with the Sephardim" (201). Leaving the question of shortness aside (I have no idea what to make of it), what we are being told is that *phenotypically* these are (Central and Eastern) Europeans. What does the blood group evidence say?

Numerous comparisons are offered in the paper. Many of the blood group frequencies (of the O, of the cDE chromosome of the Rh system, the M and N) are similar to those of Europeans, Mourant's book being the source of comparative data. In addition, Ashkenazi blood group data are compared with all of the available data on other Jewish communities, both individual oriental communities and the Sephardim. In most instances, the authors note divergence, especially with reference to oriental Jewish communities. Occasionally they note similarities, mostly with the Sephardim. In one instance they note a similarity with Moroccan and Tunisian Jews, and with respect to one "unexpected" result, they also note that a similarly high frequency of a "North European" chromosome (of the RH system) is also found "in some oriental Jewish communities." "The question was then raised as to the origin of this chromosome in these [oriental] communities" (202).

The picture that develops is rather inchoate: for the most part, differences were marked, especially with oriental communities. Some similarities were found between the Ashkenazim and one or more Jewish communities. But what interests me is the *difference in analytic practice*. Why, in contrast to every other paper, does each of the other Jewish communities appear here? One could propose that the "Other" Jews are being evaluated against an Ashkenazi (genetic) norm. But I think the significance of the comparisons moves in the opposite direction. There is a "problem" regarding the origins of the Ashkenazim, which needs resolution: Ashkenazi Jews, *who seem European*—phenotypically, that is—are the normative center of world Jewry. No less, they are the political and cultural elite of the newly founded Jewish state. Given their central symbolic and political capital in the Jewish state and given simultaneously the scientific and social persistence of racial logics as ways of categorizing and understanding human groups, it was essential to find other evidence that Israel's European Jews were not in truth Europeans. The normative Jew had to have his/her origins in ancient Palestine or else the fundamental tenet of Zionism, the entire edifice of Jewish history and nationalist ideology, would come tumbling down. In short, the Ashkenazi Jew is *the* Jew—the Jew in relation to whose values and cultural practices the oriental Jew in Israel must assimilate. Simultaneously, however, the Ashkenazi Jew is the most dubious Jew, the Jew whose historical and genealogical roots in ancient Palestine are most difficult *to see* and perhaps thus to believe—*in practice*, although clearly not by definition.

It is impossible to derive any single history of "the Jews" on the basis of the blood group data that Gurevitch and his team collected in Israel's early decades. To paraphrase M'Charek (2005), different genes different populations make. In light of a project intended—sometimes explicitly, sometimes implicitly—"to prove that superimposed on the obvious variation between Jewish communities and tribes, common genetic factors might be found" (Falk 1998, 601), there was only one gene that functioned as a "good marker" (M'Charek 2005): the CDe or the so-called Mediterranean chromosome.

The first mention of the CDe chromosome as a shared marker among the various Jewish communities was in the 1957 paper on Moroccan and Tunisian Jews. As noted, Moroccan and Tunisian Jews exhibit "a typically Mediterranean high level of CDe (53.4 and 56.09% respectively), as found in all Jewish groups so far examined" (Margolis, Gurevitch, and Hasson 1957, 67; see also Goldschmidt 1963, 19). Ranging from 60.5 percent in Jews from Persia to 41.5 percent in Cochin Jews, these values are relatively high. As Gurevitch and his colleagues explained, these high levels of CDe are found in "all Jewish communities studied by us, [and that] suggests the common origin of *the Jewish people*" (Margolis 1957, emphasis added). That is the one and only mention of "the Jewish people" in any of these publications.

Insofar as "[genetic] markers are actually involved in what genetic diversity is" (M'Charek 2005, 24), by the early 1960s the CDe was the only genetic sign available from the blood group data that enabled the researchers to draw a circle around a population called "the Jewish people." It was argued that the CDe chromosome differentiated Jews from various non-Jewish, non-Mediterranean populations with whom they (used to) live and who tend to exhibit, by way of contrast, high rates of different chromosomes of the RH system. (The CDe did not differentiate Jews from other Mediterranean populations, however.)[39]

As indicated in the papers and exhibits before the 1961 international conference of human geneticists in Jerusalem, the search for "good markers" was ongoing. Beyond the blood group work of Gurevitch and his team, which was considered the best genetic evidence for studying population origins at the time, researchers sought other polymorphisms (fingerprint patterns, haptoglobin and transferrin types) and disease mutations (thalassemia, G6PD deficiency, "defects in color vision") that might provide evidence of a shared origin for the disparate contemporary Jewish populations now resident in Israel (see Goldschmidt 1963; Kalmus et al. 1961; Sheba 1960). This search for the good marker reads like an almost random quest for genetic signs that would prove that a single, historic Jewish population does indeed exist. And it was against what seemed to be a relatively haphazard search for an answer to the question of Jewish origins—to the question of whether or not Jews "are indeed a race"—that Helmut Muhsam, a demographer by training, attempted to construct a systematic method to answer the question once and for all. "The Genetic Origin of the Jews" (Muhsam 1964) marks a significant departure from Gurevitch's earlier studies: "the Jews" as a single category operates as *the* object of analysis. Muhsam's goal was to devise a method rigorous enough to identify a distinct Jewish population descendant from an original Jewish race on the basis of blood group data.

Homo israelensis

Helmut Muhsam began his paper by articulating the problem in the then current state of research on Jewish populations: there is no resolution to the question of "whether the Jews should, or could, be considered as one single race" (1964, 36).[40] And the reason that question had not been resolved, he argued, lies in methodological obstacles: while a "wide variety of morphological characteristics" have been studied, there has been no systematic approach to the data. Muhsam set out to develop just such an approach. Before doing so, however, he cautioned his readers that *this* race might not be quite like other races:

> That this racial entity—if it exists at all—would present much higher internal variability than is commonly found in "a race," is beyond doubt, in view of the diversity observed in the Jewish people. But because of the well-established distinction between any Jewish eidah ["community"] and the surrounding gentile population (to which we shall refer . . . as the "genetic environment" of the eidah), the diversity within the Jewish people does not render the discussion of racial unity of the Jews completely futile. (1964, 37)

Diversity should not be read as a sign of the absence of a Jewish race, Muhsam argues. For both mathematical and historical reasons, a "flexible criterion" (41) must be used in order to establish a common genetic or racial origin.

Muhsam designed a method to evaluate two sets of relationships: first, the genetic relationship between the contemporary "eidoth" and their respective genetic environments, and second, the genetic relationship of each to the "original Jewish race" (38): "if the various Jewish eidoth resulted mainly from an amalgamation of the original Jewish race and their respective genetic environment . . . then any such eidah should occupacy [*sic*] an intermediate position between the original Jewish type and its environment" (38). Mathematically speaking: if X_0 represents the original Jewish race, X_1 the eidah, and X_2 the genetic environment, then $X_0 \leq X_1 \leq X_2$ or the inverse (38).[41]

Muhsam's project faced serious obstacles. Most centrally, "almost nothing is known of the original Jewish race—if there ever was one" (38). In addition, with regard to many of the contemporary eidoth, who "wandered through several countries during their history, the appropriate genetic environment with which they assumedly amalgamated, is not known either" (38). But Muhsam was not deterred: with regard to the relevant genetic environment, he relied on the "only assumption which can easily and consistently be made," which is to consider the genetic environment of any eidah to be its present one. As for the problem of virtually no knowledge of an original Jewish race: "It will finally emerge that the absence of information on the original Jewish race is a fortunate circumstance rather than a shortcoming" (39).

We are never actually told how that absence of information turns out to be an advantage. Nevertheless, Muhsam developed a mathematical model for studying the relationship of current eidoth to their present environments and to the past race from which he presumed they had descended. He further specified the assumptions and methods of his study: First, the results will be more reliable if one uses all traits about which data are available, rather than single characteristics. Second, the model will only work if one simplifies the analysis to "a certain type of attribute and to one process of amalgamation: we shall consider attributes which are completely determined by one gene and miscegenation will be considered as the process of amalgamation" (40)

Relying on statistical techniques, Muhsam set out to represent the relationship of contemporary eidoth to the original Jews in a visual form. Representing the "hypothetic Jewish race" as the "central point" in space and each genetic environment as "peripheral points," the "intermediate points"—the genetic characteristics of contemporary eidoth—should form straight lines joining the central point with the peripheral points (40). Put another way, if "all of our assumption [*sic*] are true, the relationship between the genetic origins of all the eidoth, these eidoth and their genetic environments is represented by an n-dimensional star like structure" (40).[42]

Relying on data on thirty-six communities and their respective genetic environments in Mourant's 1954 book, *The ABO Blood Groups*, the "empirical data" do not bear out Muhsam's expectations: "The general picture exhibited . . . can be considered to be largely the opposite of the starlike structure, which would be expected if all eidoth stemmed in fact from a common origin: the rays of our star seem rather to diverge into all directions than toward a single point or, at least, a limited area" (49). Those results, however, hardly deter Muhsam from clinging to his original hypothesis. As Raphael Falk puts it, what follows is "five pages of discussion . . . in which the author refuses to take No for an answer" (Falk 1998, 601). According to Muhsam,

> [The results make] it very unlikely that all Jewish eidoth included in this study stem from a common genetic origin. But at least three different explanations of this *apparent irregularity* can be offered, each of which alone is sufficient to account for the deviation of the empirical data from the expected starlike structure, without contradicting the basic assumption of a common origin of all Jewish eidoth. (1964, 50, emphasis added)

The choice of the wrong genetic environment,[43] a nonrepresentative sample of the gentile group who "mixed" with the Jewish population,[44] or "any selectivity on the ABO locus"[45]—each of those "extraneous facts" (52) could have distorted the results. There are, in addition, numerous other possibilities, the most important of which is that "there never was a single Jewish race, or in any event, even before the Jewish people was dispersed among the various Gentile populations, it did not present a genetic unity" (52). While that is a possibility that "should not be disregarded"—there were twelve Jewish tribes that left ancient Palestine at different times—Muhsam argues it is not very likely that "there were genetic-racial differences between the 12 tribes or that genetically selected groups left Palestine at various periods for distinct destinations" (53). Muhsam thus turned to a "modified" version of this "hypothesis of a highly diversified origin." While there was a single, original Jewish race, there was also "secondary amalgamation." Jewish groups of the diaspora were originally "isolates" that lived separately even one from another. But at later historical moments, they reintegrated, biological speaking that is. In turn, those new population mixtures subsequently also merged with their respective genetic environments. He concludes that, while the current data does not support his original hypothesis,

> In view of various historical facts such as wide-range migration and the formation of isolates for many generations, the simple model may not be able to explain the relationship between each eidah and a hypothetical common genetic origin. Neither the existence of this common origin nor that of some kind of relationship should be considered to be ultimately denied by our analysis. It is hoped that the analysis of additional traits, taking full advantage of the possibility to extend our model into a multi-dimensional attribute space will throw further light on the problem. (1964, 53–54)

For these researchers in population genetics, the fact of Jewish genetic diversity confronted them again and again as a problem no longer at the level of racial phenotype but now with reference to blood group and, as was also a widespread scientific practice at the time, on the basis of disease mutation data. But evidence of diversity was never enough to undermine the original hypothesis that there was an original Jewish race or population from which current Jewish communities descend. In practice, researchers relied on other lines of evidence and other methodological commitments to compensate for the inability of the biological data to produce an acceptable result—from the perspective of the biblical stories and the history of Jewish origins, homeland, and exile taken for granted in modern Jewish nationalism—to produce an answer regarding their questions about Jewish origins that "made sense." As indicated in Muhsam's conclusion, it is not that history can be read or confirmed by the evidentiary terrain of population genetics alone. In light of the counterintuitive results that his analysis produced, the logic must be reversed, at least in part: this novel form of evidence, the evidence drawn from work on blood groups in population genetics, must be interpreted in light of already "known" historical information. I turn to discuss Chaim Sheba's keynote address before the Second International Symposium on Human Genetics held in Jerusalem in 1971 to show that Muhsam was not alone in making this move. Sheba called upon geneticists to turn to existing historical documents in order to interpret the evidence at hand.

Chaim Sheba was a prominent political and scientific figure in the Yishuv and the early years of statehood, culminating in his directorship of the Ministry of Health and subsequently his leadership of Tel Hashomer hospital, which he transformed into an important center for the study of genetic diseases (Bondi 1981, 302). More generally, under his direction, Tel Hashomer hospital (subsequently renamed the Sheba Tel-Hashomer Hospital in his honor) became a center of medical outreach into the *maabarot*, where Jewish immigrants from Middle Eastern and North African states lived, and into the remaining Palestinian villages (ibid., 171). He was also a key proponent of medical selection vis-à-vis new immigrants to the state: he feared that all the healthy Jews were moving to other countries, and the old and the weak coming to Israel at "the expense of the Jewish people" (152). He worried that the medical system was not equipped to deal with the burden and range of diseases immigrants were bringing with them, and was concerned that, given the unsanitary conditions of the camps, the diseases brought by immigrants would spread to the population as a whole. And he was anxious that if too many physically and mentally ill immigrants came to Israel, there would be deleterious genetic consequences for the population as a whole (152–53). But Sheba's interest in medicine and genetics went beyond issues of public health and national biological fitness. A committed nationalist, so too was he interested in biological data as a source for reconstructing Jewish history (303), and "Jewish history" did incorporate Jewish communities for whom, in other contexts, Sheba could hardly contain his disdain.[46]

At a conference on the "Epidemiology of Cancer and Other Chronic Diseases in Migrants to Israel," Sheba gave a lecture on "Jewish Migration in Its Historical Perspective" (1971).[47] Beginning with an autobiographical account of his medical education and Jewish upbringing in Europe, his immigration to Palestine and his clinical experience there, Sheba explained that he gradually came to be interested in population genetics "and its place in clinical medicine and public health" due to evidence of differences in disease incidences among different Jewish communities of Palestine (ibid., 1334). But his interest in population genetics did not end there: clinical experience and the specific lessons on Jewish diversity he learned in Palestine ultimately led him to become "deeply enamored of what I love doing—putting Jewish migration into its historic perspective" (1335). And before a room full of human geneticists and medical doctors, that was the task of his keynote address.

What most struck Sheba on coming to Palestine was the fact of Jewish phenotypic diversity: Yemenite, Kurdish, Iraqi, and Ashkenazi Jews, for example, exhibited different incidences of different diseases. How to reconcile this observation with the fact of common Jewish origins?

> It is my task to try and sort out for you who these migrants to Israel are and leave it to the experts to present the facts and figures. In such a situation one is faced with the problem of deciding whether it is the influence of the environment or of the genetic make-up or, as in most instances, an interplay of both which could be held responsible for the remarkably different frequencies of cancer and chronic diseases, which occur in the various exiles returning to Israel (1335).

And he began that task of sorting out who the migrants are by noting: "The Bible is the major source for the genealogy of the Jewish People (1335). Beginning with the story of Abraham and his journey to the promised land, mentioning along the way fellow-travelers (the Phoenicians, for example), Sheba cautioned: "It would . . . [be] . . . dangerous to decide on the genetics of a group by one marker only, as the Gileadites did to the Ephraimites and killed every person who could not pronounce 'Shibolleth.' Aware of such a fateful mistake in each and in every group returning with the 'ingathering of exiles,' gene frequencies of different mutations were investigated and compared to established ratios" (1336). On the basis of that data, Sheba drew a picture both of specific Jewish communities and their histories and of the Jewish people as a whole. For example, how might one explain that both Libyan Jews and Armenians have the highest gene frequencies for Familial Mediterranean Fever (FMF)?

> The Libyan Jews and the Armenians . . . must have left the common Semitic gene pool in and around the 8th century B.C., most likely from the Kingdom of Israel. They most probably left as refugees into Libya, separated by the

> Western Desert from the two superpowers at war in those days, Egypt and Mesopotamia; . . . the exiled ten tribes, driven out of Samaria and by the Assyrians, could well be an early, and therefore important component of part of today's Armenians. (1336)

Sheba proceeded to support his claim that Armenians are descendants of the ancient Jewish tribes with evidence from ritual practices, physiognomy, and so forth (1336–37). Through a similar logic, he tried to explain the presence of the frequency of cases of Dubin Johnson syndrome (DJS) in Iranian Jews:

> The Iranian isolate with one DJS case for every 1,200, as against 1:40,000 in the Iraqis, can only be explained by the fact that the deportation of King Joachin (of the Tribe of Judah) in 597 B.C., with his entourage including Kish, four generations prior to Mordechai and Esther (of the Tribe of Benjamin), must have taken with them all there was of this mutation into the Iranian exile. (1338)

Sheba also asked: why do "the European (Ashkenazi) Jews practically lack G6PD deficiency" (an X-linked genetic mutation) among other deficiencies commonly found in other Jewish communities (e.g., DJS and the thalassemias)? Relying on a reading of Josephus' *The Jewish Wars,* Sheba argued that the Jews exiled into Italy were only males, sold into slavery or "thrown to the lions in the arena." Those who survived married slave girls who were "imported from north of the Alps, in other words they were Japhethites and not part of the Semitic gene pool" (1338). He continued:

> Thus under the conditions of slavery the Semite males had to marry, and obviously convert, the Japhethite females. Therefore the core of those Jews who moved into Europe in the paths of the various invaders of Rome from North of the Alps did not, in most cases, have wives of Semitic origin; and thus these women did not carry the mutant gene for G6PD deficiency to their male offspring at all. (1338)

In his interpretation of the biological data, Sheba sustained the historic authenticity of Ashkenazi Jewishness by abandoning the Middle Eastern origin of its maternal line.

Over and over again, Sheba turned to "history"—to the Bible, to Josephus, to archaeological findings, to "documents of medieval travelers who searched for the 10 lost tribes (assuming that the travelers were remnants of the two tribes of Judah and Benjamin)" (1340)—in order to give an account of how it was that specific "Semitic genes" got into certain contemporary Jewish populations and not others, or at least in notably different frequencies. He never entertains the possibility given *his own biological evidence* that these different communities might have different, non-Hebrew ancestors:

> The Hebrews had a geographical concentration in this country between 1200 B.C. and 70 A.D. But already in the 8th century B.C., *parts of the body of the people* broke off taking some mutations with them. Communications being as they were in ancient times, and restrictions being imposed on the Jews by their captors, many isolates could never exchange genes (except in sporadic cases). The variety of mutations known to exist among Jews perplexed us, in view of our uniformity in language and religion, until we began to look at the genetic disorders in terms of our history. (1339–40)

Although Sheba relied solely on a history of "isolates" rather than on a history of isolates and then "secondary amalgamation," as had Muhsam, the logic remains the same: read the genetics through the "known" history and the "problem" of Jewish genetic diversity is washed away. And in an admonition to look to history that is worded much more strongly than was Muhsam's, Sheba warned his audience of doctors, geneticists, and epidemiologists of the dangers of not doing so:

> Whoever wishes to dig deeper into the genetic, and even environmentally influenced "*Homo israelensis*" of today, has no alternative but to spend part of his time in reading the Old Testament and then a wealth of information accumulated in Aramaic and Latin (such as the "Jewish Wars" of Josephus Flavius) as well as in the middle ages in the glorious period of Spanish Jewry. Here is an opportunity for the humanities to meet with the natural sciences, not only to add background to the scientists, but also *to prevent geneticists, epidemiologists, and clinicians from drawing conclusions which disregard already existing data*. (1340, emphasis added)

That already existing data is textual data. For its proper interpretation, genetic data depends on older, humanistic sources. The truth of Jewish origins lies elsewhere. It resides in longstanding traditions, in biblical texts, in existing "historical" sources. And as such this search for the biological evidence of Jewish unity—of a sustained Jewish Difference—is an experiment that can never end (Galison 1987). And yet the experiment was reiterated over and over again, whether by Jewish scholars who sought to specify the racial character of the Jews in Europe and the U.S. at the turn of the twentieth century or by Israeli scientists who sought to specify the genetic distinctiveness of Israel's Jewish citizenry in the early decades of statehood, even as those iterations occurred within partially distinct scientific epistemologies and in important ways radically distinct social, institutional, and political contexts.

The effect of that constant reiteration, and in particular its effect within the context of the Israeli state, was complex. Yes, the work of Israeli population genetics reiterated an anxiety that *edot ha-mizrach* could not really be Jewish kin. And it did so in a context in which medical management together with other forms of social and political intervention and control produced and reinforced not just the difference but the inferiority of these non-European Jewish communities (for the most part, Middle Eastern and North African). At the same time, however, the work of population genetics produced oriental Jews *as fellow Jews*: every time researchers studied the biology of "the Jews," they performed the fact of Jewish national unity—understood in the long shadow of racial thought and Zionism, as these were articulated in the early twentieth century—despite the fact that no biological evidence of that unity ever materialized through their work.[48]

By the late 1970s, Israel's "oriental Jewish communities" had become the Mizrahim, and for the most part their kinship with Ashkenazi and Sephardic Jewry had emerged as an unquestioned biological and historical fact (see Bonné-Tamir 1980). The State of Israel had expanded the known Jewish world even if, as I discuss in the following chapter, within the grammar of biological sciences designed to track origins and descent, Mizrahi Jews will never be a category as legible to—as "authentic" within—the grammar of this scientific practice as is the category of the Ashkenazi Jew.

The field of human population genetics has always been bifurcated into two distinct (if often overlapping) kinds of projects, those that seek to understand the "Family of Man," to return to Donna Haraway's words (1989), and those that seek to specify particular (national) populations. As I discussed in chapter 1, a standard history of the shift from race science to population genetics has long privileged the universalizing side of this work: that we are all one species and that in studying biological diversity at the population level, what we are seeking to understand is the processes of *human* evolution and *human* unity. But there is a whole other history that needs to be emphasized far more than it has been to date. Charting biological *difference* has remained as central to the practices of population genetics as has been the desire to chart human evolution and unity. As is evident in the case of the research that I have discussed here, Israeli population genetics was a biopolitical project of relevance to—even if not seamlessly directed by—the interests of a newly founded state and the struggle of its various elites (political, military, scientific) to produce a Jewish nation that it presumed already to exist. The data generated by the work of Israeli researchers proved as important to the international field of human population genetics as to its Israeli equivalent, but the investments of each of those scientific communities in the data were not precisely the same. Even as data on the biology of Israel's Jewish population was understood to be crucial to the management of the vitality, health, and disease of the state's citizenry, for Israeli researchers so too was knowing the genetics of the Jews a practice of nationhood. And insofar as this research was driven by the political unconscious (Jameson 1981) of the Zionist state, there were virtually no population genetic studies of about 12 percent of Israel's citizenry: those Palestinians who remained after 1948 (see Kirsch 2003, 645).[49] Moreover, the most obvious of comparisons could never be made: if this was a project to demonstrate Jewish origins in ancient Palestine, why not compare the data on Jewish populations with data on Palestine's remaining indigenous residents, its so-called "Israeli-Arab" citizens? For Jewish nationalists in the Israeli state in these early decades, the possibility of that kinship was unimaginable. It was completely out of bounds. For Israel's Arab citizens, in contrast to its "oriental Jewish communities," the effect of research on the biology of the Jews was to produce nothing but their absolute Otherness.

CHAPTER THREE

Know Thyself

> Despite being one of the World's most intensively studied groups, there is still only sparse documentary and archaeological evidence from which the histories of the many diverse Jewish communities of the Diaspora can be constructed. There is also considerable uncertainty concerning the extent to which those communities may be descended from the Jewish people of antiquity. As a consequence the genetics of living Jews (and their present and previous neighbours) has been used to illuminate both the communities' demographic histories and their origins.
>
> —Bradman et al. 2004, 89

As evidenced in these opening lines to an article summarizing the results of recent studies of Jewish origins, the questions being asked by genetic anthropologists are not new: What are the origins of today's Jewish communities? Are contemporary Jews descendants of the "Jewish people of antiquity"? Are the different Jewish communities related to one another or not? The kind of evidence, however, is novel. No longer an examination of racial phenotype, of blood groups, or other now "classical" genetic markers, this is molecular genetics.

As was true in earlier scientific efforts to study the biology of "the Jews," researchers are confronted by the fact of biological diversity. But, in their practice of genetic history, they understand biological diversity differently from those who worked according to prior scientific paradigms. The whole analytic "problem" posed by biological diversity has been washed away. From its inception, population genetics posited both the existence and the importance of diversity within population groups. That commitment characterized its departure from race science, which presumed a priori absolute differences between natural kinds. Nevertheless, as I demonstrated in chapter 2, in the actual practices of Israeli population genetics in the 1950s and

1960s, diversity remained a hurdle to be overcome.[1] For population geneticists in the postwar period, as much as for race scientists at the turn of the twentieth century, evidence of biological diversity was a problem that had to be explained away if the existence of the Jewish nation—or the Jewish race—was to be legible and true.

In the practices of anthropological genetics, the existence of intrapopulation diversity is a fully assimilated norm. The fact of genetic diversity among Jewish groups demands no explanation. The goal of the molecular studies is to identify genetic continuities *despite* the admixture (race mixing in an older parlance) that has occurred over the generations. Relying on mitochondrial DNA and the Y-chromosome—the two genetic systems that are inherited lineally, from one's mother (by daughters and sons) or from one's father (in the case of sons)—allows one to "see" kinship and origins without presuming purity. Researchers point out that different "genetic systems" reveal different genealogies and genetic kinships: "Demographic histories of human communities can be extremely complex, affecting *in different ways* the distribution of genetic diversity *in the three separate genetic systems*: the paternally inherited NRY [non-recombining arm of the Y-chromosome], the maternally inherited mtDNA, and the autosomal genome that is inherited from both parents" (Bradman et al. 2004, 89). As Lisa Gannett and James R. Greisemer argue regarding research on blood groups in the 1950s and 1960s, biological groups were "constructed by the privileging of certain traits over others—different choices, different kinds of 'races,' different 'racial groups'" (Gannett and Greisemer 2004, 158). Today it is not just meta-critiques of the biological sciences that recognize the shifting boundaries of population groups. Researchers acknowledge that the choice of different genetic systems can generate different outcomes in sorting human genetic diversity into specific population groups.

It should come as no surprise that diversity is not a problem for molecular genetic attempts to read the history of the Jews. Within the context of a multicultural political imagination that has emerged as axiomatic in many liberal nation-states, diversity is a social and political and not just a scientific norm. Over the next two chapters, I explore the scientific and social conditions of possibility within which historical narratives are questioned or made plausible, or, quite simply, *made*, particular human collectivities and individual identities are fashioned, and a specific kind of scientific ethic and scientist (or knowing subject) emerge. What epistemological and political shifts had to occur in order for genetic history to emerge as a social field privileging such practices of knowing the self? What are the ways that we think and "argue now" (Anderson 2006) that make possible the embrace of genetic historical facts by so many persons and publics today? And what specific practices and understandings of the self is the work of genetic history generating?

The genetic historical studies I analyze are situated squarely within the history and grammar of multiculturalism, as a transnational political form and in its particular American articulation(s). From scientists who study "themselves" to genetic genealogists who seek to discover their familial and ethnic origins, this is a scientific and social quest for one's authentic cultural self. It is a project committed to investigating and discovering evidence of a sustained Difference. In contrast to all the serious ethical debates that have surrounded medical genetics—What are the dangers of breaching the species barrier or of creating life? What are the consequences of identifying genetic proclivities for diseases or pathologies in populations or individuals?—genetic history, and especially the commercial market in genetic ancestry testing I discuss in chapter 4, is saturated by a language of self-exploration and fun. That remains true even when, as in the case of efforts to reconstruct African American roots, it is simultaneously argued to be of significant political import (Gates 2007). To borrow Stanley Fish's term, anthropological genetics is "boutique multiculturalism": multiculturalism not interested in confronting the difficult—the potentially impassable—problems of recognizing and "tolerating" radical difference (Fish 1997; Povinelli 2002). Instead, for many it is a matter of "lifestyle" (Fish 1997, 384)—the quest for self-knowledge and self-definition as a practice, a "politics" in and of itself.[2]

Nevertheless, even as it enables boutique multiculturalism, the historical genetic imagination is not characterized by a radical rupture from the earlier forms of Jewish self-identification and articulations of the Jewish collectivity that I explored in the previous chapter. Therefore, I examine a matrix of contradictory assumptions and commitments, some legacies of the early to mid-twentieth century and others that emerged of late, out of which a distinctly genetic historical self materializes: There is the persistence and power, even in a Jewish diasporic imagination, of a Zionist articulation of Jewish peoplehood, that is, that the Jews are a biologically linked collectivity who originated in ancient Palestine and that that biological-qua-historical origin matters a great deal. More broadly, there is the persistence and power of the belief that biological data tells Jews something fundamental about who they are, collectively and individually. Nevertheless, there is simultaneously a commitment to and a celebration even of the fact that Jews are who they are because of the choices that their ancestors made over and over again. In other words, there is the simultaneous commitment to the authentic self as a *parvenu*, as someone who makes herself. Moreover, genetic historical projects of self-knowing and self-exploration are most often not tied into collective political projects—the Zionist movement, the work of building the nation-state in Israel—as were previous projects of Jewish biological self-definition. (See chapter 5 for a collective political project for which genetic history's working objects are appropriated, albeit one still squarely situated within contemporary forms of identity politics.) Knowing the self exists in the domain of individual self-exploration, even if that desire to know one's individual self simultaneously builds an understanding of an authentic Jewish collectivity. This is identity politics not nationalist politics per se—albeit it is a diasporic identity politics that rearticulates rather than breaks from the Jewish nationalist imagination that Zionism first articulated and that lives on in and underwrites the Jewish character of the Israeli state.

If that is one tension evident in the projects and practices of Jewish genetic history, there is a second tension that runs through the heart of its "epistemic procedures" (Knorr Cetina 1999), that is, how and by whom knowledge is made. Through a reading of public representations of the work on Jewish origins, I argue that an image of the detached expert—the hallmark of a "real," objective scientist—sits side-by-side that of a Jewish scientist interested in knowing himself. This is a coupling of scientific objectivity with identity politics, of expert detachment with personal attachment that likewise characterizes other projects in genetic history as well as broad domains of biomedical inquiry in the U.S. today. I point to that juxtaposition not to call into question the reliability of the scientific work. There is no reason to assume one cannot produce reliable accounts of a phenomenon in which one has personal investments, whether as a Jew or as someone with a family history of heart disease or breast cancer. Instead, I am interested in the joining of the detached expert and the attached Jew in the person of the knowing subject as a new configuration of "a scientist" that is felicitous today, a consequence of the fact that multiculturalism has saturated domains of inquiry that reach well beyond the humanities and the social sciences, those fields most often accused by their critics of having been overtaken by identity politics. This novel configuration of the "Scientific Self" (Shapin 2008), I argue, is crucial to work in genetic history because of the ethical work it is seen to do: It allows one to produce knowledge about the biological difference of, in this instance, the Jews and yet maintain one's distance from the specter and dangers of race science. In our multicultural age studying "the self" is increasingly providing an alibi for studies of group-based human genetic diversity in the biological sciences, helping to fashion such studies as politically safe, at times even as politically valuable, and often it seems as simply "fun."

The Choice of Jewishness

Even though the mid-to-late 1990s witnessed a proliferation of Y-chromosome studies of Jewish origins, studying Jewish genetic history via the patriline presents a problem. Priestly status is and always was passed down from fathers to sons, but since Talmudic times (circa 200 B.C.E. to 500 C.E.) Jewishness has been determined on the basis of maternal descent (see Cohen 1999). In the twentieth century, various streams of Judaism began to recognize patrilineal descent as a legitimate ground for membership in the Jewish world. Nevertheless, matrilineal descent remains the primary and the only universally agreed upon principle of transmission in the Jewish world and in the Jewish state. Mitochondrial DNA, therefore, the genetic line that tracks the maternal lineage, is considered to be particularly salient when investigating the genetic history of the Jews (Goldstein 2008).

Mitochondrial DNA emerged as a working object for anthropological geneticists long before the Y-chromosome. As argued by Marianne Sommer, in order to carry out comparisons of intrahuman (as distinct from interspecies) genetic variation, scientists needed "regions of DNA that showed high enough mutation rates to function as 'carriers of the very recent history of human diversification'" (2008, 509). That scientific object appeared in the mid-1980s when Rebecca Cann, Mark Stoneking, and Alan Wilson published their work on mitochondrial DNA (mtDNA) (Cann, Stoneking, and Wilson 1987).

Mitochondrial DNA is not found in the nucleus of the cell. Outside the cell nucleus but within the cell membrane, mitochondria are found in the cytoplasm. The source of energy for the cell, the mitochondrion is the only cellular structure other than nucleus that has its own genome. In its genome, composed of about 16,500 base pairs of nucleotides (chemical components), there are only thirty-seven genes. The rest are "control regions" that "control the way in which mitochondrial DNA copies itself during cell division" (Sykes 2006, 103). As explained by Bryan Sykes, parts of the mitochondrial control region can function regardless of the precision of the nucleotide sequences: "The vital consequence for us of this tolerance in the DNA sequence of the control region is that when a mutation happens it doesn't affect the performance of the mitochondria at all" (103).

Like the Y-chromosome with its large expanse of noncoding DNA, the control region of mitochondrial DNA, having no known function, is believed to enable scientists to chart origins and descent. When Rebecca Cann and her colleagues first reported having deciphered the mitochondrial genome, it was considered the "perfect epistemological object" for the purpose of mapping intrahuman diversity: "It was relatively simple to analyze; it evolved rapidly, due to the lack of an efficient repair system; and it was inherited strictly maternally and therefore was free of the complexities caused by recombination" (Sommer 2008, 510).

Since the 1980s, there have been numerous efforts to study the Jewish maternal line. Those studies produced evidence for two different accounts of the Jewish matriline: First, that Jewish women from different communities displayed genetic similarity, indicating a common genetic origin. And second, that there has been a large amount of admixture with the communities among whom Jewish women (and men) lived, thus rendering women from different Jewish communities more genetically similar to their "host" populations than to other Jewish women (Thomas et al. 2002, 1412). In the late 1990s, Neil Bradman and Mark Thomas, of the Centre for Genetic Anthropology at University College London, in cooperation with colleagues at the Centre and elsewhere, designed a study to clarify the history of the mtDNA gene pool in Jewish communities. They did so by comparing data on mtDNA in Jewish communities with mtDNA data on non-Jewish women of their "host" communities, on the one hand, and with the Jewish Y-chromosome evidence, on the other. That Y-chromosome evidence, as I discussed in chapter 1, had "confirmed" a shared origin between Jewish and non-Jewish Near Eastern populations (1412).[3] Does the maternal line tell the same story?

Replicating the logic that guided the studies of the Cohanim, researchers began with the following hypothesis: "*If* the system of matrilineal inheritance of Jewish identity has been strictly followed, *we could expect* it to be reflected in systematic differences in the pattern of mtDNA and Y-chromosome genetic variation within and among Jewish populations" (ibid., 1411, emphasis added). Given the matrilineal principle, fewer women than men should have married into Jewish communities. In other words, if the matrilineal principle were upheld, fewer Jewish men would have chosen to marry non-Jewish women. Genetically speaking, that means Jewish women across a broad range of Jewish communities should share more haplotypes (a set of linked mutations) with one another (they should display *less* genetic diversity) than should Jewish men. That, however, is not what the researchers report. The mitochondrial evidence did not support the belief in the shared origins of today's Jewish women in ancient Palestine. Their findings flew in the face of a longstanding narrative regarding the origins of the Jewish diaspora. Nevertheless, as I illustrate, by embedding choice and agency into a reading of the data, a different kind of origin and a different basis for historical and cultural authenticity are produced out of the mitochondrial evidence.

The researchers did not find shared "modal" (most common) haplotypes among the different female Jewish populations.[4] Instead, they found evidence indicating that most of "the Jewish groups *formed independently* around (at least) eight small, distinct nuclei of women" (ibid., 1417, emphasis added).[5] Contrary to expectations, there was far more genetic diversity in the mtDNA gene pool than was found in Israelite Y-chromosomes (that is, there was much more "genetic distance" among Jewish women than among Jewish men).[6] Moreover, in most cases researchers were not able to determine geographic origins for the mtDNA haplotypes (1416–17). Consider the example of Georgian Jews, who had the highest frequency of its mtDNA modal haplotype (51 percent), the most extreme example of a "founder effect" (that today's female population is descended from a single or just a few related maternal founders) identified in the study.

The Georgian Jewish modal haplotype was found in only one non-Jewish Georgian individual. When the revised Cambridge Reference Sequence mtDNA database was searched (the standard reference for identifying the geographical origins of mitochondrial sequences), it produced a match in Syria (two individuals out of 69) and in Iraq (one individual out of 116). In addition, there was one directly derived type (one or two mutations removed) present in two Georgians. Derived types were also found in the North Caucasus (two out of 208 individuals in total), in Turkey (one out of 218 individuals), in Armenia (one out of 191) and in Sicily (one out of 90). Those matches are but isolated findings, however. None of that data lends insight into the geographic origins of the Georgian Jewish maternal gene pool. As David Goldstein explains in his popular book on the genetic history of the Jews, statistically speaking finding one in a hundred is the equivalent of finding zero in a hundred (Goldstein 2008, 30). "For the Georgian modal haplotype, there is therefore no clear indication of provenance, although an indigenous origin is certainly possible, given the data" (Thomas et al. 2002, 1416).

The researchers drew three main conclusions from the mitochondrial data, the first two of which do not square with traditional stories of the origins of the Jewish diaspora: First, most Jewish communities were founded not just by relatively few women but by different and *unrelated* women. In other words, as seen via the maternal line, today's different Jewish communities are not phylogenetically related to one other.[7] Second, none of these communities seems to have an origin in ancient Palestine.[8] And third, following the founding of these maternal lineages, the communities became and remained endogamous.

The mitochondrial evidence contradicts what is commonly believed about the origins of contemporary Jewry, that Jewish *families* were expelled from or fled persecution in ancient Palestine and that women together with their men founded the communities of the Jewish diaspora. As David Goldstein sums up the results of this mitochondrial study, although the research cannot tell us "exactly what is behind the different demographic histories,"

> the results are consistent with a story of Jewish men, perhaps traders along the Silk Road or the Arabian Peninsula, traveling long distances to establish small Jewish communities. They would settle in new lands and, if unmarried, take local women for wives. The communities might have been augmented by additional male travelers from Jewish source populations. . . . These communities began with Jewish males and, presumably, non-Jewish women, and not many at that. (Goldstein 2008, 98)

Once begun by Jewish men, however, these communities closed in upon themselves, with men marrying very few women from the outside (Thomas et al. 2002, 1417). (If not, researchers would have found far more genetic diversity—in effect, more founding mothers—in the mitochondrial gene pool.) Subsequent generations of Jewish women were and are descendants of these original founding mothers. As summed up by Nicholas Wade of the *New York Times*, "even though the founding mothers of most Jewish communities were not born Jewish, *their descendants were*. 'It's precisely that custom that allows us to see those founding events,' Dr. Goldstein said" (Wade 2002, emphasis added).

Conversion of course is permitted in Judaism.[9] Insofar as the convert is given the name "son" or "daughter of Abraham," s/he is "adopted into the family of Abraham and assigned a new 'genealogical' identity" (Boyarin and Boyarin 1993, 317). Nevertheless, the conversion of *individuals* means something quite different than the claim that Judaism's "founding mothers" did not originate in the ancient Middle East but were converts to Judaism. That tale tells a very different diaspora story: a migration of men who settled elsewhere and married local women. Those women presumably converted to Judaism *and then remained true to their (new) faith*. This is perhaps less a story of "origins" than one of "beginnings," to borrow Edward Said's distinction (Said 1975, xxi).

Whereas for Said, "origins are divine," beginning—and beginning again—is historical. Origins are passive, beginnings active: "X is the origin of Y," but "the Beginning A leads to B" (ibid., 6). "The beginning is the first point (in time, space, or action) of an accomplishment or process that has duration and meaning. The beginning . . . is the first step in the intentional production of meaning" (5). The fact of Jewishness here can be read through just such a notion of beginning, of an *accomplishment*: even if Judaism's founding mothers did not originate in ancient Palestine, once they married Jewish men, once *they converted to Judaism*, the history of their Jewishness began. The Jewishness of Judaism's maternal line was created and then it was sustained by the genealogical practice of passing down Jewishness from mothers to their children. That intentional act produced the reality that, even though the founding mothers were not Jewish, *their children were*, as David Goldstein put it. Whatever the origins of these women, the fact of Jewishness, once begun, emerged over the generations as "visible" in the mitochondrial record.

Origins continue to matter greatly here, however, even if not in Said's sense of the divine. Where did Jewish communities originate? Do they all come from the same place and derive from the same source population? Those are the questions being asked. Nevertheless, there is more than one kind of origination recognized in this scientific work, one made possible by genetic history's entanglement with a larger contemporary cultural logic of personhood and identity. In contrast to the work of Jewish race scientists at the turn of the twentieth century, purity does not have to commence or be sustained from the point of exile from ancient Palestine to prove the biological coherence—the racial status—of contemporary Jews (see chapter 2). These women *chose* Judaism in places other than Palestine, they were *parvenus* and then they remained true to their faith. And that is why their children and their daughters' children and their daughters' daughters' children were and are Jewish. Over the generations the choices of those founding mothers, and the choices made by Jewish women and men in every subsequent generation to be true to the principle of passing Jewishness down the maternal line, have left a trace in the genetic system passed from mothers to their children.

The facts of mitochondrial DNA and its control region extend the logic of junk DNA. But they also do more than that. Within this genetic historical imagination, it is not just that biology functions as evidence of origins *if one has been "true"* to one's traditions or faith—that cultural practices leave and sustain a biological trace. Choices—the agency of particular women—create a meaningful, bounded, and enduring human group: an origin and a culture.[10]

There is an ongoing scholarly debate regarding the epistemological commitments of genomics and post-genomics. Is this biological determinism all over again, a recreation of race science on new evidentiary terrain (Duster 2003)?[11] Or is the logic of genomics and post-genomics structured by the concept of risk, demanding that its subjects act responsibly, that they make the right choices in order to live long and healthy lives (Rose and Novas 2005; Heath et al. 2004)? Insofar as a particular science with its distinctive epistemological commitments makes imaginable and possible certain ways of living and being in the world, I take a brief detour through that debate in order set the stage for considering the forms of politics and personhood specific to genetic history.

For scholars who warn of the political dangers of the new genetics, the specter of racial determinism and its stepchild, eugenics, looms large. Scientific and popular literatures propagate the "gene myth" (Hubbard and Wald 1993), scholars have argued, granting biological reductionism a renewed authority. Genomics is providing simple answers to complex social problems such as alcoholism, mental illness, and crime (Duster 1998; Lippman 1991; Nelkin and Lindee 1995).[12] We are witnessing the resurgence of race science and eugenics in new (genomic) clothes.

Others have argued that the new genetics operates with a different logic. The Human Genome Project (HGP) aimed to identify the genetic bases of diseases, a goal further pursued by the HapMap project, which was designed to study genetic diversity at the population level. Within the logic of the HGP and the scientific projects that followed, identifying genetic mutations—and finding them in individuals through genetic testing now or as imagined in some techno-genomic future—is not the same as arguing that someone has or even necessarily will develop a particular disease. Genetic information only rarely tells a patient or prospective parent that she or her progeny will develop a particular illness, and even then, the time of onset and the question of severity remain uncertain. This style of reasoning generates uncertainty, scholars have argued (Davison 1996; Rose 2001). Post-genomic medicine is driven by an epistemology of the "*potentiality* for illness," "the *risk* of future pathology" (Sunder Rajan 2006, 23; see also Rabinow 1992; Taussig et al. 2003; Novas and Rose 2000; Rose and Novas 2003; Lakoff 2005). Post-genomic medicine abandons an earlier binary of the normal and the pathological (Canguilhem 1989; Canguilhem and

Delaporte 1994). It transforms "healthy" persons—all persons—into "patients-in-waiting" (Sunder Rajan 2006, 23–24).

Many social theorists have argued that the notion of risk is central to neoliberal economies and their attendant techniques of rule and forms of citizenship (Rose 1999; Beck 1992; Calhoun 2006; Giddens 1999; Knorr-Cetina and Preda 2005; LiPuma and Lee 2004). If one examines the epistemological logic of post-genomics in light of its intersection with broader political rationalities and economic forms characteristic of the neoliberal age, it becomes clear how the epistemological commitments of post-genomic medicine diverge from the belief in biological determinism that anchored race science.[13] The post-genomic subject is the neoliberal citizen-subject governed by risk. Moreover, she is that (neo-)liberal "rational, calculating creature . . . whose moral autonomy is measured by . . . [her] capacity for 'self-care' . . . [an] individual who is fully responsible for her/himself " (Brown 2003).

In relation to questions of health and illness, responsibility tends to involve making the decision to act prudently to avoid the "diseased" outcome indicated by one's genetic profile (see Rose 2001, 2007).[14] Within the epistemological commitments of genetic history, the political grammar of responsibility and choice is configured differently. There is the responsibility to be true to one's faith, to one's traditions, as I have argued in my reading of junk DNA (see chapter 1), an argument that extends to the control region of mitochondrial DNA: biological markers are read as indicators of cultural practices. In addition, a political rationality that privileges responsibility for oneself is evident in the act of recognizing origins in "events" more accurately described as beginnings. For Judaism's "founding mothers," Jewishness may have been an accomplishment achieved in places other than ancient Palestine. But that origin-as-accomplishment does not diminish the truth—the authenticity—of the story of "the Jewish diaspora." Identity *is* an accomplishment. Populations are visible to the molecular gaze because of a sustained and repeatedly chosen fidelity to one's traditions. And those choices, that responsibility, that fidelity is what testifies to—what produces, in fact—a *meaningful* human collectivity, a "culture" that is.

Nevertheless, individual choice and human agency do not replace the significance of more standard origin stories, either in general or with respect to the specific question of the origins of Judaism's "founding mothers." We are not *just* subjects of our own will and responsibility. If authenticity is a matter of being true to oneself (Taylor and Gutmann 1994), genetic data is emerging as a grid of intelligibility against which one's collective and individual authenticity is read. In the case of Jewish origins, evidence of Jewish peoplehood is sought against the background of a now longstanding commitment, first articulated by Zionist activists and scientists over a century ago, to the idea that contemporary Jews are direct descendants of an ancient people that originated in ancient Palestine and to the belief that that fact can be—indeed, will be—substantiated on the basis of biological data, once it is properly read and understood. That is the conception of Jewishness that has been sought and reproduced in what is now over a century of the near continuous study of the biology of the Jews. And that origin story, which need not be couched in the Divine, is continually recuperated and reiterated in a variety of ways in genetic historical research today. I begin with the choice to privilege Y-chromosome evidence in narrating Jewish history.

In an article published in November 2001, Harry Ostrer, a professor and Division Director of Human Genetics at the Langone Medical Center at New York University, argues that "a genetic link among Jewish groups" is supported by "genetic polymorphism studies" (Ostrer 2001, 891). He begins his article by stating what he represents as historical fact: "The Jewish people originated in the Middle East during the Bronze age. For more than 2,000 years, Jews have been a migratory people, establishing communities throughout the Middle East and the Mediterranean basin" (891). As Ostrer explains, "the designation of who was a Jew was prescribed by religious law as one whose mother was a Jew. Entry into the community was possible through religious conversion, but was probably rare. As a result, Jewish identity has been continuous up to the present" (891).

What then does the genetic evidence show? Recognizing that "older single-locus studies" have been divided "on whether Jews have had significant admixture with non-Jewish populations, including possible mass conversions," Ostrer explains the limitations of such studies, including the possibility of alleles (genetic mutations) under selection. By way of contrast, "although by religious law Jewishness is a maternally transmitted trait, studies of mitochondrial and Y-chromosomal polymorphisms provide strong evidence for both matrilineal and patrilineal transmission, and many generations of endogamy" (891). In other words, genetically speaking, patrilineal transmission and not just matrilineal transmission—evidence of a sustained endogamy—has characterized Jewish history. But what about the question of origins? What about the origin of the Jewish people in the Middle East during the Bronze Age? Ostrer focuses on the Y-chromosome evidence: Jewish populations share thirteen common Y-chromosome haplotypes with non-Jewish Middle Eastern populations, "indicating that the original Jews might have arisen from local peoples and are not the offspring of a single patriarch" (891). He then discusses the discovery of the Cohen modal haplotype and explains that it may have originated during the "First Temple Period" in Jewish history (thereby presuming the biblical account to be historical and not mythical), rooting Jewish origins in the ancient Near East here via the priestly line. There is no specific mention of mitochondrial studies, whose historical conclusions in 2001 remained "divided."[15] It is the Y-chromosome evidence that tells the story of "a genetic link among Jewish groups," as the section on the "maternally and paternally transmitted trait" is named (891).

Privileging the Y-chromosome as the basis for telling the story of "the Jewish people" or "the Jewish diaspora" is even more apparent in public accounts and understandings of the research. Take the example of Jon Entine's book *Abraham's People*. A journalist, a former producer of Tom Brokaw's *NBC News*, and a visiting fellow at the American Enterprise Institute, Entine summarizes the results of the Cohen study as follows: "In genetic terms, the Hebrew Bible is the story of the Y chromosome. Although Jewish tradition has traditionally emphasized the importance of the female lineage . . . the male lineage was the tribal standard of the ancient Middle East" (Entine 2007, 65). The Y-chromosome emerges as the standard for his historical narrative as well. Assessing the evidence on the origins of Ashkenazi Jewry ("Converts or Abraham's Children?"), Entine writes:

> By the late 1990s, geneticists were finally in a position to unravel the mystery of the Jewish genome shaped by these periodic bottlenecks. Spurred by the Cohanim studies, an all-star roster of geneticists from the United States, Europe, South Africa, and Israel proposed drawing a definitive picture of *Jewish ancestry* back to the time of the diaspora that began in 586 BCE and even stretched back to the time of Moses. The scientists compared nearly 1,400 Jewish and non-Jewish *males* from around the world—twenty-nine populations in all. Their findings . . . surprised and delighted biblical literalists: Jews could trace their male lineage back to biblical Palestine. (2007, 210, emphasis added)

After summarizing the mitochondrial evidence, Entine notes that if Thomas's explanation "holds, based on the matrilineal determination of Jewishness, many of the founding mothers of Askhenazi Jews may be the descendants of righteous converts or maybe even women living as Jews who never converted—by Israeli law, not Jews at all" (ibid., 218). But he immediately recuperates Jewish origins in ancient Palestine via the male lineage:

> Perhaps not every Jew is descended solely from the ancient populations in Judea and Samaria, as the Bible suggests, but *most Jews* do share a common ancient ancestry. *Most Jewish males* appear to have originated in the eastern Mediterranean. After being expelled from the Middle East, and after diaspora stops along separate routes in Italy and Asia, Jews trickled into Europe. They brought with them some wives, but more often than not, they coupled with local women (220, emphasis added).

"Jews," Entine reports, "have received these studies confirming their relative purity with a combination of enthusiasm and concern. They reaffirm the central threads of Jewish history and identity: *diaspora Jews are ancestrally connected to each other and to biblical Israel,* a belief that binds religious, nonpracticing, and even atheist Jews" (220–21, emphasis added).[16] (The "concerns" are a consequence of the specter of anti-Semitism raised by producing Jews as a population with a shared biology.)

The results of Y-chromosome studies have circulated widely as evidence of the truth of Jewish origins in ancient Palestine. A reader of the *Financial Times* of London cited that same evidence in response to Tony Judt's review of Shlomo Sand's book, *The Invention of the Jewish People,* as I mentioned in opening the book. To quote the letter writer once more: "To simplify but not distort, the DNA record (*at least on the male side*) shows that Jewish communities everywhere, east and west, Ashkenazi and Sephardi, are more closely related to one another than to the non-Jewish populations they live among" (Silverman 2009, emphasis added). It is evidence regarding the descent of men that apparently proves wrong Sand's argument that contemporary Jews are not descendants of a people exiled from ancient Palestine over two thousand years ago (Sand 2009).

If the Y-chromosome evidence has circulated widely and has been received by and large as straightforward fact, the mitochondrial evidence has not experienced the same reception. Unexpected results are a lot harder to assimilate. I asked one of the lead researchers what happened in such cases. "Very straightforward," he said. "If people like what they hear, they shout from the rooftops. If they don't, they ignore it. It doesn't enter their discussions." His account is a bit too cut and dried. Genetic evidence based upon studies of the two linearly inherited genetic systems, mitochondrial DNA and the Y-chromosome, has generated a shift in narrative form by privileging the patriline as the basis for narrating Jewish history. Nevertheless, his comment carries a significant resonance of truth. The public response to the Cohen studies and to the subsequent Y-chromosome studies was unconditional acceptance. A second researcher told me that one of the reasons he thinks certain people have been so willing to participate in genetic studies was the Cohen study results: some people participate for "religious reasons," wanting to prove "the biblical tradition, that it isn't unsubstantiated myth." Had the Cohen study not "confirmed" that tradition, many would have been far more wary of participating in future projects, fearing that "the results would fly in the face of what they had been taught."

The results of the studies of Jewish male origins were widely accepted, almost unconditionally. *HaAretz*, the *New York Times*, and a variety of other newspapers in the U.S., the U.K., and Israel as well as the U.S. Jewish press reported that Y-chromosome evidence confirms what was always known—or believed—about the origins of the Jewish people (Wade 2002). As summed up in the title of one *HaAretz* article, which reflects a distinctly Israeli spin, "And Who is the Most Similar Genetically to the Jews? Palestinians"[17] (Traubmann and Suni 2000). Or as Dr. Lawrence H. Schiffman, at the time Professor of Hebrew and Judaic Studies at New York University, told Nicholas Wade of the *New York Times*:

> The study fit with historical evidence that Jews originated in the Near East and with biblical evidence suggesting that there were a variety of families and types in the original population. He said the findings would cause "a lot of discussion of the relationship of scientific evidence to the manner in which we evaluate long-held academic and personal religious positions," like the question of who is a Jew. (2002)

Even the researchers seemed willing to accept the conclusion of a shared Near Eastern ancestry long before the evidence could be argued plausibly to be conclusive: In the initial Cohen studies, the hypothesis that the Cohen modal haplotype was a *possible* signature of the ancient Hebrew population was highly speculative at best (see chapter 1). Nevertheless, studies were launched to "test" the Lemba's Judaic origins on its basis (see chapter 5). Only one scholar (Zoosman-Diskin 2000) called into question the validity of the claim of the CMH as a signature of ancient Hebrew ancestry, and his criticism was widely dismissed or simply ignored. Michael Hammer and several colleagues published a paper in 2009 that reassessed the results of the initial *Nature* papers on the Cohanim. The researchers reaffirm their initial conclusion that the CMH originated in the Near East prior to the dispersion of ancient Jewry, but most contemporary self-identified Cohanim descend not from a single lineage but from "a *limited number of* [independent] paternal lineages" (Hammer et al. 2009, 707).

The study of Judaism's "founding mothers" met with a markedly different response. Nicholas Wade went back to Schiffman for comment: "Most of those founding narratives do not have strong historical support," Schiffman told him. "The new genetic data could well explain how certain far-flung Jewish communities were formed. But he [Schiffman] doubted that it would account for the origin of larger Jewish communities that seemed more likely to have been formed by families who were fleeing persecution or making invited settlements" (Wade 2002). I am not arguing the mitochondrial results were rejected out of hand. Another prominent Jewish studies scholar found the historical implications "by no means implausible." According to Shaye Cohen, "the authors are correct in saying the historical origins of most Jewish communities are unknown. Not only the little ones like in India, but even the mainstream Ashkenazi culture from which most American Jews descend" (quoted in Wade 2002). Nevertheless, as reported in the press, the mtDNA results generated controversy, not clear historical conclusions or facts. And as one biological anthropologist put it in an interview, "if we are willing to take positive results at face value, what about negative results?" Negative results pose the greater challenge, she said: "What are the standards for disproof?"

The difficulty of accepting unexpected results is not a characteristic of the public domain alone. The results of Thomas and his colleagues' mtDNA study did not achieve the status of scientific consensus either. As one researcher told me in an interview in 2004:

> The hard thing we found was that those communities were founded by a relatively small number of women, and then there was a very tentative suggestion that those women who were founders were, at least in some cases, more likely to be of local origin rather than Middle Eastern origin. I believe that Doron [Behar] has published a paper arguing against that [Behar et al. 2004]. So the issue about the identity of those women, of were those women of Middle Eastern origin or not, that's the most controversial.

According to John Entine, there is a "friendly war over those results between the London and the Haifa researchers" (2007, 218). Reestablishing the genetic origins of the Jewish maternal line in ancient Palestine is an ongoing goal of research on the mitochondrial line. That nationalist commitment is not so easily abandoned in favor of the self-made, in "exile" that is, Jew.

Doron Behar (of Haifa's Technion) and colleagues have published two subsequent studies of the mtDNA lineage, one on the Ashkenazim (2006) and a second on "non-Ashkenazi" Jews (2008). The researchers rely on different molecular techniques to "count the founding mothers, and where possible date and localize their origin" (Behar et al. 2008, 2). In so doing, they challenge the results of the paper published by Thomas and his colleagues a few years before. I begin with the results of the study on Ashkenazi women that, the researchers argue, indicates a shared, and most likely, Near Eastern origin.

As reported in "The Matrilineal Ancestry of Ashkenazi Jewry" (Behar et al. 2006), the maternal gene pool displays a strong founder effect. Contrary to the findings of the earlier study (Thomas et al. 2002), which argued there is no evidence that only a few women founded the current Ashkenazi population, Behar reports that "four ancestral women" account for "fully 40% of the mtDNAs of the current Ashkenazi population (~8,000,000 people)" (Behar et al. 2006, 490–91).[18] But Behar and his colleagues wanted to do more than just "infer the actual number" of maternal founders. They wanted to figure out the "temporal and geographic origins" of these maternal founders. They doubted the results of Thomas's study that argued that in all likelihood Ashkenazi women are descendents of a diverse set of founding mothers who were not of Near Eastern origin. By way of contrast, Behar and his team report "a striking overrepresentation" of specific haplogroups: K and N1b. The K haplogroup accounts for 30 percent of maternal lineages and the N1b 10 percent. (Haplogroups are clusters of haplotypes that share one or more ancestral mutations. Haplogroups are increasingly used to root haplotype trees so as to control for the problem of assuming particular haplotypes are genealogically related when their similarity may be a result of "convergence," that is, the same mutation happening more than once in relatively short historical time spans.) A subclade (subgroup) of haplogroup K (K1a1b1a) accounted for 62 percent of the Ashkenazi K mitochondrial DNAs (~19.4 percent of the contemporary Ashkenazi population). And K1a1b1a is a subclade, the authors argue, whose "sister lineages . . . can be found in Portugal, Italy, France, Morocco, and Tunisia" (ibid., 492). Its "phylogenetic tree is [thus] of a wider Mediterranean presence and origin" (498). Taken together with the phylogeography of N1b—that it is "virtually absent in Europeans but appears at frequencies of ~3% or higher in those from Levant, Arabia and Egypt" (489)—Behar and his team conclude that there is a "likely" Near Eastern origin for the maternal gene pool of Ashkenazi Jewry. They argue that their technique for analyzing the mitochondrial line allowed for "the detection of a small set of only four individual female ancestors, likely from a Hebrew/Levantine mtDNA pool, whose descendants lived in Europe and carried forward their particular mtDNA variants to 3,500,000 individuals in a time frame of < 2 millennia" (493).

Ashkenazi Jewry, however, were not the only focus of Behar and his team's work. After all, Thomas and his team had included women from various other Jewish communities in their study. As such, Behar's team carried out a second study of Jewish maternal descent that focused on a population that, they explain, is much less studied.[19] "Contemporary Jews, whose number is estimated at 13 million, can be divided [in]to Ashkenazi and non-Ashkenazi, which are each in turn comprised of numerous different constituent communities" (Behar et al. 2008, 1). While "Ashkenazi refers to Jews whose recent ancestry over the past millennium traces to Central and Eastern Europe," the authors write, the "geographically much more widespread non-Ashkenazi Jewish communities are also culturally more diverse, and are comprised of Jewish communities that have continuously resided in the Near and Middle East and in North Africa and in different geographic locations to which Jews fled or to which they were deported including the Iberian expulsion in 1492–1495" (ibid.). Replicating the problems of classification that characterized Israeli population genetics in the 1950s and 1960s (see chapter 2), Behar and his team struggle with nomenclature: "These communities also share similar religious rituals, probably due to their presumed common historical origin from the descendants of the much earlier Babylonian exile. As a result of common ritual practices, they are sometimes collectively referred to as the Sephardic (Spanish) or Mizrahi (Eastern) Jews" (ibid.). Sephardic really better refers to those who fled Iberia; Mizrahi does not generally include Italian and Yemenite Jews. "Non-Ashkenazi," therefore, is a better name. "Cumbersome as it is, [it] encompasses here all non-Ashkenazi communities, currently estimated to comprise about 5 million individuals" (ibid.).

As indicated by the very choice of name, this Jewish collectivity remains a residual category for biologists studying the population structure of "the Jews" (see chapter 2): in contrast to the Ashkenazim, this is not a culturally or historically well-defined community, and, according to Behar's study, it is not a community recognizable biologically either, at least not along the mtDNA line. It exists only in relation to what it is not, that is, *not* Ashkenazi. And that fact becomes only more evident from the molecular data itself. Behar and his colleagues report three patterns among non-Ashkenazi mtDNA:

> Unlike the previously reported pattern observed among Ashkenazi Jews, the numerically major portion of the non-Ashkenazi Jews, currently estimated at 5 million people and comprised of the Moroccan, Iraqi, Iranian, and Iberian Exile Jewish communities showed no evidence for a narrow founder ef-

> fect, which did however characterize the smaller and more remote Belmonte, Indian, and two Caucasus communities. The Indian and Ethiopian Jewish sample sets suggested local female introgression, while mtDNAs in all other communities studied belong to a well-characterized West Eurasian pool of maternal lineages. (Behar et al. 2008, abstract)

A biologically diverse population, non-Ashkenazi Jews are particularly difficult to study: "The non-Ashkenazi Jewish communities studied herein pose an analytic challenge to population geneticists." They are "spread across a broad and diverse phylogeographic area." There are multiple patterns, and they are not related one to the other. "The attempt to estimate the size of the founder effect in a large number of separate communities, each prone to different fluctuating demographic events during their histories, the latter often relying on oral narratives rather than archival or better substantiated historical records, is intrinsically more complex than studying phylogeographically less complex cases, such as the Ashkenazi Jews" (ibid., 9–10). This quotation raises an important question that is never asked: why study "the non-Ashkenazi" as a *single category*—"a population"—at all?

Evident in the work of Behar and his colleagues is the fact that origins *in the ancient Near East* (for the maternal line) do matter. These studies sought to disprove the results of Thomas's earlier work. More specifically, based on rather thin genomic evidence, Behar and his colleagues argue for a "Near Eastern origin" as "likely" for the founding mothers of the Ashkenazi community. But of the 40 percent of the population whose mtDNA displays a strong founder effect, only 10 percent (the N1b haplogroup) seems to have a *possible* Near Eastern origin. They might have a geographic origin that ranges anywhere from Arabia to Egypt. Of the remaining 30 percent of the founding lineages, at best about two-thirds show a circum-European-North African Mediterranean phylogeography. As David Goldstein writes "the ultimate origin of their four founding mothers is unclear" (2009, 97).

The desire to reestablish the origins of the Jewish maternal line in the ancient Near East is evident in Behar and his team's studies of mtDNA. But more than the specific commitment to the biological-qua-historical truth of that nationalist origin tale is evident in this work. As a more general principle, if genealogical descent—if a shared gene pool—defines the truth of (historic) identities, then "a population" with shared and legible genetic origins will necessarily emerge within a genetic historical grammar as the authentic one, and in this instance that means "the Ashkenazim." From a different analytic and historical perspective, however, Mizrahi Jews, the so-called non-Ashkenazim, are not best described as a "large group of *separate communities."* They are not best defined as a community united by *"common ritual practices"* but of no clear or unified phylogeographic ancestry (Behar et al. 2008, 1, emphasis added). The Mizrahi Jew is a category of identification, the Mizrahim a political community, forged in the Israeli state in response to the discrimination experienced by Jews who immigrated to Israel in the 1940s and 1950s from, for the most part, the Arab world. In non-genetic terms, the geographic origins of "the Mizrahi Jew" is very clear: it is an artifact of a political practice of *self-definition*—an appropriation of their designation as *edot ha-Mizrach* and a resignification of the meaning of the category itself—that emerged in the Israeli state in response to and as a mode of resistance against an Ashkenazi political establishment and cultural elite that defined the "new Hebrew" and that tried to re-create the "oriental Jew" in its image (see Shohat 1989; Chetrit 2010; Eyal 2006; Shenhav 2006). In short, the Mizrahim are a product of modern and not ancient Israel. That kind of a population and practice of identity-making genetic history can give no account of: no evidence of its existence is to be found in the genome.

Like race science and population genetics of old, these genetic historical investigations of Jewish origins and unity can never live up to their positivist commitments. Despite experimental designs, the original hypothesis that *there is a "population"—a race, a people—of "Jews"* that traces its roots to ancient Palestine can never be undermined. For contemporary Jewish communities, the truth of Jewish unity and identity lies elsewhere. It resides in biblical texts and in religious beliefs and practices. It resides in cultural and political commitments. It resides in the very fact of the Jewish state. The biological evidence is incapable of undermining the original hypothesis.

Nevertheless, there are different cultural and political imaginations within which scientific arguments and the "histories of the Jews" get inscribed, and those discourses portend different forms of scientific ethos, politics, cultural practice, and conceptions of the self. Vis-à-vis the contemporary practices of genetic history, there are different kinds of origins recognized as grounding identity claims. There are also different kinds of scientific and analytic choices to be made: mitochondrial DNA or the Y-chromosome? On the basis of what kinds of evidence is the "history" and "origins" of the Jews to be told? But, as I show in what follows, there is also a tension or ambivalence over the status and significance of genetic historical facts themselves. In what ways might origins or beginnings matter? Put another way, how does this work matter, or does it matter at all to the question of who is a Jew? I consider those questions through an analysis of the intersection of identity politics and scientific virtues in the quest for Jewish origins.

Self-Definition

In *Science in Action,* Bruno Latour talks about the Janus-faced nature of science. On the one hand, scientists claim that nature is prior to science. Nature resolves scientific disputes. On the other hand, scientific practice demonstrates that nature is an outcome of what scientists do. Nature—as consensus, as a "black box" emerges when scientific disputes have been resolved (Latour 1987). I use Latour's notion of a Janus-faced practice to think about the ambivalence and incoherence evident in the discourse of and about genetic history. On the one hand, genetic history is just another historical science. It is a scientific pursuit designed to provide insight into unresolved questions about, in this instance, Jewish history. It has nothing to say about "identity"—in the sense that "identity" indexes criteria for membership, i.e., who is a Jew or who is a Cohen or, for that matter, what it means to be a Jew or a Cohen. On the other hand, genetic history tells us something about who we are, really, to borrow Charles W. Mills's turn of phrase (1998, 41). These projects are driven by a desire to know-thy-collective-self, which, as I show in the next chapter, is generating a desire to know thy-individual-self via tests offered by genetic ancestry-testing companies.

Scholarly papers are replete with the language of detachment and objective scientific pursuit. As explicated in "Threads to Antiquity," the sex-specific studies of Jewish origins have enabled researchers to tease "out two important threads of Jewish history," a male origin in the ancient Near East and "a pattern of localized, independent, female defined foundation events in Diaspora communities" (Bradman et al. 2004, 96). More broadly, in "undertaking these studies we have: (a) evaluated genetic data to test alternative hypotheses based on non-genetic data; and (b) used analysis of genetic data to formulate hypotheses that can be evaluated using other disciplines" (96). Testing alternative hypotheses, tracing threads of Jewish *history*. That is all genetic history does. As David Goldstein writes in *Jacob's Legacy* (2008), after recounting the story of a genetic study of the Bnei Menashe that found no evidence of Judaic roots in this community that believes itself to be a Lost Tribe, "some commentators took the opportunity to say, essentially, 'Ha—told you they weren't "real" Jews.' . . . In my view, this is a perfect example of when we are best served by ignoring DNA—if living, breathing people want to embrace Judaism, genetics should have nothing to say about it one way or another" (2008, 89).

In popular accounts and in interviews that scientists give, the scientific nature of this work is emphasized over and over again, "the scientific" standing for "'impartiality, detachment, disinterestedness and a willingness to submit to evidence'" (Daston and Galison 2007, 379) by those expert enough to do so. There are unexpected results that researchers explicate and stand by: that Judaism's founding mothers did not originate in the ancient Near East (Thomas et al. 2002); that the Ashkenazi Levites do not originate in the ancient Near East and may be evidence of the partial truth of the Khazar hypothesis—a long dismissed theory that Ashkenazi Jewry is descended from the population of the Khazar Empire that converted to Judaism and ruled the Northern Caucasus in the seventh through tenth centuries (Behar et al. 2003; see Koestler 1976 for the original hypothesis). The ability to recognize unexpected results performs the very impartiality of the research. It testifies to the fact that these are scientists who are willing to *submit to evidence*.

Being willing to submit to evidence—that is a longstanding image of the scientific self. Scientific work, as various scholars have demonstrated, has always presupposed a certain kind of self (see Weber 1958; Daston and Galison 2007; Shapin 1994, 2008), and popular accounts of work in genetic history stage that self for the reading public. For example, David Goldstein writes:

> We know that mitochondrial DNA and Y-chromosome variation reflect, in some way, female and male genetic history, but how should we use variation in these systems to assess the effects of female-defined ethnicity in Jewish populations?
>
> One of the hardest things to get across to students in genetic history is that there is no right answer to a question of this sort. In fact, if a student working in my lab asks, "What should I do with the data?" the only honest answer I can give is, "Look for something interesting." . . . This is not to say that genetic history is nothing but creative interpretation. Genetic history that is researched accurately and intelligently depends on increasingly sophisticated tools developed in allied disciplines, most notably population genetics. The analyses borrowed from these disciplines must be carried out correctly, and they must be interpreted with caution and a sense of fairness. This is no easy trick, and anyone unable or unwilling to come to grips with the algebra and detailed conceptual structure of population genetics is well advised not to venture into genetic history.
>
> Because we don't know the history at the outset, however, we don't know which analyses will be instructive and which irrelevant. Hence, there is noth-

ing to do but get to know the data and see what they have to say. (Goldstein 2008, 90)[20]

Goldstein sketches the kind of person a genetic historian needs to be. The student of genetic history must be open minded and curious: there is no simple answer to empirical questions, no a priori way of knowing the history one may find. She must be well trained in the "sophisticated tools" (complex mathematical tools) of allied disciplines, in particular of population genetics (what I refer to as anthropological genetics). And she must be willing to learn *from the data*. The student of genetic history is the cautious, knowledgeable, and curious expert. And she—or he—is a knowing subject who values meticulousness perhaps above all else:

> [Neil Bradman] had the good sense to hire Mark Thomas, a postdoctoral fellow at Cambridge University. Nothing enters or leaves the Thomas lab without a complete recording of time, place of origin, and contents; freezers are accessed only at particular times, laboratory buffers can be made up only under his supervision, and on and on—every "i" dotted and every "t" crossed. Beyond meticulous, Thomas is a scientist's scientist. . . . Once, when someone accused our group of sample manipulation, I had not a second's concern. In this collaboration the samples had all been received, stored, and inventoried by Mark. If anyone on the planet could document precisely what had happened to each and every sample he had ever handled, it was Mark Thomas. (Goldstein 2008, 12)

Popular representations of the work—and the workers—of genetic history are replete with descriptions of care and detachment, of the willingness to suspend a priori judgment in order to be able to submit to evidence (see *The Lost Tribes of Israel* 2000; Vickers 2010a, 2010b). So too, of course, are scientific papers framed in the rhetoric and distance of objectivity. Historical evidence—oral and documentary traditions—is being evaluated on the basis of newly available molecular data and techniques. Hypotheses are being tested, data read, and conclusions reached. That is what the scientific work of genetic history involves.

This rhetoric of scientific objectivity and detachment and accounts of the specifically scientific self and life, however, sit side-by-side other kinds of narratives that invoke more intimate relationships with and personal attachments to the research at hand. I am interested in those narratives of attachment not in order to question the credibility of the research. Rather, I am interested in explicating a novel configuration of the scientific self that has emerged in particular domains of natural scientific research, most prominently but not exclusively in the United States, and the kind of political and ethical work that novel configuration of the knowing subject does.

Consider the origin stories of the first Y-chromosome research project. As I mentioned in chapter 1, Karl Skorecki's inspiration for a genetic study of priestly descent came to him one day as he sat in synagogue:

> My mind was drifting perhaps from some of the liturgy and prayers, and during the synagogue ceremony, members of the priest tribe or Cohanim are called up for particular contributions to the service. So I was sitting there, and another member of the congregation was called up as a Cohane, or Jewish priest, and his origin was from North Africa. And my origin in terms of where my parents came from is from Eastern Europe and Poland, and we are both Cohanim, or priests. So I thought to myself at the time, well what might we have in common other than the fact—other than this tradition that we have? So this led to the notion that perhaps we could find, somewhere in the human genome, a similarity. Well, the part of the human genome that's also passed from father to son is of course the Y-chromosome. (*The Lost Tribes of Israel* 2000)

Meanwhile in London, Neil Bradman (who identifies himself as a Levite) was having an inspiration of his own: his son, who wanted to spend the summer in Israel with his girlfriend, was looking for a senior thesis project. Bradman suggested his son collect samples for a study of the Jewish priestly lines (the Cohanim and the Levites) and that suggestion proved more than a passing inspiration (interview). A businessman who decided after a long and successful career to return to the academy and pursue a Ph.D. in human genetics, Bradman used his private wealth to found the UCL Centre for Genetic Anthropology and to cover many of the costs of the research on Jewish genetic origins—including "salaries in the lab," as one former student told me.

Nearly all of the main researchers have a personal narrative about how and why they got involved in the studies of Jewish origins. The following is David Goldstein's published account:

> My entry into Jewish genetic history was both personal and professional.
>
> For a long time, when asked how I came to study the genetic history of the Jews, I had ready rather pat answers about the intersection of population genetics, human evolution, and the oral and written traditions of Jewish civi-

> lizations. Those subjects have fascinated me for twenty years and remain at the heart of my intellectual life. But with bits and pieces of this book in front of me that recapitulate the better part of a decade's worth of work, it is also clear that my evolution as a genetic historian is more personal than simply the interaction of genetics, evolution and history. This is for me in part a personal story. (Goldstein 2008, ix)

Goldstein narrates his trajectory into Jewish genetic studies while a graduate student at Stanford University. He migrated from an interest in producing mathematical models to make sense of the evolution of ploidy (the number of sets of chromosomes in any given species), to an interest in what *did* happen (population/anthropological genetics) rather than what *could* happen (theoretical population genetics), and finally, to his cooperation with Luca Cavalli-Sforza and Marcus Feldman in their work on microsatellites. Having worked on developing a good way to "use microsatellites to understand how long ago human populations separated," he received an email from Neil Bradman (ibid., ii):

> By that time . . . I had earned the moniker Mr. Microsatellite and was an imminent transplant to Britain, nervously starting my first job at beguiling and frustrating Oxford University. In his email, Neil Bradman introduced himself to me, told me what he was doing, and extended a cordial invitation to Shabbat dinner at his home. He signed himself "Neil Bradman (of the Jews)." I had no way of knowing it then, but Neil Bradman of the Jews was to become one of my closest research partners over the next five years. (ibid., 13)

Goldstein then writes: "I suspect that had I no Jewish heritage, this work would likely have never led me into Jewish genetic history" (ibid., xii).[21] For his part, Michael Hammer told a reporter for a Jewish newspaper "one reason he began the research was his curiosity about his own Jewish roots." The other reason was an interest in why Ashkenazi Jews have high frequencies of rare diseases (Garifo 2000). And like those original projects, subsequent projects have been launched because of individual curiosity, even if not always that of the scientists themselves. For example, I was told by a researcher that the impetus for studying Kafkhazi Jews came from a specific individual. This person contacted him, telling him that he was convinced that Kafkhazi Jews are descendants of Khazars. The researcher asked the man if he would like to participate in a study, even though they will never be able to prove descent from Khazars: there are no "verification" samples.

The individual helped organize the study—"it was his obsession." The man got his friends involved.

As I mentioned above, I point to these personal motivations—to these attachments—not to call into question the validity or rigor of the scientific work. Nor do I doubt the sincerity of a scientific ethos of objectivity and intellectual openness. Rather, I want to explore the merging of that scientific epistemology, ethos, and self with an ethics of attachment. I do so in order to think about the context within which we consider that amalgamation a plausible and felicitous scientific self today and to consider what kind of work it does in securing the authority of specific kinds of knowledge. How is it that scientists present their results as reliable, objective, experimentally determined and themselves as disinterested scientists and yet simultaneously and openly declare personal attachments to the projects, their desires to know something about themselves, as Jews, as Cohanim, as Levites? How is it that they can declare—*and be*—both of those things at one and the same time?

In earlier eras, it is worth noting, a scientist's presumed Jewish identity—let alone his/her explicit self-identification as a Jew—was often understood to undermine the credibility of his or her scientific claims. As Mitchell Hart has pointed out with respect to scientific studies of "the Healthy Jew" in the late nineteenth and early twentieth centuries, "when Christian authors write and speak in positive ways about Jewish health and hygiene, the power of such narratives to convince may very well be greater than narratives produced by explicitly identified or identifiable Jews. Non-Jews are given an authority precisely because they are non-Jews." The flip side of that authority, of course, was the Jewish scientist who was not considered credible because of his or her Jewishness, "evidence of excessive bias and subjectivity" (2007, 22).

In *Objectivity* (2007), Lorraine Daston and Peter Galison historicize the concept of objectivity. They demonstrate how objectivity does not subsume all of scientific epistemology, and they do so by tracing the historical development of three different styles of knowing from the early nineteenth to the mid-twentieth centuries. In exploring the epistemological commitments of each distinct way of knowing, Daston and Galison emphasize that a history of scientific epistemology is simultaneously always also a history of the scientific self, that is, of the specific subjectivity that a scientific formation promotes of what or who a scientist is:

> If objectivity was summoned into existence to negate subjectivity, then the emergence of objectivity must tally with the emergence of a certain kind of

> willful self, one perceived as endangering scientific knowledge. The history of objectivity becomes, *ipso facto*, part of the history of the self.
>
> Or, more precisely, of the scientific self: the subjectivity that nineteenth-century scientists attempted to deny was, in other contexts, cultivated and celebrated. (Daston and Galison 2007, 37)

In the case of genetic history, scientific epistemology does not demand the denial of a form of subjectivity cultivated and celebrated elsewhere: social attachment (as identity politics) and expert detachment are not understood to be in conflict. What is the broader context within which this "novum" (ibid.)—the joining of "judgment" (the scientist as expert) with attachment (the scientist as Jew)—is recognized as a plausible scientific self?[22] And what work might this scientific self do?

These studies of Jewish origins are but a small set of genetic genealogical studies carried out within a broad, flourishing field of research: studying population-based diversity in the aftermath of the Human Genome Project. Launched in 1991, the Human Genome Diversity Project aimed to produce a map of contemporary genetic diversity in order to study how "current diversity had evolved, and how people and genes had spread over the world" (M'Charek 2005, 2). Alain Pottage has argued that population genetics, best exemplified in the Diversity Project, "has come to style itself as a form of 'sincere' science, conscious of the broader ethical implications of its activities. The sincerity of population genetics is advertised by its basic normative programme, which is to establish a historical geography of the human family that begins with the egalitarian and [seemingly] apolitical substance of genes rather than a divisive and asymmetrical category such as 'race'" (Pottage 1998, 760; see also Reardon 2005).

That self-styled "sincerity"—specifically, a self-styled antiracist project (Reardon 2005)—met a maelstrom of opposition. In focusing on "indigenous peoples" who were considered to be the "'treasure keepers of original information'" (M'Charek 2005, 2), indigenous activists and their supporters dubbed the Diversity Project the "Vampire Project": primarily U.S. and European scientists were taking genetic material from indigenous groups and giving nothing in return. Scientists were accused of extracting "indigenous" genes as "raw material" for commercial profit (ibid., 3). As the battle over the HGDP ensued, its social and ethical aspects came under increasing scrutiny, ultimately leading to the development of an "ethical protocol" that sought to protect the rights of the source populations (14). In Pottage's words, "these protocols allow indigenous peoples to be represented 'as partners in the work rather than merely subjects of it'" (Pottage 1998, 761).

It was not just for the Diversity Project that ethical imperatives came to loom so large. The Human Genome Project was already the largest ethical project in history, with approximately 5 percent of an annual budget of about $350 million allocated to exploring "Ethical, Legal, and Social Issues" (Stevens 2002, 111). "'Life,'" Paul Rabinow argues, "is problematic today because new understandings and new technologies that are involved in giving it a form are producing results that escape the philosophical self-understanding provided by both the classical world and the Christian traditions" (Rabinow 1999, 16). With the possibility of manipulating DNA, of breaching the boundaries between species, and of creating life, there is "a leitmotif among scientists, intellectuals, and sectors of the public . . . [that turns on] redeeming past moral errors and avoiding future ones . . . ; a heightened sense of tensions between this-worldly activities and (somehow) transcendent stakes and values" (ibid., 17–18). "Forth-rightly secular moderns," Rabinow argues, have become obsessed with the ethics of life (18).

If ethical concerns are pervasive in the practices of and debates about genomics and post-genomics, they take a very specific form in anthropological genetics. Ethical debates are not focused on the status of life, on the dangers of transgressing the boundary between nature and culture that are permeable in ways hitherto unseen. Instead, the ethical concerns that anthropological genetics raises emerge within the context of decidedly postcolonial and multicultural political critiques. For the Diversity Project it was the specific specter of colonialism that loomed large. Unlike race science at the turn of the twentieth century or population genetics in its earliest instantiations, an "ethical requirement" is built into the work of anthropological genetics today, and it is inscribed (however inadequately; see Pottage 1998) in protocols or rituals of informed consent. And that is a legacy not just of Nazi genocide but of the history of colonialism and racism, at home and abroad. But that ethical requirement to obtain informed consent—a requirement of all human subjects research—hit a wall when the consent of an individual was no longer seen to be enough. Once inscribed into a long history of European/North American colonial racism and capitalist extraction, the Diversity Project's efforts to collect DNA samples were argued to be a pillage of "the patrimony" of indigenous *groups* for the scientific and commercial benefit of scientists and communities elsewhere. *Individual* consent thereby proved inadequate to solve the ethical problem.

The Jewish genetic history studies are a subset of these studies of human diversity. They are practices of knowing the self.[23] And as practices of knowing the self, for such studies it has been easier to avoid the political maelstroms that have been generated by scientists studying "Other" human subjects.[24] Studying the self—biological research projects in which the subject and object of research are represented as or taken to be the same—provides an ethical and political alibi for those who fear the specters of race and its history of violence. Studying the self, moreover, is a "solution" to the "problem" of epistemology, that is, of *who* can create authorized knowledge about *whom* in our multicultural age.

The "self-study" nature of studies of Jewish genetic history is not unique in the biological sciences today. Identity politics has met post-genomic medicine as well: as various scholars have demonstrated, the emergence and commodification of race as a valid category in biomedical and pharmacogenomic practices are driven in part by self-identified minority doctors, researchers, and communities. "Many actors at the center of this ethnography," Duana Fullwiley writes of her work on population admixture studies in medical genetics laboratories, "focus on race as a function of their personal identity politics as scientists of color. This is to say, they are driven not by racist notions of human difference, but by a commitment to reduce health disparities and to include 'their' communities in what they describe as the 'genetic revolution'" (Fullwiley 2008a, 695). In 2002, when the Centers for Disease Control and Prevention appropriated $747,472,000 for "minority health" (Kahn 2004), which *required* researchers to include race as a variable in the various domains of post-genomic medical research, the decision received widespread support among minority physicians (Fullwiley 2008a). In 2001, the company NitroMed gained support from both the Association of Black Cardiologists and the Congressional Black Caucus for federal approval of BiDil as a drug to treat heart disease in African Americans, the first racially tailored drug. (Researchers argued that BiDil was effective particularly in African American patients suffering from heart disease; for critical perspectives, see Kahn 2003, 2004; Sankar and Kahn 2005.) And in 2003, Howard University announced its intention to build the first DNA database on people of African descent with a view toward "jump-start[ing] an era of personalized medicine for black Americans" (Kaiser 2003, 1485). More broadly, citizens have made demands on the state for federal recognition of racial difference, including in matters of health; and medical researchers (many of them self-identified as minority researchers) pursue population-level differences to redress the imbalances in medical research and clinical practice (Ziv and Burchard 2003; Risch et al. 2002; Fullwiley 2008a).

Identity politics has met the world of the natural sciences. If identity politics represents an effort toward self-definition (see Brown 1995), that practice now extends to broad domains of medical and biological research. In fact, self-definition and self-study may have emerged as so common, expected perhaps, that David Goldstein declares (humorously) that "you don't have to be Jewish to . . . become one of the world's foremost experts on the anthropology of the Jews" (2008, 41; referring to Tudor Parfitt).[25]

Just as different scientific epistemologies cultivate different kinds of scientific selves, so do distinct historical moments and socio-political horizons generate different social histories of truth (Shapin 1994). As Steven Shapin has argued, "we cannot understand how our various scientific and technological knowledges are made, and made authoritative, without appreciating the roles of familiarity, trust, and the recognition of personal virtues" (2008, 2). Trust, familiarity, personal virtues—they take many forms in the production of scientific and technological knowledge today. In particular domains of biological and biomedical research, and especially in the context of a U.S. polity in which identity politics is so taken for granted in the public domain, "familiarity" and "trust" are in large part grounded in—seen to inhere in—one's "cultural self." Thus, scientists, at least scientists operating in particular domains of human subjects research, no longer need to deny their "identities." To the contrary, they proclaim them publicly, as they narrate their reasons for getting involved in specific research projects, and as they demand redress for imbalances in medical research and clinical practice. In contrast to earlier eras of Jewish biological self-studies, scientists today frame their investment in the work in terms of *personal* attachments, curiosity even. They are not engaged in arguments—public political arguments—over prevailing anxieties regarding Eastern Jewish immigration to Western European nation-states or to the U.S., and they are not involved in intra-Jewish arguments about the grounds for or validity of Jewish nationalism.

In *The Way We Argue Now* Amanda Anderson engages critically with central ethical and epistemological commitments of "what is broadly known as 'theory' across several humanities and social science disciplines" (2006, 1). The politics of identity has entered into the epistemological and ethical groundings of knowledge claims, she argues. Philosophical engagements that have sought, within this critical tradition, to undo the presumption of "stable, if embodied and situated identities" (ibid., 28) have maintained that absolute critical distance, that is, *critical detachment*—the possibility of the view from nowhere—is but a "God's eye trick" (Haraway 1991).

I want to highlight the ways in which those assumptions regarding the politics of identity and the production of knowledge have saturated domains that extend far beyond various forms of poststructuralist scholarship in the humanities and social sciences. As I illustrated above, the politics of identity has entered into biological inquiry as well: contemporary research into human population diversity grapples with a similar set of political challenges. And the response is not so far afield. It is not just literary theorists or cultural critics who own their "identities" as they produce academic knowledge. So too do researchers in various domains of biological research, including genetic history. The way "we live and know now" (Shapin 2008, xiv) is saturated by a multicultural imaginary.

By way of contrast with literary and cultural critics, however, for researchers in genetic history—and other domains of medical and biological research I reference above—declaring personal attachments to the research at hand is not understood to have any *epistemological* consequences: it is not seen to affect the substantive claims generated by their research. These scientists do not claim to be producing "situated knowledge" (Haraway 1991). Scientific knowledge is still understood to be the view from nowhere: scientific *facts*, generated by the knowing subject who embodies scientific virtues—that is, who knows how to evaluate and submit to the weight of evidence—retain their claims to distance, to detachment, to truth.

What then are the ethical and political implications of the coexistence of claims to attachment and detachment cultivated by certain practitioners in the biological and medical sciences today? We end up with knowledge that is factually true and, apparently, politically benign—or even good. We end up with an ethical or "sincere" science (Pottage 1998) that is presumably shorn of the specter of race science and eugenics, and of genocide and slavery that still haunts so much research into population-level human biological diversity. And we end up with scientists—and other experts—we can trust because they are one of us. (I discuss this at more length in chapter 4.)

A distinctive kind of politics of identity, however, is emerging through the practices of genetic history. Identity is *a project*, as Nikolas Rose has written of biological citizenship (Rose 2007). But if identity is a project—the reiteration of the choice to be true to one's traditions and to one's men, or making the choice to become what one has always already been once one receives the results of one's ancestry test—it is constrained by the epistemological authority of specific kinds of biological facts. Within the practices of genetic ancestry testing as well as in particular political projects to recognize communities as (Lost) Jews, as I show over the next two chapters, certain kinds of choices emerge as legible and credible and as desirable and even inevitable in light of genetic historical evidence. As I elaborate in the next chapter, choice is an ideological effect produced in the space(s) where genetic history's epistemological commitments and proliferating genetic archive intersect with the circulation of the discipline's claims and the commercialization of its working objects. Together, the science of and commerce in genetic history are generating novel ways of knowing and becoming who one really is.

CHAPTER FOUR

The Politics of Identity, Inc.[1]

A few days after *Nature* published the findings of the first Y-chromosome study of the Jewish priestly line (Skorecki et al. 1997), Karl Skorecki was fielding phone calls from Orthodox Jews who wanted to be tested to prove descent from the biblical figure of Aaron (Cohen 1997). "We're in a dilemma," Michael Hammer told a reporter. "We can't prove or disprove very easily if someone is a Kohen from this data. Do we want to? If we don't, will somebody else come along, a genetic-testing company, and do it, or would the Orthodox rabbinate hire some company to do it for them? Are we in a situation where someone could patent this as a genetic test" (ibid.)?

It was not long before some of what Hammer predicted came to pass: a genetic-testing company located in Houston was founded, and it offers customers the opportunity to test themselves for Cohen descent. And with the launch of the commercial market in ancestry testing, it is not just Cohen ancestry tests that are being sold: consumers can take tests to "find out" if they have Jewish or Native American ancestry; they can purchase tests that purport to identify the ethnic group in Africa to which their enslaved ancestors belonged. And with the birth of a new technology called Ancestry Informative Markers, they can find out what percentages of European, African, Asian, and Native American ancestry they have.

In this chapter I further elaborate the relationship between genetic history and the politics of identity through a focus on the commercial market in genetic ancestry testing in the United States. More specifically, I examine the ways in which explorations of selfhood, the politics of multiculturalism, and the commercialization of (biological signs of) identity articulate in the market for ancestry testing. Testing companies constitute the most concrete and direct "network" (Latour 1987) through which genetic history's epistemic things and truth-economy circulate in the public domain.

On the one hand, these ancestry-testing companies are prime examples of "entrepreneurial science" (Shapin 2009), that is, of "profit-making companies whose products are more accurately defined as new intellectual goods than as material artifacts" (ibid., 230). These companies are becoming, particularly in the United States, a feature of "the way we live and know" (ibid., xiv). On the other hand, genetic ancestry-testing companies are part and parcel of a global phenomenon, what Jean and John Comaroff have named the "identity economy" (Comaroff and Comaroff 2009, 2). Genetic genealogical-testing companies are start-up enterprises privately owned and, for the most part, not publicly traded. They offer their products—scientific techniques and facts—to individual consumers with the promise of providing knowledge about the individual self. The commodities bought and sold through genetic genealogy companies are always, already a part of us. It is *our genetic material* that is taken from us and then sold back to us in the commodity form. And that commodity is believed to reveal facts about our truest selves.

Discovering our truest selves is a desire that characterizes our multicultural age, what Charles Taylor has described as the quest for authenticity (Taylor 1992). That desire to know oneself, of course, takes various forms, is rooted in different histories, different forms of politics, and various conceptions of the self (e.g., as indigenous, as racial, as ethnic, as gendered, as sexed). In the case of Jewish projects of genetic self-fashioning as they manifest in the contemporary U.S., the politics and the fantasy of the immigrant nation stand center stage.

From its very beginnings, as Matthew Frye Jacobson has argued, multiculturalism was rooted in far more than the Civil Rights movement and black nationalist politics, as is often assumed. It was, he writes, also "a fantasy about immigration and America—about 'success,' and subsequent reveries of the second, third, and fourth generations at mid-century" (Jacobson 2006, 15). By the early 1960s, the melting pot ideology of the U.S. state seemed to be receding: in 1963 John F. Kennedy made a presidential visit to Ireland, marking "the emergent public language of 'enduring links' between Americans and their many homelands" (ibid., 15). The "ethnic revival"—white ethnics reclaiming their immigrant roots—was in part a response to the Civil Rights movement that "heightened whites' consciousness of their skin privilege," making it not just "visible but uncomfortable." In turn, Civil Rights and later, black nationalist politics provided a language through which to produce an "identity that was not simply 'American'" (ibid., 2) but *ethnic* American.

During the bicentennial, the U.S. was celebrated as an "immigrant nation"—in Hollywood films, in the heritage industry, in public celebrations of the bicentennial that represented the Ellis Island immigrant as the archetypal citizen. And with Alex Haley's *Roots* as its model—a narrative that circulated not as a specific account of the story of African Americans but as a tale of universal value to all ethnic groups (Jacobson 2006, 42)—"personal" ethnic heritage quests emerged as a widespread cultural practice, and they often took shape in "a newfound passion for genealogy." In the process of reconfiguring themselves as ethnic Americans, those children of Jewish, Italian, or Irish immigrants whose parents had "become white" were now a little less so. They were now not *that,* or not *quite* white (ibid., 21).[2]

There is an extensive historical literature on the political and social dynamics through which American Jews "became white" (Brodkin 1998; Goldstein 2006; Jacobson 1998; Rogin 1996). Paralleling the path of many other immigrant communities in the U.S. over the course of the early to mid-twentieth century, Jewish immigrants aspired to "fit in," to become just another American within the American melting pot. By the mid-to-late 1960s, however, their children began to "find" their Jewishness once again. Part of a general trend among the children of European immigrants, "pluralism," that is, the celebration of one's ethnic difference rather than the desire to assimilate to the American "melting pot," became the norm. By the turn of the new millennium that now "multicultural" understanding of American citizenship, what Jacobson calls the "hyphen-nation," was ever more entrenched. According to Eric L. Goldstein, writing at the turn of the millennium, many American Jews are caught in a bind: "Exhilarated, on the one hand, with the possibility of achieving success and acceptance of a kind unimaginable even twenty years ago, some are also afraid that the Jews may cease to exist as a distinct group. As a result, many Jews fret over the chance for 'Jewish survival,' aware of the costs of assimilation even as they pursue it" (Goldstein 2006, 211). In response to that prevailing anxiety, Goldstein argues, there is evidence that younger Jews are becoming increasingly uncomfortable with their (self-) designation as simply "white" (236).

Talk about "Jewish survival" has generated a lot of debate, scholarly and public, about what it means to be a Jew (e.g., Sklare and Greenblum 1967; Freedman 2000; Goldstein 2006; Cohen and Eisen 2000). As I demonstrate in this chapter, genetic ancestry testing, its results, and the kinds of practices that it elicits in those who take the test provide a particular kind of answer to the question of what it means to be a Jew (or to be an any other "X"). And the answer it provides allows individuals, at one and the same time, to embrace the liberal virtue of self-creation—of self-definition—and to assert the fact of a true and enduring Jewishness, biological signs of which Jewish individuals carry around within.

Genetic history is born of an age in which new subjectivities and new politics are emerging on the terrain of rapid developments in the biological sciences, biotechnology, and biomedicine (Rabinow 1999; Rose 2007). But what its molecular evidence generates, grounds, and authenticates are narratives of origins, kinship, and history. Those narratives are grounded in biological facts, but they are not represented as biological narratives. As I argued in chapter 1, genetic history is not governed by the biological determinism of race science: one isn't a Jew *because one shares particular haplotypes*, a specific biological nature, even if, as I show, upon receiving the results of ancestry tests, people do sometimes declare, "I discovered I am a Jew." Rather, being Jewish—and being a member of any other "ethnic or geographic" group (as the categories of classification are named by ancestry-testing companies)—requires that one actively embrace that "ancestry," that one learn about and fashion oneself according to its cultural or religious principles, thereby transforming ancestry into identity or selfhood.

That is not to say genetic history has no power, however. As genetic historical data and techniques circulate through the marketplace of genetic genealogical-testing companies, the quest for self-knowledge is fashioning a determinism all its own. Genetic historical evidence does not make me a Jew (or an any other social "X"). In other words, being Jewish is not a "heritable *trait*" (Jacobson 1998, 176, emphasis added). Nevertheless, through biological *signs*, I can discover truths about myself that I did not know. Upon discovering those truths I may desire, I may *choose* to learn more about—and to become—that which I always already was. As evidenced by the practices that emerge from genetic-ancestry testing, via a complex dialectic between scientific truths and individual choice, molecular facts can determine who we, collectively and individually, actually are.

The Birth of Genetic Genealogy

A few years following the publication of the 1997 *Nature* paper on the Cohanim (Skorecki et al. 1997), Bennett Greenspan, an entrepreneur who wanted to know whether launching a genetic genealogical-testing company was a plausible idea, contacted Michael Hammer for his advice. Greenspan had read about a study that used the Y-chromosome to determine whether or not Thomas Jefferson fathered a child with Sally Hemings, his slave (Foster et al. 1998). He had also read about the Cohen studies. Might the Y-chromosome be used for genealogical purposes? As a genealogist himself, he wanted to know if the Y-chromosome could be of any use to his own—and other genealogists'—personal quests:

> I was trying to find someone to prove that my cousin in California was related to someone with the same last name in Buenos Aires, and they had the same oral traditions that we did, we just couldn't prove our genealogies were the same—and so, for that reason I went to [Hammer] because he was the only American professor of either of those two [Cohanim] studies. . . . I reached out to him and I said "Look, would you do this DNA testing?" . . . He said, "Look I'll bottom line the science so you don't make mistakes, and you deal with the general public." . . . It was a good division of labor. In April of 2000, we founded genetic genealogy . . . the concept was genetic genealogy, the name of the firm was Family Tree DNA. (interview)

Family Tree DNA was the first genetic genealogical-testing company established in the U.S. And it has become a thriving venture: Family Tree DNA's gross revenues grew from $2.6 million in 2004 to $12.2 million in 2006 (Bradford 2008). In the words of Nicole Bradford: "Whoever said it is pointless to live in the past didn't try to build a business on it" (ibid.). Family Tree DNA is part of a growing market niche. According to Henry T. Greely, by 2005 there were at least sixteen companies and one nonprofit (the Genographic Project) that sold genetic genealogical services to the public in the U.S. (Greely 2008). By 2007 according to a report by the American Society of Human Genetics, approximately thirty companies offered genetic ancestry tests (Hudson et al. 2007). This is "a major potential market" for genomics that is outside the domains of medicine (Greely 2008, 210).

Greenspan started his company by approaching a few scientists and asking them, "How can we do this testing? How do we get the source information?" With the help of mtDNA and Y-chromosome experts (Theodore Schurr of the University of Pennsylvania for the former, and Michael Hammer for the latter), an initial database was compiled on the basis of publicly available data. From there Family Tree DNA built its own database from the samples collected from its clients. It now boasts the largest commercial genetic genealogical databases of any company: "As of September 14, 2009, we have a total of 262,471 records!" the website reports (familytreedna .com, accessed 9/12/2009). And the size of a company's database is considered critical, by entrepreneurs and consumers alike:

> Whether your goal is to verify your genealogy or to discover your deep ancestral origins, the size of the testing company's database is of fundamental importance. To get the best and most accurate answer, you want to compare your results with as many others as possible, which means choosing the company with the largest and most robust database. Family Tree DNA is in the lead—Our databases are several times larger than all the others' combined! (www.familytreedna.com)

Family Tree DNA also claims to have the largest "Jewish comparative databases . . . containing records for Ashkenazim and Sephardim, as well as Levites and Cohanim."[3] And if size makes a difference for matters epistemological, so too does it make a difference for matters commercial: emphasizing the size of one's database is a marketing technique for genetic genealogy-testing companies. And as is true across the industry, Family Tree DNA's databases are proprietary.

FTDNA samples are analyzed at Michael Hammer's laboratory at the University of Arizona. More specifically, the samples are run by the laboratory's commercial arm.[4] In addition to a research laboratory, Hammer directs a laboratory that provides genotyping services for university-based researchers and private businesses. For samples sent by university-based researchers, the laboratory charges cost. "In the case of FTDNA we can charge market value. And that profit then goes back into the university. . . . [It] can be fed back into buying equipment, hiring positions, and so forth" (interview). Hammer's lab provides similar services for African DNA, an African genetic ancestry company started by Henry Louis Gates Jr in cooperation with Bennett Greenspan. In addition, the commercial laboratory runs all the sequencing data for National Geographic and IBM's Genographic Project, a project carried out in cooperation with Family Tree DNA.

The respective databases, however, are kept separate. Family Tree DNA's genealogical database is different from the Genographic Project's database, and both are maintained separately from Michael Hammer's research database and samples. To quote one scientific consultant: "Any project involving both a commercial element and a research element needs a firewall." That is a matter of ethics: an individual's sample cannot be passed around without consent. In addition, it is an epistemological matter: the databases are not compatible. Family Tree DNA's database is "dominated by family genealogical studies," not by samples from "indigenous communities, classified on the basis of specific cultures and languages, the kinds of samples and information that an anthropological project [such as the Genographic Project] requires" (interview). In addition, genetic genealogical-testing companies get "pressure from their own clients" to offer higher resolution tests. Thus Family Tree DNA's commercial database is an archive of different and often higher resolutions. The Genographic Project might only need seventeen short tandem repeats (STRs) for its purposes. FTDNA now provides customers "with twenty-four STR tests and north" (for the Y-chromosome), one consultant told me. In fact, Family Tree DNA now sells full mitochondrial sequences for the maternal line.[5]

Commercial ventures have different incentive structures than do research labs. Therefore, Family Tree DNA also owns its own laboratory. It is what Greenspan calls "a discovery lab." As he elaborated in an interview, "we're looking for new markers, we're looking for markers that have not been published by other scientists, that a very very very very small percentage of our general customers might be interested in." Hammer's lab has no reason to run those tests, but Family Tree DNA does. Small profit margins make a difference for the commerce in ancestry testing. Greenspan recounted the following story in order to make that point:

> There's a branch of the tree of mankind called F on the male side. And we had an astrophysicist that said, "You know I've been looking and I've been reading the papers and I think that the [haplogroup][6] F that you've been stating that these 45 or 50 men [have] out of 140,000 in your database—I think they're F4. And I'd like someone to run this one sample." I have his permission to run it against F4. That's never been run before. It's been discovered at the University of Arizona, but [even though] . . . we have a commercial relationship with them . . . they're not going to run a single marker for a single guy. You know they're not interested in doing that. On the other hand, we are potentially interested in doing that, if we can make a couple dollars and we can add to science. And so the guy paid us 150 dollars to . . . set up the test and . . . we ran this one guy to determine whether he was actually F1, 2, 3, or 4-subdivisions of haplogroup F. . . . That's the kind of stuff that we do [at the Houston lab].

Running increasingly higher-resolution tests and building databases out of them has made Family Tree DNA relevant to academic researchers, according to Greenspan. "[Doron] Behar and [Michael] Hammer are constantly telling me, 'your databases are superior to the anthropological community's'" in terms of both size and the resolution of SNP [single-nucleotide polymorphism] testing. For example, the Houston Laboratory runs the full genome sequence analyses of mitochondrial samples (the hypervariable and the coding regions; see note 4). It has compiled a mitochondrial database that is larger than any other. In addition, the Houston laboratory is beginning to service academic researchers: "We're now starting to have researchers who have come to us who want us to run for a fee their academic samples so that they can publish—because we can do it more inexpensively, more efficiently than they can do it in their own facilities because we do it every day." In short, there is a "trading zone" (Galison 1997) developing between ancestry-testing companies and university research projects and laboratories, a reliance on the technological capacities and expertise of each other even as their goals, databases, and the test resolutions each needs do not fully converge.

There is an extensive scholarship on the commercial nature of the new genetics. Donna Haraway argues that one of its signal characteristics is its integration into "the New World Order, Inc." (1997). That is, it is fully integrated into the neoliberal economy such that market logics and enterprises structure research in contemporary genomics down to its very basic science core. As Steven Shapin demonstrates in his recent book, on the one hand our visions of a prior, more pristine science that was protected from the domain of industry and profit is a distortion of the historical record: twentieth-century U.S. science has always been characterized by the marriage of science and commerce. On the other hand, he argues, nonetheless a novel configuration of that relationship has emerged over the past several decades. The amount of private money funding university-based scientific research increased from approximately one quarter billion dollars in 1980 to about two and a half billion in 2000. In addition, there has been a widespread shift in understandings of that relationship. Beginning in the late 1960s and early 1970s, not only did "research universities increasingly offer . . . themselves as handmaids to industry." So too was "helping" industry "identified as public service." Moreover, "the sorts of professorial commercial ties that were once a source of administrative unease were now enthusiastically encouraged" (Shapin 2008, 213). The 1980s and 1990s witnessed even more dramatic shifts. No longer were scientists merely consultants to or employees of industry. Scientists in particular disciplines were founding or leaving the academy to take equity shares in start-up companies, biotechnology companies included (215). For the first time, scientists faced the possibility of becoming downright rich (ibid.).

The commercialization of molecular biology began in the late 1970s.

With the invention of recombinant DNA technology and the potential—the promise (Fortun 2008)—of genetic engineering, the "cloning gold rush" began (Wright 1986, 36). In turn, as this new technology met the transformation of the economic and political environment—as it met deregulation—the distinction between basic and applied research collapsed, and the privatization of the practices of the new genetics followed suit (Waldby 2002; Haraway 1997; Rabinow 1996b; Sunder Rajan 2006).[7]

A technologically driven science, (post-)genomics is part and parcel of the domain of speculative finance. The world of biotech and pharmacogenomics is risky business: market value is hitched to a promise of treatment options that are as yet largely unrealized. "Biovalue" structures this increasingly commercialized world of molecular biology: the production of use value (in the form of medical treatments, for example) and surplus value out of vitality itself (Waldby 2002; see also Rabinow 1999, 1996b; Fortun 2008; Sunder Rajan 2006).

Anthropological genetics emerges from within these broader political-economic and scientific developments, without which its technologies for reading the human genome would never have emerged. Nevertheless, the commercialization of anthropological genetics has followed a different path. As the organizers of the Human Genome Diversity Project argued in defending themselves against charges of bio-colonialism, their research was basic science. In the wake of the Diversity Project's failure, the Genographic Project, for example, is consciously represented as having no applied or commercial interests: it is "a purely anthropological, non-medical, nonpolitical, nongovernmental, nonprofit international research project involving scientists from both the developed and the developing world" (Wells 2006, 171). "No genetic data will ever be patented from the project" (172).

The signal characteristic of anthropological genetics is neither its inherent commercialization nor the collapse of the distinction between "basic" and "applied" research. For projects in genetic history, applied science—in this instance, commercial ventures in genetic ancestry testing—has followed upon the "success" of basic research: if one can investigate a *population's* origins, if one knows the relevant markers and has or is developing the relevant databases, why not develop genetic ancestry tests as a commodity to be bought and sold? As one researcher noted in an interview, "I think [the Cohen] study was seminal in a way. We had no way of predicting where that was going to take us or anybody else. I think it was timely in the sense that people were thinking more on these lines, and the connection was made obvious by the work that we had done." According to Greenspan, the Cohen study "really started this and the Jefferson study finished it. And then it took eighteen months for me to realize the practical interest" (interview). In turn, demand—the desire for higher resolution tests by genealogists who want more specificity and certainty—has driven the development of increasingly refined tests for both the Y-chromosome and mtDNA. Moreover, commercial profit is being fed back into basic research: the University of Arizona uses the income to fund academic work. Basic science itself is increasingly reliant on profit from commercial ventures.

In addition, genetic ancestry-testing companies may be inverting the relationship between biomedical work and the other disciplinary fields that have developed on its coattails, including genetic history. On the one hand, the desire to "know thy-ancestral-self" on the basis of genetic data makes sense within the discursive matrix of a "genes are us" world, and that world has been driven by a paradigm shift led by biomedical fields towards the genome. On the other hand, genetic ancestry testing seems to be the model for the now burgeoning "personal genome-testing" companies, which offer tests not just for ancestry but primarily for disease risk, a different dimension of this "genes are us" world.

In February 2008, Family Tree DNA spun off a new company:

> DNATraits specializes in DNA testing to identify genetic disorders and susceptibility to inherited diseases and characteristics. DNATraits offers the most extensive and least expensive tests in the world, complete with a free consultation both prior to and after testing. Whether you are starting a family now or in the future, are concerned about a hereditary disease, or want to determine your risks for certain medications (coming soon!), contact DNA-Traits today and discover the science of better living.[8]

According to Max Blankfeld, who co-founded both Family Tree DNA and DNATraits with Bennett Greenspan, "this is a field that we were reluctant to get in, as genealogists normally don't like to mix genealogy and health information. However, we noticed that over the years more and more people approached us on the subject of genetic diseases or inherited conditions. This lead us to form a separate entity for this specific purpose, where we use totally different test kits, absolutely unrelated to Family Tree DNA tests and stored DNA."[9] Unlike other recently launched personal genome-testing companies (23andMe or deCODEme), DNATraits offers tests for Mendelian (single-gene) disorders. For their part 23andMe and deCODEme also offer ancestry tests. Has ancestry testing blazed the trail for personal genome companies? "Well I think that when 23andMe decided to do this," Greenspan responded,

> they looked at the success of Family Tree DNA and National Geographic in selling direct consumer DNA tests and said we're going to change the paradigm and we're going to offer it in medical venues. . . . I think the passive, nonmedical nature of what we do showed people that doing DNA testing can be good and those other companies have taken advantage . . . of the fact that the public had started to become accepting of doing DNA tests. The difference is that doing Family Tree DNA testing is fun; doing medical testing and finding that you're a carrier for a disease isn't fun.

Ancestry Tests

On November 13, 2008, the American Society of Human Genetics issued a "Statement on Ancestry Testing." It was incumbent on the society to address the issue. The statement raised epistemological questions about the scientific rigor and transparency of the products on offer, and it raised questions about "the impact of ancestry testing on people, families, communities and societies [that] traverses a wide range of psychosocial, ethical, legal, political and health-related issues" (Hudson et al. 2007, 1; see also Soo-Jin et al. 2009; Bolnick et al. 2007). Many scholars have questioned the reliability of the ancestry tests sold by genetic genealogical-testing companies. Can one actually specify an individual's so-called ethnic origins? Deborah Bolnick and her colleagues have argued that "when an allele or haplotype is most common in one population, companies often assume it to be diagnostic of that population." But they point out, given high levels of genetic diversity within any given population and given gene flow between populations, "very few alleles are . . . diagnostic of membership in a specific population" (2007, 400). A researcher involved in the Jewish origin studies was critical of testing companies in far less compromising words: "I just think it's fraudulent. There are companies that say we will test you for the Viking gene, the Jewish gene. . . . There are particular variants that are found at high frequencies in particular populations, but they are rarely found at high frequencies in only one population" (interview).

Ancestry-testing companies, however, beg to differ. According to Tony Frudakis, the Chief Scientific Officer of DNAPrint Genomics, which offers ancestry tests that calculate the percentage of one's "geographic" or "ethnic" ancestry (Native American, Asian, African, European), "DNAPrint's website and print publications clearly explain that the bases for human-derived notions of 'race' incorporate genetics as well as geography, religion, culture, and even socioeconomics." While companies can provide probabilistic calculations of their customers' racial ancestry, in other words, they do nevertheless take "great pains to explain . . . that both genetics and race are imperfectly correlated with geography" (Frudakis 2008, 1039).

Family Tree DNA offers information regarding ethnicity on the basis of Y-chromosome and mtDNA tests.[10] "DNA testing," they state on their website, "can show: if two people are related; your suggested geographic origins; if you could be of African ancestry; your deep ancestral ethnic origins."[11] With regard to Jewish ancestry, Family Tree DNA informs its potential male customers: "Males can test their Y-DNA to determine the origin of their paternal line, including the possibility of Jewish and Cohanim ancestry." More specifically, the website states that Y-chromosome testing can provide information about the following: "if two people are related; your suggested geographic origins; if you could be of Jewish ancestry; if you could be of Cohanim ancestry; your deep ancestral origins" (www.familytreedna.com, accessed 9/13/2009).

Not all consumers are looking for ethnic or geographic origins.[12] Many are genealogists in search of more specific information regarding familial relationships, and to aid that quest Family Tree DNA recently added a new product, the "Family Finder," that "helps you find family across all your lines, up to 6 generations back, by checking hundreds of thousands of points in your autosomal DNA, and comparing your results with others in the Family Finder database" (www.familytreedna.com, accessed 12/13/2010). (I will return to the ways in which ethnicity and family are entangled in genetic genealogy.) Moreover, unlike companies like African Ancestry or its rival, African DNA,[13] neither pinpointing someone's specific ethnic origins nor calculating the percentages of "bio-geographical ancestry"—what percentage of one's ancestry is Native American, African, European, or East Asian—is the primary commodity on offer at Family Tree DNA. According to Greenspan, Family Tree DNA is the company most favored by "serious genealogists": it is most favored by people seeking to establish family trees and to identify specific family members.

Nevertheless, Family Tree DNA does provide customers with information about their "ethnic and geographic origins" both "recent" and "deep," and that is a common reason why people test. One person who started a post on the web forum titled "Ancestral Heritage Curiosity" wrote: "What I hope to accomplish with a DNA test is not so much to tie my Family Tree to someone else's (although that would be nice) but to get an idea of my ancestral heritage." Of presumably German-Irish descent, she is one of eleven children, and "several of us have a darker than normal complexion." "Can a DNA test help me with this? If so, which test would be the best?" After a long back and forth with various people on the web forum regarding which tests might be most helpful, another person wrote: "A lot of Irish people are dark haired and dark featured. That might be one of the reasons. But other than that DNA testing might find the answer, might reveal an area of origin you didn't know about" (accessed 7/27/2009). A second customer started a post "Jewish heritage questions!!" "All my life I've believed myself to be Jewish because of maternal Jewish ancestry. Lately, I've decided to try my hand at becoming religious and have discovered the hard way that being ethnically Jewish does not mean that you are halachically Jewish. . . . To be Jewish under Jewish law, your mother needs to be born Jewish or you have to convert." He had discovered recently that his grandmother's father was Jewish but his mother's mother's ancestors were not. After months of trying to trace his mother's maternal line via names, he posted in the hope of finding someone who might know whether or not "these names are Jewish in origin." In addition, "I'm in the process of getting my DNA tested by Family Tree, and while that may personally confirm I'm Jewish, it does not mean anything to a court of Jewish law." He asked, "What mtDNA haplotypes are most indicative of Jewish ancestry?" To quote one response: "As the other posts have noted certain subclades [i.e., subcategories] of [haplogroups] K and N are common among Jewish women. However haplogroups, indicators of early human migrations, such as H, T and J are also common although they are not at all unique to the Jewish population."

In the summer of 2009, I decided to test my mitochondrial line. I took the HRV1 test (the first hypervariable region), which provides information on relatively deep ancestry. I did so for two reasons: first, I wanted to receive the results page so that I could better understand what information was being provided to customers about ethnic ancestry. Second, I wanted to engage with others in Family Tree DNA's online forum who had taken the test in order to better understand what was at stake for those who take ancestry tests and how they respond to the results. Within a week, I received several emails from customers whose samples "matched" mine and who wanted to know if I am Jewish or have any known Jewish ancestry. Each suspected they had Jewish ancestors and were hoping to substantiate that suspicion by finding matches with self-identified Jews. One woman explained, "I'm searching my family tree to find out who my ancestors are and also to find relatives who also have Jewish ancestry. We are possibly secret Jews or Anousim or sometimes called Marranos or Crypto-Jews from Spain." She elaborated in a subsequent email: "I tested because I wanted to find new relatives. I am very active in genealogy and tracing the roots of my family. I have mtDNA matches in 160 different countries worldwide, which includes Norway, but the biggest majority come[s] from Germany, England, Ireland, Wales, France. I also suspected that I might have Jewish ancestors so I wanted to learn what I could about them." Explaining her suspicion that she is of Jewish ancestry, she wrote: "I have been for many years drawn to Israel and Jews. I discovered that my birth family had some Jewish practices though being Jewish was never once mentioned. I do have some matches who say their ancestors were either Ashkenazi or Sephardic. I have sent emails to every mtDNA match and several have said they have Jewish ancestors, they are Jews now or suspect that they might be."[14]

My response to the results of my mitochondrial test was quite different than that of the woman I quoted above: I could not understand how or why those results could tell me anything meaningful about either kinship or ethnicity. Given the time frame of the HVR1 test, the likelihood that I am related to the above woman within what Family Tree DNA calls "a genealogical time frame" is quite remote. As explained in the documentation I received with my test results, "An exact match on Hyper Variable Region 1 means you share a common female ancestor, but in only 50% of cases did this common ancestor live within the last 52 generations"—approximately 1,300 years. "Matches of the mtDNA have more Anthropological significance as the time frame for a common ancestor could go beyond the Genealogical time frame." In other words, the test may give you matches that are more relevant for understanding one's human ancestry than one's familial connections.[15] Moreover, I learned that my mitochondrial line is a common and very old European lineage, something that I considered neither here nor there. But as I learned from subsequent email exchanges, my response was not typical of those who take the test.

How then are ethnic identifications made? They are not a matter of incontrovertible fact. Ethnic identification is perhaps better conceived of as a statistical common sense generated within the context of a specific database. According to Greenspan,

> If we have a bunch of Native Americans who have tested and a guy with an oral paternal tradition of being Native American and he matches people [Michael] Hammer and [Theodor] Schurr have obtained by going on reservations, then I am in good shape. If have STRs, if match exactly or one off or the same last name, or an oral tradition of being Native American then I am

comfortable they are Native American. There are a lot of Native American wannabes, and I tell them they are not. (interview)

Greenspan went on to explain, "we have people who *feel* that they're Jews, or *feel* that they're Native Americans, and we test for it and we can't find it. Now that in itself doesn't mean that they don't have any Native American genes or any Jewish lineage. It's just on the Y-chromosome or the mitochondria, which happens to be a very clear, unambiguous roadmap . . . we can't see it" (interview).

Family Tree DNA is not identifying so called private markers. There are (virtually) no genetic markers found exclusively in one population and therefore, as Bolnick and her co-authors have argued, ancestry tests "cannot pinpoint the place of origin or social affiliation of even one ancestor with exact certainty" (2007, 400). Ethnic identification is a matter of finding a "match." And that match depends on the database with which one compares the sample (see Bolnick 2007; Hudson et al. 2007).[16] As several exchanges on Family Tree DNA's online forum have pointed out, a lot of people get back low-resolution results (i.e., tests that rely on fewer markers) that indicate Jewish ancestry. One person created a post, "Confused, possibility of Ashkenazi Ancestry?": "My paternal ancestry comes from mainly Germany but my results (12 marker) are showing a strong connection to Eastern Europe. In a lot of the comment sections of the results there's 'Ashkenazi' and if I'm not mistaken that's Jewish. I just want to know . . . if there's a strong chance that some of my ancestors were Ashkenazi Jews or not?" He posted his results: Haplogroup R1a1 and the list of matches (with other customers who have tested) identified by the Family Tree DNA database. One responder noted that as far as she knew, R1a is one of "the most common haplogroups in Eastern Europe," but is also "common in Germany and Scandinavia. It is also found in about 10% of Ashkenazi Jews." She then cautioned: "Very many Ashkenazi Jews have tested [at Family Tree DNA], so they often appear in matches." She suggested he take higher resolution tests in order to resolve the question (accessed 7/27/2009).

Each genetic ancestry test is read against a body of data that serves as the control group: Does your marker match or nearly match haplotypes present in the existing samples of Native Americans, for example? Does it match haplotypes in the existing samples of (Ashkenazi, mostly) Jews? In the case of tests offered by African DNA, which "ethnic group" in contemporary Africa—as represented in the company's genetic database—does your DNA marker match? In explaining the logic of ethnic identification, Greenspan told me the following story. There was a Polish guy whose mother was adopted. This was 2002. Family Tree DNA still had a small database. But since the Polish man matched two Jews, Greenspan told him "it is clear you descend from Ashkenazi Jews." The customer was intrigued by Greenspan's claim because he knew that his mother was adopted in 1941, in Poland. By 2004, Family Tree DNA found five more matches in their mtDNA database on the HVR-1 region but now not just with Eastern European Jews but with non-Jewish Europeans as well. So the Polish man upgraded to a HVR-2 test. He matched Jews and non-Jews. He reached out to a match living in Holland and asked her if she is Jewish. The Dutch person said she did not know. Her mother was adopted. Family Tree DNA then ran a whole mitochondrial sequence for the Polish man for free. They found that he matched one Jewish sample and was one mutation away from another Jew. More generally, they had found six mutations in the coding regions of mtDNA that separated the Jewish cluster from non-Jewish clusters. That provided "100 percent proof" that the Polish man "came from a Jewish genetic gene pool," according to Greenspan. By 2009, Family Tree DNA had identified more matches between the Polish man and Jews. The Dutch woman, however, clustered with the English group.

Assigning ethnicity is nothing more than "guilt by association," to borrow Greenspan's felicitous phrase. Say it is December 24, he explained, say you are driving your car and you go into a neighborhood with no Christmas lights on. Either there is a power failure or they don't celebrate Christmas. Say you see furry black coats and funny hair. "When I have five people who match and all say they are Ashkenazi Jews, and now I have a non-Jew who also matches and whose father was adopted . . . then most reasonable people will accept" that he is Jewish (interview).

If ethnicity is a significant part of what is on offer at Family Tree DNA, what are the consequences? What are the effects of "discovering" that one "has" Jewish or any other "ancestry," as promised on Family Tree DNA's website? As Deborah Bolnick and her colleagues have argued, even though ancestry testing is often referred to as merely "recreational," "the tests can have a profound impact on individuals and communities" (2007, 318). The "Statement on Ancestry Testing" of the American Society of Human Genetics emphasizes this point: "Knowledge about genetic ancestry—if undesirable or unexpected—can elicit a range of psychological responses including shock, disbelief, denial, anxiety, anger, fear and other well-known reactions to unwanted news. It can also lead to the reshaping of individual or group identity" (Hudson et al. 2007, 7).

In an insightful ethnographic study of genetic ancestry testing among African Americans seeking ethnic roots in Africa, Alondra Nelson provides a rich account of ancestry testing. Describing the product on offer by African Ancestry, Nelson explains:

> A hypothetical root-seeker . . . may learn that her mtDNA traces to the current Mende people of Sierra Leone and that her Y-DNA test, for which she submitted her brother's DNA, traces to the Bubi group of present-day Equatorial Guinea. African Ancestry analysis might thus be regarded as ethnic lineage instruments through which undifferentiated racial identity is translated into African ethnicity and kinship. (Nelson 2008b, 254)

This may seem to provide strong evidence that the practice is "essentialist," Nelson writes—that it presumes identities are inborn and passed down through the generations, here via DNA. But her ethnographic research reveals something quite different, she argues. DNA is but one contributing element to the "potential transformation" of subject positions. "The cogency of African Ancestry's testing is derived significantly from social sources that shape how facts are anticipated, interpreted, and mobilized by roots-seekers"(2008b, 254–55). For example, the results of the mitochondrial DNA tests of Bess, one of Nelson's informants, indicated two possible origins: she could be "linked to the Kru of Liberia plus/or Mende-Temne of Sierra Leone"(259). Which of the "two possible ethnicities" (ibid.) was most compelling to Bess? "'My sister was married to a man from Sierra Leone . . . she replied obliquely, intimating that she would likely travel to the natal home of her deceased brother-in-law" (ibid.). This story captures a key aspect of what Nelson refers to as the "interpretive arc of ethnic lineage testing" (ibid.). Ancestry testing "occasions 'biosociality' between African communities and their diasporas" (261). Nevertheless, she argues: "Affiliations that incorporate biogenetic facts may nonetheless be the 'families we choose'"(263).

Nelson's analysis is correct: test results must be interpreted. They must be incorporated into narrative structures, without which they remain meaningless, incomprehensible in fact. The Polish man mentioned above could assimilate the possibility of his Jewish ancestry because he knew his mother had been adopted in 1941 in Poland. The woman who emailed me about our match could interpret her genetic kinship with others of known or suspected Jewish ancestry as evidence of her own Jewish ancestry because of an already existing narrative of herself as someone with "signs"—cultural and emotional—of being a lost Jew. And to take one more example: A man, let's call him F, tested his ancestry in three different companies: Ancestry By DNA, Omnipop, and DNA Tribes, all of which use a set of genetic markers in order to supply customers with percentages or fractions of ethnic/racial ancestry. (The technology is called admixture mapping). In his instance, Ancestry By DNA reported he is 77 percent European, 22 percent Native American, 1 percent sub-Saharan African, and 0 percent East Asian. "Can someone tell me how I can get ABDNA results like I did but Omnipop and Tribes is telling me I am African American?" F gets a variety of answers. One person explains: "As for Tribes it matches your profile against other populations in its database. If it doesn't have the population then it picks the closest. My Mum is 50 percent English 50 percent unknown American and yet she pulls no matches to the British Isles with Tribes!" A second person responded: "Just because your Tribes score says Africa certainly does not mean African in the cultural sense. Rather, East and North Africans have strong tries to Asia Minor and Europe respectively, as seen throughout history." F then posted his "thanks": "I am a little calmer this morning. Since the ABDNA test confirms what my genealogical work has found, I am putting more ancestral confidence in that test."

Genetic genealogical web-chat forums are replete with examples of such interpretive work. One person responded in the following manner to the man mentioned earlier who was confused about the possibility of having Ashkenazi ancestry: "The highest probability of your origins seems to be from [presumably non-Jewish populations in] Czechoslovakia and Poland, not Ashkenazi. What is your family myth of origin?" Moreover, web-chat forums are replete with examples of ancestry tests and the fulfillment of desires: the desire to be Jewish, the desire to be Native American, the desire to find *some* "minority" ancestry: "I just updated my 15 marker and found some interesting things when comparing to the latest populations. My majority is NW European, but it is the minority race that most of us are interested in." As Matthew Frye Jacobson has argued, in the age of American multiculturalism, many white ethnics no longer desire to be *all that*—or just—white (2006).

In spite of all the interpretive work involved, however, I pause at naming these families that we choose (Weston 1997). We do miss a lot when we argue that ancestry testing involves "succumb[ing] to a population-based genetic essentialism" (Brodwin 2002, 328). Nevertheless, I want to highlight the epistemological power of genetic facts. Doing so makes more evident the ways in which their very facticity makes certain choices and certain kinds of affiliations available and truthful, and others less so. I elucidate the dynamic relationship between the authority of biological facts and the interpretive work of self-fashioning and self-recognition in order to sketch the form determination taking shape within the field of genetic history. As the previous chapter demonstrated, a fundamentally liberal belief in and commitment to choice and agency are inscribed within what kind of facts are recognized as evidence of "origins." As I have also argued, that liberal belief in agency (sometimes configured as choice) is inscribed within genetic history's reading of junk DNA and the control region of the mitochondrial genome (see chapters 1 and 3). As I demonstrate here, so too is it inscribed in the very market rationality of the commerce in genetic ancestry testing. But insofar as liberal fantasies always exist within structural constraints, I want to analyze the juxtaposition of "choice" and "agency" with the epistemic authority of scientific facts. Given the epistemological power of genetics—and not just genetic history but genetic knowledge more broadly in our genomic age—genetic history is providing evidentiary grounds on which certain choices and interpretations and certain practices of self-recognition become possible, legible, desirable, and truthful. The "choice" of Jewishness or of any other ethnic or national affiliation becomes available at the crossroads of genetic history's particular epistemological power and the commercial network through which its epistemic objects and truth claims circulate in the public domain.

The Marketing of Choice

In the marketplace of ancestry testing, one's first choice is not interpreting the results of ancestry tests. It is deciding whether or not to test. Greenspan believes that compared to other populations Jews are more open to genetic ancestry testing: Jews started in DNA early, "in the 1970s with enzyme testing." Educational campaigns about Tay-Sachs, an inheritable and ultimately fatal genetic disorder, provided a "good education" about genetics more generally. By the time genetic genealogy came along, Greenspan argued, Jews were "more receptive to DNA testing" (interview).

But that does not mean that most Jews are willing to take genetic ancestry tests, not even most Jewish genealogists. By the summer of 2008, genetic genealogy was a lot more present at the annual meeting of the International Association of Jewish Genealogical Societies in Chicago than it had been several years earlier during the annual meeting in Washington, D.C. While more of a footnote in earlier years, at the Chicago conference DNA was a ubiquitous subject. Conference members were carrying tote bags with "Family Tree DNA" logos on them. A sign by the Illinois Jewish Genealogy Society's table offered a special discount for Family Tree DNA testing. Greenspan estimated that approximately 15 percent of participants at the 2008 meetings were buying ancestry tests from Family Tree

DNA. That was a large increase from the estimated 3 percent who did so the first year Family Tree DNA showed up at the meetings and offered the test (interview).

There are many reasons individual genealogists do not test, not least of which is the cost. I asked Greenspan where most of his customer base resides. "A lot of the market is U.S. based," although they do have a relationship with a company in Switzerland, their "European office."[17] "More and more Israeli Jews are coming in as customers. . . . I have tested hundreds of Israelis." But Israel is "not a rich country. The test is one hundred to three hundred dollars" (interview). This is a luxury that presumes disposable wealth.

Even for customers who have already run some tests with Family Tree DNA, cost remains paramount. As discussed on web-chat forums and as I was told in numerous interviews, there are many "upgrades" people decide they cannot afford. Collect the DNA now by purchasing low-resolution tests, the argument goes, and wait until the technology gets cheaper—or "I win the lotto," as one person said—to order more tests. But collecting the genetic material *now* can be of crucial importance: "I have a record of mom's cousin when he was alive, and now he has passed away. Imagine if I waited." Or as one person posted on a thread titled "Jewish Test" on Family Tree DNA's web forum, "I urge you to test your grandfather for the sake of preserving his DNA. The DNA of all family elders is irreplaceable!" (accessed 8/4/08). From the perspective of genealogists, this is a project in familial salvage genetics.

Expense is but one common explanation for not testing, however. For many genealogists there is considerable confusion about what kinds of information one gets out of ancestry testing and about whether or not it will be of any genealogical significance. After one session on genetics and disease risk at the annual meetings of the International Association of Jewish Genealogical Societies in Chicago in 2008, one person approached the speaker after her talk and asked about genealogical DNA testing and genetic predispositions to disease and genetic disorders. Ms. Sitron, the speaker, insisted that these were two separate issues and that the kind of results that "Bennett" gives does not offer any direct medical insights. But the questioner explained that she had used Bennett's DNATraits service and had received news that she carried one gene for a blood disorder. She was confused if this meant that she herself was "infected," "affected," or simply a "carrier." She said that she was so distraught over this that she nearly did not come to the conference. The speaker took her by the hand to go talk with Greenspan about this and get the matter sorted out. To give one more example, another genealogist explained:

> I'm interested [in ancestry testing] but hesitant. . . . I have friends of friends who are having a whole full scale [test], like the BRCA test. Do you know about that? That there's a higher incidence of breast cancer among Ashkenazi women? So you can do . . . genetic testing that does the medical aspect. This is more the anthropological. . . . I'm sort of interested but yet hesitant. . . . It's like, how much do you want to know and how much of a role do genes play, how much is lifestyle? . . . Science will play out. We'll understand some of these tests. We'll understand what percentage is the genetic component. (interview)

After moving seamlessly back and forth between genealogical and medical testing, she continued: "With the whole genealogy genetics, it has a certain appeal, particularly for those of us who hit roadblocks with traditional forms of research. [But] just because you match someone genetically doesn't mean you'll have anything in common with them otherwise."

Others decide not to test because there are genealogical things they would rather not know. "I would like my husband to [test] because that will track at least his father's father's father," one woman said in an interview. "They also claim to be Cohanim, so that would also be interesting to see how far back we could get with that." She laughed, "I know he won't take that test." Why not? "Why would anyone who believes that he's a Cohen take a test that might show he's not?"

> I would imagine that for some [the Cohen test] would seem to be validation. But for most, I would imagine it's a question of the father and what you grew up with and what you're told. . . . Science is constantly changing you know, the interpretation of results changes accordingly. It's not exactly like taking a test to see if you might have an inherited disease, for example. . . . This is sort of putting a piece of your identity into question that perhaps is very important to your sense of who you are. I'm just thinking of my father-in-law. . . . He was all about being a Cohane. . . . It was a huge part of his identity and so I'm not sure that he would have accepted a result that told him otherwise.

And there are those genealogists who refuse to test simply because it would ruin the fun. It would destroy the hard work that makes genealogical research what it is. One person told me, "genetics threatens genealogy. . . . It threatens to cut out the research project. For example, I am meeting with someone later today named E (my maiden name). We are going to look at family trees and figure out if we are related." That whole project of sharing trees, sharing photos, and so forth, the "beauty of the process could get lost" with genetic testing. It is the "uncertainty that makes genealogy fun" even if "it is certainty" that one "is chasing after."

Choice is built into the very commercial structure of this enterprise. Ancestry tests are a commodity I can buy or choose not to buy. Moreover, given the logic of the commodity form, my sample belongs to me: my information can only be shared with other customers if I have signed a consent form. I can have my information removed from the database at any time. I can choose to have my sample destroyed. And these are proprietary databases: neither my sample nor my information will ever be passed along to other companies (say, insurance companies) or to the state. As one person put it, one hurdle of genetic genealogical testing is "first you have to convince your relatives that their DNA will be used *only* for what they consented to and nothing else" (interview).

Given the many testing companies in the market, my ability to choose extends beyond the decision of whether or not to test. I can choose which company to test with. And as is the case for many genetic genealogists, I can choose to test with more than one. Genealogists consider various criteria when deciding where to test: the size of the database, the presumed reliability of the results, an assessment of the company's "expertise" and credibility. According to one genealogist, she chose Family Tree DNA partly because of its collaboration with the Genographic Project: "with National Geographic choosing them, another vote of confidence." And for many, there is a question of "trusting" Greenspan. Family Tree DNA is well known for being "there" for its customers. Greenspan makes himself available to discuss results. As one genealogist told me, she chose Family Tree DNA because of "their care. They are very communicative. They have good follow-up. When I had questions, I emailed them, phone calls. They always responded." And quite crucially, Greenspan is "one of us": Jewish, a genealogist, very present at the meetings of the International Association of Jewish Genealogical Societies. If you tell anyone at the IAJGS meetings that you are interested in DNA, someone said in an interview, the response tends to be: "You need to talk to Bennett." According to one informant, "well it's because he's been part of this Jewish club for a number of years now. Now the thing is getting bigger, but he's kind of the guru." And as often seems to be the case, Greenspan took a personal interest in this man's case: "He's the one that's looking at my three markers, because it's an interesting case for him. . . . They're trying to figure out, what's the possibilities. He doesn't even know." If the "the virtues of familiar people" matter to the "rationally calculative worlds where late modern finance meets technoscience" (Shapin 2008, 270), then Greenspan's interest in and presence for his customers, and his "familiarity" as a fellow genealogist and a fellow Jew, have made Family Tree DNA the company of choice for many a Jewish genealogist.

If there is no single reason why people choose not to test, neither is there a single reason why people purchase genetic genealogical tests. Consumers wish to find out various kinds of things other than ethnic or ancestral origins. Some wish to confirm or disprove that "a relative" she has discovered via traditional genealogical research methods is indeed a blood relative. Others are engaged in more of a fishing expedition. They turn to genetic testing when classical genealogical research has "hit a wall"—the most common turn of phrase for describing the genealogical value of genetic testing. One genealogist told me that genetic ancestry testing might be particularly helpful in figuring out her husband's family history: "On his dad's side, I can go back three generations but I haven't been able to get back to Europe. DNA support could be helpful." I asked her what she is hoping to find out: Where he came from? The region? Specific family connections? "The region is the weakest link. People moved back then. We don't believe it, but people moved." Recently Family Tree DNA found a 67 STR (single tandem repeat) match with someone from the Ukraine on her mother's side. But her mother's side is from Lithuania going back to the 1700s. "The fellow said 'Wow, there is certainly a chance our ancestor's son—or daughter—went to Ukraine.' Two days ago, I passed him my family tree which he gave to his sister to see what she can find." What is important is "not the region as much as making connections based on our own research. My goal? I know this is my cousin, now it is up to us to show that past." Another informant explained, "the reason genealogists do [genetic ancestry testing] is because we're trying to connect with people . . . trying to find other relatives, to find people who might be related, who might be from the same town, and maybe they know more about the history of it. But you could never find them otherwise [i.e., without DNA.]"

If an individual quest, genealogical research—traditional and genetic—is nevertheless a collective project: genealogical research relies on pooling information and creating databases and websites and archives to which people can turn. There is JewishGen, a website that documents available sources, connects genealogists with others of the same name, and produces a computer disk of "The Family Tree of the Jewish People," which compiles all the family trees that individual genealogists contribute to the public database with a view to elucidating more and more familial connections among Jews. In its "Jewish database" Family Tree DNA provides the genetic equivalent. In addition to saving genetic material for future tests before a family member dies, one genealogist recounted in an interview, "I see the benefit for everyone": the larger the database, the more likely it is that someone will find a match. A frequent commentator on the Family Tree DNA's web forum responded to someone asking about a "Jewish Test"—i.e., what tests might be most useful in revealing whether he has Ashkenazi or Sephardic ancestry:

> Test results will be ACCURATE but may not be CONCLUSIVE for or against your hypothesis of Jewish ancestry. But all tests [*sic*] results are useful (to know that your maternal grandfather's direct paternal line is not Jewish is just as informative as knowing that it is . . .) And all DNA test results are useful, if posted on public databases, as they add to the public store of DNA information. (accessed 8/4/08)

And that seems to be a widely held view.

> The DNA is kind of the newest idea [in genealogy] and it's really still in its infancy. . . . I think they've been testing actively some of these companies for six to eight years now. . . . They're beginning to get some more reasonable numbers. Of course it requires that we get a large database. So what our society, JGS of Illinois, now has just started . . . is kind of formally encouraging our members to participate, to submit their DNA for testing. And we have a group plan, and you get a little discount for Family Tree DNA, which has most of the Jewish data, most of the [Jews who test] . . . are in that database. (interview)

He recognized the complications of encouraging individuals to test but nevertheless saw its larger, collective value:

> We're encouraging people to spend money, but we don't want them to get their hopes up, like "Oh this is it, you're going to swab your cheek and your family tree will come out." That was the biggest concern, that this is very experimental and there's a very low probability of finding someone. But we're trying to also further the resources of the genealogical community, and the more people that participate in it, the next person is going to have a better

chance. . . . We're not a big club. So if we get the Jewish community invigorated on this, maybe we can find where we're from and rebuild our family circles.

As those family circles expand and incorporate more and more individuals, the familial begins to "add up" to a larger collectivity: "The Jewish People," and as a subset of that population, "Ashkenazi Jewry"—a genealogical community also increasingly "discernible," according to Greenspan as well as various scientific studies, on genetic lines.[18] (The Jewish database held by Family Tree DNA is overwhelmingly Ashkenazi, as is the U.S. Jewish population.) There was a long and increasingly hostile exchange on Family Tree DNA's online forum about whether Jewishness has anything to do with genetics and whether it makes any sense to list "Jewish" as a criterion for identification on Family Tree DNA profiles. It is "not fair," one person wrote, other "religions" are not listed in the databases. It is misleading to identify haplogroups as being "Jewish." "Elizabeth" replied:[19]

> My parents, my grandparents, my great-grandparents, and even my 5th great-grandparents, all who were Ashkenazi Jews living in Eastern Europe, are who define my culture, my family history and my ancestry. So your comments about what haplogroups are or aren't "real" Jews are completely offensive and unwarranted, and they need to stop, immediately. If you want to bash Jews, do it elsewhere, NOT on this board.

In defense of her original statement, "Elise" responded: "I think you misunderstood my point. I wasn't bashing anyone's devoutness or great grandparents. My point was that religion, especially if it is only one religion, shouldn't be mixed with science. I could put that I am an mtdna H1 Hare Krishna, but that would be irrelevant or misleading to anyone else who is H1."

Elizabeth:

> No, you completely miss the point that being Jewish is about MUCH more than religion. My great-grandparents immigrated from Belarus. They were *not* Belarusian and I do not have Belarusian ancestry. I am Jewish, they were Jewish, our ancestry is Jewish. I can convert to Christianity and I would *still* have Jewish ancestry.
>
> When you understand the distinction between being culturally Jewish and practicing Judaism as a religion, as well as the distinction between

having Jewish ancestry and descending from the ancient Hebrews/Israelites, only then will you begin to understand my outrage at your statement that an H1 [haplogroup] being Ashkenazi is "nonsense."

The commitment that Jewishness is "not merely" a religion is a long-standing one, a key element of the Zionist rejection of Jewish assimilation in the late nineteenth and early twentieth centuries. As I demonstrated in chapter 2, prominent scientists and political activists who sought to make Judaism more than that, more than "just" a religion, turned to biology. Over time, the fact of a shared biological (-qua-historical) origin in ancient Palestine emerged as taken for granted in the nationalist ideology of the Israeli state. Contemporary practices of genetic genealogy reiterate, they perform—*and they provide empirical evidence for*—that (ethnic nationalist) understanding of Jewish identity, even as they also provide evidence for two related but not identical groups: the Jews as an ancient, all-encompassing genealogically based peoplehood and the Ashkenazim as a distinct genealogical subset of contemporary Jewry (and the numerically dominant one in the U.S., as is often pointed out).

Genealogy is practiced as an individual pursuit of family history and lost relatives. At the annual meetings of the IAJGS there is little *direct* discussion of a collective Jewish identity or of the relationship between Jewish identity and genetics. And with respect to genetic tests and genealogical knowledge, many people insist on the need to distinguish "biology" from "Jewishness." DNA is helpful for "filling in the gaps." That is all. Moreover, there is often rather vociferous objection to the implication that kinship is *only* biologically based. For example, following a talk by Colleen Fitzpatrick (August 2008), during which she discussed the case of someone whose DNA test results appeared to be at odds with his genealogy, someone in the audience loudly protested that this did not mean that he's not part of the family. The speaker heartily agreed. She added: this was likely the result of a "non-paternity event." There was some general commotion during this exchange, with audience members supporting the person who voiced the position that the subject in question should still be considered part of the family. However, the audience member who first made this claim followed up by saying, "Look, it's just a lab test!" The speaker took this to mean that the lab test might be wrong, and explained that the odds of this were very low, but that even if the results came back differently a second time, this should hardly make a difference. The families that matter may well include individuals who are not *biologically* related.

Nevertheless, the practice of genealogy builds upon, reinforces, and performs a collective, descent-based sense of a collective self, even in the absence of explicit discussions of "Jewishness." Moreover, through ancestry testing, the descent community is rendered visible. Jewish peoplehood and Ashkenazi Jewry are rendered empirical facts within a new regime of truth when they emerge as scientifically legible and genetically true. Individuals are often led to genealogy by commitments to their Jewishness—whether the sense of always having identified strongly as Jews or the desire to recuperate a past and relatives lost to the Holocaust, a constitutive "event" in American-Jewish life (Novick 1999).[20] Does your genealogical work have anything to do with the question of Jewishness for you, I asked one woman? "Oh yes. I always knew and believed and embraced being Jewish. I loved going to religious school. Back then girls were rarely bat mitzvahed. I insisted. . . . Being Jewish was always important to me." Another woman, someone who worked at a Jewish genealogy center, told me:

> These are people who are obsessed with tracing their ancestors; they are attached to wanting to know. The further back, and the higher the level of certainty, becomes very exciting for them. . . . [For genealogists] it is exciting to see all these Jews going back and back and back. It gives them a stronger sense of being Jewish. In that sense, continuity is important.

She remarked that people have long used "genetic analogies to talk about ancestry." But all that talk of "blood" was metaphorical not literal. "The risk of population genetics is that it might transform what might have been a metaphor into something really real." The danger is that you could forget that Jewish identity is complex. "You could forget the element of choice." And yet "sometimes that which is chosen is essential or inalienable; and that stuff that one has, one can get away from." For many, exploring family connections and histories can result in a deeper sense of their Jewishness. For others, it can result in the "discovery" *that one is Jewish*. One genealogist told me that a person has to be pretty open minded to do genealogy. "You never know what you might find out. I found out I am Jewish."

While the search for specific familial connections that leads some to test their or their male relatives' Y-chromosome and mtDNA, others want to find out things they just do not know—because there are no family records, as is the case with testing for African ethnic origins (Nelson 2008a; Gates 2006 and 2008), or because no one is interested in their questions, knows anything about the family, or is willing to talk about the past. Deborah, as I will call her, came from a small family and "almost by coincidence" became interested in making her family tree. As far as she knew, her mother's father was Cajun and her mother's mother was Irish. What she found out was quite a surprise: her mother's line was Jewish, something her mother claimed not to have known. Whether or not her mother was telling the truth she will never know. What she does know is that in her family Jewishness "had migrated underground" (interview).

Deborah discovered her Jewish ancestry via traditional genealogical research: the death certificate of a long lost great-aunt, newspaper clippings about a funeral for another relative. She took that information to an Orthodox rabbi who declared her "Jewish by birth and by Jewish law." By the time she took the mtDNA test she ended up "confirming something I already knew." Nevertheless, the confirmation, she told me, was meaningful. There was another person who matched with her group. He had been adopted and his mother had been adopted. "For him it was almost a shock."

What is the significance of such "shocks?" What effects might they have? Alondra Nelson is right: there is no simple biological determinism here. One doesn't suddenly become Jewish or anything else as a result of genetic genealogical tests. Being Jewish means more than just carrying genetic material read as an indicator of ancestry—both for oneself (as I show below) and, of course, insofar as group membership requires that others recognize and accept one as a member of the group (see chapter 5). Nevertheless, instances when people find out things they did not already know provide analytic insight into the epistemological assumptions and power of genetic ancestry testing in ways that perhaps classical genealogical discoveries (a specific cousin I didn't know, a specific region or town in Europe to which my Ashkenazi family can trace roots) do not. One must interpret the results. One may well embrace and identify with those scientific facts and the genealogical implications they purportedly entail. And if embraced, one must translate them into cultural practices and connections. In the case of Bess discussed by Nelson, one must choose with *which of them*, Kru or Mende/Temne, to identify. But for most people who test—people for whom the epistemological authority of genetic ancestry testing is often already a given—test results are not so easily dismissed, ignored, or simply interpreted as irrelevant or incorrect.[21]

Let me dwell for a moment on the case of Bess: she could quite easily be *neither* Kru *nor* Mende/Temne. As one biological anthropologist put it, "what they are telling you when they tell you about your ancestors, is that 'where you come from is where the haplotype you have has the highest frequency.' But that may not be where you come from."[22] Upon receiving those results, Bess did not say, "well, I could be a descendant of an individual who happened to carry this marker but was neither Kru nor Mende/

Temne." Like many others who receive their genetic ancestry test results, she did not consider the possibility that these facts could quite simply be *false*. Instead, she began a journey of self-fashioning in the terms of the facts African Ancestry provided: You are *either* this *or* that.

The determining character of genetic genealogical testing is evident in precisely such moments: in the aftermath of test results genealogists tend to follow the trails laid down by the genetic facts. Not everyone chooses to test, but those who do by and large presume that there is meaningful and credible knowledge to be had. And upon discovering some unknown "ethnic or geographic origin," individuals tend to do more research on—and they tend to develop an *interest in*—the ancestry they now know that they have always had. The genealogist who discovered unknown Jewish roots on her mother's line became a practicing Jew. She raised her children Jewish.[23] She subsequently discovered that her husband's presumed Scottish roots—"his name is Macfarlane, for God's sake"—were in fact from "Norway," or so she found out from a Y-chromosome test done by Family Tree DNA. "I started doing more research about Vikings and did a lot of reading on Scottish history" and "I found out that a lot of Scottish ancestry, prior to being Scottish, was Scandinavian." Meanwhile, her daughter-in-law—a Filipino woman who had always thought her father was of Spanish ancestry—had a test "come back Chinese." "She's actually more of a native person. It took her at least a year to accept that."

Unlike phenotypic characteristics that were (presumably) visible on individual bodies and that marked *and made* one's racial difference, the genetic historical facts of ethnicity do not attach directly to individual persons. Rather, ethnicity becomes a quality of the DNA, which we, as individuals, carry within. In that partial displacement—from the visible body to the DNA within—one's true self can be at one and the same time a fact of biology and an "object of choice and self-construction" (Comaroff and Comaroff 2009, 1). One can be of Jewish *ancestry* and yet not or *not yet* (fully) Jewish.

Ancestry and identity are distinct terms and concepts in the world of genetic genealogy. Ancestry does not make one an X, even as it provides the facts in relation to which particular practices of self-recognition and self-fashioning become possible, especially for those who discover things unknown. Ethnicity is to be discovered in the DNA, which gets woven back into an individual's sense of his or her own personhood via a dialectic between the genetic facts and the practices of self-fashioning he or she subsequently adopts. Take the following comments from one exchange regarding Jewish ancestry on a Family Tree DNA online discussion forum:

> If you go on to do 37 or 67 markers you may find some cousins, but whether or not they self-identify as Ashkenazi *I don't believe R1a* [a haplogroup] *is Jewish. It is European.*
>
> The Khazars, who became (Ashkenazi) Jews by royal decree, *were not DNA-Semitic*, but attracted Semitic-Jewish immigration and inblending thereafter. That may be one explanation for non-J1 and J2 haplotypes occurring in eastern European Jewish families.
>
> However, after 1200 years, it is still theoretically possible to have totally Khazar *non-Semitic DNA* and a millennium or more of family Jewish religious observance. And the converse, a definitely Jewish, Cohanim DNA modal but a (at least recent) non-Jewish family history. But so rare!

In a posting titled "Semitic People," another genetic genealogist wrote, "About 30% of all Jews and [a] small percentage of Arabs are J2 haplogroups. However, J1 is the only haplogroup that is Semitic in origin." Someone responded: "I did not know that only J1 was regarded as semitic and J2 was not regarded as semitic. Does anyone have further information on this study?" (accessed 7/27/09). Someone else weighed in to point out that the term "Semitic" appears nowhere in the paper cited in the discussion thread.

Haplogroups (such as R1, J1, J2) are argued to be identifiable in ethnic terms. (More accurately, *some* haplogroups are presented as identifiable in ethnic terms. Others are seen to be too geographically widespread to be reliable ethnic markers.) *We* are all mixtures of various ancestral populations, but anthropological genetic and genetic genealogical databases presume there were *ethnically pure—"original"—populations* sometime in the past that can be identified in our DNA. The presumption is that there were "source parental populations," as Duana Fullwiley refers to them (2008a, 704), and that those source parental populations—carriers of *ur*-ethnicities—identify where or who we descend from. Genetic markers are taken to be empirical signs of those original ethnicities that we, as individuals, carry within.[24]

Can genetic testing really identify ethnicity, I asked one person who has done many genetic genealogical tests? Yes. "It can tell you that. Telling you the countries of origin. It opens doors. It opens imaginations. It gets you to do more research. You know you are from this region. . . . My friend and my boss, she's got Egyptian, African American, Native American ancestry—she's got tons of fun things in her genes. . . . It's a wonderful thing." And customers increasingly sign their web posts according to test results that identify percentages of "ancestry" in their genomes: "53% European, 40%

Sub-Sahara African, 7% East Asian," one customer's signature line writes of his "mother's autosomal breakdown."[25]

One anthropological geneticist told me that testing companies and the National Genographic Project have a "real public science" component to them. "Many people are becoming technically sophisticated in their dealings with the gene," something that becomes quite clear reading through chat forums and blogs as well as conducting interviews. There is a wide range to how well individual customers understand the workings of DNA and the mathematical models on which ancestry testing depends and, therefore, their own test results. Nevertheless, there is clearly a small and growing group of serious amateurs, people who have learned quite a bit in their efforts to navigate their own genetic genealogies. How did you learn so much about DNA, I asked a genealogist. "I read *a lot.*" And then by corresponding with people in "my group"—her "matches" at Family Tree DNA who have formed a group to work together to discover more about their shared ancestry—and by asking a lot of questions. "What does this mean? What does this not mean? You have to hone how you talk about it." (See Epstein 1995 on the role "lay expertise.")

For Bennett Greenspan, going into haplogroup testing was a conscious effort to move his customers' interests beyond a genealogical time frame:

> In 2003, we launched a new database at Family Tree DNA where we said, "This is your known haplotype because we've tested you and this is probably the haplogroup that you descended from." And so we started explaining what we call deep anthropology, or as you know, your deep, real anthropology to genealogists. . . . Some people absolutely loved it and some people just didn't get it. . . . But over the years, we've even seen our customers produce bumper stickers that say, "What is your haplogroup? No? Let me Guess!" (interview)

Haplogroup identification increasingly appears in the signature lines on Family Tree DNA's online forum: "Y-DNA: L2, mtDNA: U1a" one customer signs off. And "group projects" are increasingly organized along haplogroup lines, for both mitochondrial and Y-chromosome tests (see familytreedna.com, Group Projects).

The identification of selves by virtue of one's haplogroup might well signal the emergence of novel "biosocialities," that is, the formation of social networks and subject positions on the basis of shared DNA markers (Rabinow 1992). But at the level of what they are taken to mean, such practices of identification may not be generating anything all that new: identifying one's haplogroup seems to just push the question of nation and ethnicity back into the deep ancestry of human groups. To return to what Family Tree DNA tells its customers they can learn from their Y-chromosome tests, knowledge about one's "deep ancestral *ethnic* origins" is on offer (www.familytreedna.com, emphasis added). Ethnicity and *nationality*—through Family Tree DNA's listing the "countries of origin" of their customers—are inscribed into the very way in which "matches" are identified in test results, both "recent" and more remote. In explaining the chart detailing matches with my mtDNA test results (remember, a "match" indicates a 50 percent possibility of a shared ancestry within the past 1,300 years), Family Tree DNA informed me that it provides information on:

> Each country from which you have matches
> The number of people you match for each country and comment combination
> Any additional information your matches provided about their origins
> The total number of people you match from that country
> The total number of people who have reported this as their country of origin
> The percent of the people we have tested from this country who match you

The document then instructed me on how to read the chart: "You match 8 persons out of 489 people from Austria, this is 1.6% of the population tested from Austria." (In other words, I have a 50 percent probability of having a shared maternal relative over the past 1,300 years with eight persons *from Austria*, as if that national identification carries any meaning that far back in time.) And such national explications frame Family Tree DNA's discussion of haplogroup origins as well: "Here you will find the haplogroup assignments and countries of origin of your matches who belong to the same major haplogroup as you. The purpose of this page is to provide you with the haplogroup information of your matches listed on your mtDNA Ancestral Origins page." The countries of origin with which my haplogroup matched include: Austria, Bulgaria, Czech Republic, Denmark, Egypt, and so forth—reading back into the deep historical record the contemporary "national order of things" (Malkki 1992).

In the practices of genetic genealogy, haplotypes and haplogroups do not function as evidence of genetic diversity within *the unity* of the human species. They are epistemic objects that transform species-level diversity into population distinctions. And in genetic genealogy's epistemological logic, a "population" is rendered commensurate with the modern cultural and political notions of an ethnic and/or a national group.

The Self Within

In *The Empire of Love,* Elizabeth Povinelli explores the distribution of "the liberal, binary concepts of individual freedom and social constraint" across the globe and draws a distinction between the autological and the genealogical subject. If the former is fashioned in terms of "discourse, practices and fantasies about self-making, self sovereignty, and the value of individual freedom," the latter identifies subjects governed by the social constraint of "various kinds of inheritances"—kinship, tribal, racial, religious (2006, 4). The autological subject or Enlightenment Man required its Other—not capable of self-making, held by social custom, and, it is important to emphasize here, held by his or her biological fate. Primitive man was also man of nature and not just a man of social restraint. He was European man's racial other. As Nancy Stepan has argued, in the late nineteenth century citizenship was premised on being free, rational, and thus, a European man. It was on the basis of their determination by nature that women and racial others were excluded from its domain (1982, 1998).

The distinction between autological and genealogical selves and societies was always a liberal fantasy. It was always a projection. Neither in the political order of the nation-state, nor in terms of our membership in families and other social worlds, have we ever been simply parvenus. But that fantasy or, more accurately, a double-speak about it is both evident in and *constitutive of* the practices of genetic genealogy. People can embrace ancestry tests because of the ways in which they can simultaneously assert the importance, the meaningfulness, of their biological ancestry—of "the genealogical grid" (Povinelli 2006)—and yet simultaneously fashion genetic genealogy as a project of self-making. How did you "feel about the results?" genealogists often ask one another about unexpected results. I can *decide* to do research on my (unexpected) Native American or Jewish or Norwegian or Mende ancestry. I can choose to forge ties to relevant communities. I can choose to adopt cultural practices. I can choose to become a practicing Jew. But all those choices are made possible—they are made authentic—by a genetic genealogical grid that provides me with a truth about who I always was, a genealogical grid with an epistemological robustness—and increasingly so—that makes those claims plausible to others, and quite crucially, potentially plausible to the communities that one may wish to join (see chapter 5). As John and Jean Comaroff have argued, "cultural identity, in the here and now, represents itself as . . . two things at once: the object of choice and self-construction, typically through the act of consumption, *and* the manifest product of biology, genetics, human essence" (2009, 1). The industry of genetic ancestry testing—an industry that draws upon the expertise of a science that embeds human agency (and consciousness) *and* essence into its very epistemic objects and interpretive practices—is socially and politically felicitous precisely because of the ways in which it presumes and promotes that paradox.

The paradox between choice and determination, Steven M. Cohen and Arnold M. Eisen have argued, characterizes contemporary understandings of Jewishness, at least among the majority of American Jewry whom they identify as the "moderately affiliated" Jew (2000, 9): "Anxiety about the Jewish future has led to increased interest among scholars, communal leaders, and lay people alike in the factors which shape, nourish, and sustain Jewish commitment." Eisen and Cohen suspected, however, that one could not accurately measure Jewish affiliation or identification on the basis of *public* criteria alone: synagogue attendance, membership in Jewish organizations, support for the State of Israel. Instead, they found that on the whole American Jews "construct their Judaism mostly in the private sphere of family, friendship and reflection" (2).

Characteristic of the "private" Judaism that Eisen and Cohen identify is the prevalence of a discourse of choice. Following Weber's argument about the disenchantment of the world, they propose that the "'Sacred canopy' . . . no longer overarches existence, and so the demand to choose and re-choose identity . . . is inescapable. Nowhere have these processes been more evident than among Jews" (7). And quite important, nowhere have these processes been more evident than among moderately affiliated American Jews today. On the one hand, virtually all of their interview subjects emphasized that they *chose* Judaism—that, often as adults, they became more interested in Judaism, more committed to their Jewish identity, that they made *the decision* "to be a serious Jewish self" (9). But such autological self-descriptions stand alongside genealogical commitments: David and Molly, for example, who were "fully aware (and happy with) the fact that they have chosen Judaism believe nonetheless that it was theirs all along. It is a birthright which they have voluntarily claimed, a given which they have autonomously elected to receive" (22). This paradox—"choosing chosenness" (22)—stands at the center of American Jewish life, they argue, a commitment to "tribalism" and to the self-made Jew at one and the same time. Moreover, returning to the question of a prevailing anxiety among scholars and communal leaders with respect to a Jewish future in the U.S., Cohen and Eisen argue that because one's Jewishness is given a priori, because it "resides in the *self* and is independent of the course one's life takes" (23, emphasis added), one cannot become more or less Jewish. "One does not sacrifice any quotient by opting for less. . . . One does not sacrifice any quotient of Jewishness by marrying a non-Jew." As a quality *of the self*, "the children of an intermarriage will automatically be Jewish for the same reason, as will their children" (23). Jewishness is a criterion "of blood" (10).

Genetic ancestry tests that enable one to reconstruct lineal descent lines have particular resonance at this moment of anxiety about loss: if Jews keep intermarrying, what Jewish future will there be? More generally, in the era of American "multi-racialism"—and for African Americans, given the long history of masters fathering children with their slaves—lineal descent allows us to recuperate, to find, and to sustain the so-called identities or ethnic affiliations that we believe we have "lost" or that we fear we may well lose in the near future. While technologically different than Y-chromosome and mtDNA testing, admixture mapping—the assessment of one's "percentage" of African, European, East Asian, and Native American ancestry on offer by companies other than Family Tree DNA—is also felicitous in an age of anxiety about "mixed" marriages and cultural loss. It is simultaneously, and paradoxically perhaps, the perfect technology for those who seek to recognize and celebrate their own multi-racialism and for those who want to find a trace of a particular ancestral line. I may be mixed and increasingly so. I can either celebrate that mixture or I can discover—I can *still know and document*—a specific heritage. My truest self—my *genealogical* self, whether read via particular descent lines or through admixture maps—can never be lost.

The Specter of Race

In "Diaspora: Generation and the Ground of Jewish Identity," Daniel and Jonathan Boyarin state that group identity (as a process of *self-construction*) can be either "a product of a common genealogical origin" or it can be generated with reference to "a common geographical origin" (1993, 305). Both of those traditions exist within Judaism, they point out, and stand in tension with one another. While the latter has "a generally positive ring" in contemporary social theory, the former tends to be regarded pejoratively. The Boyarins argue that while the more obvious reason for the contemporary denigration of genealogy as a grounds for identity is the history of modern racism (305–6), there is also a second, prior source: a disdain for genealogy that characterizes the letters of Paul, which "lie at the fountainhead of Christianity" (306). In brief, the Pauline letters substituted "an allegorical genealogy for a literal one": through baptism one would be born anew into the family of Christ. In Christian theology, "the physical connection of common descent from Abraham and the embodied practices with which that genealogy is marked off as differences are rejected in favor of a connection between people based on individual re-creation and entry de novo into a community of common belief" (307). But "descent from a common ancestor" is essential to the construction of a Jewish identity, they write, a "myth" of descent that operates "on the semantic field of the body" and not "on the semantic field of status through land" (329).

The Boyarins were responding to Walter Benn Michaels (1992), who argues that, in its celebration of "culture," identity politics never really did displace race: the only principle by which I might need to "learn about *my* culture" because it is something that I have "lost" is a genealogical one. Without the principle of descent, how could I possibly have lost something I never actually had? Race, Benn Michaels argues, remains the underbelly of identity politics. The genealogical principle, the Boyarins respond, is not reducible to race. Genealogy has always been cardinal to the Jewish tradition, and that was true long before the emergence of the concept of race.

In principle the Boyarins are right. Not all genealogical principles can be reduced to the question of race. Kinship orders, for example, are not racial orders. They are certainly not necessarily racial orders. Nevertheless, given the history of race science and racial thought, given the history of slavery, eugenics, and genocide, it is hard to imagine how the genealogical principle can remain unaltered. Moreover, ever since prominent Jewish social scientists and Zionist activists recast Jewishness in the language of race and biological descent, the meaning of Jewish "genealogy" has been fundamentally transformed, at least for the vast majority of Jews around the world who never rejected Zionism, either in cultural or in political terms. More specific to the practices of genetic ancestry testing and its parent field, genetic history, insofar as they involve the search for *biological* evidence of shared ancestry, the specter of the race concept—and of race science—is forever hovering on the horizon. This is a biological science born of the history and legacy of race science and racial politics, even as it construes the meaning of biological distinction in different terms, that is, as "mere signs" of an albeit authentic and presumably meaningful ancestry. And insofar as it operates within the genealogical descent grid and the historical specter of racial thought, substantive biological claims about group-based differences, *even ones that present themselves as historical and cultural arguments*, are not far afield.

In early 2006, Gregory Cochran, Jason Harding, and Henry Harpending published a paper on the "Natural History of Ashkenazi Intelligence" in the *Journal of Biosocial Science*. As recounted by a reporter for the *Los Angeles Times*, Cochran had long pondered the following puzzle: "Why are European Jews prone to so many deadly genetic diseases?" (Kaplan 2009). This fact "offended Cochran's sense of logic: natural selection, the self-taught genetics buff knew, should flush dangerous DNA from the gene pool. Perhaps the mutations causing these diseases had some other beneficial purpose. But what?" One night—apparently, at 3 a.m. no less—he came up with the answer: "The 'faulty' genes . . . make Jews smarter" (Kaplan 2009). Cochran emailed Henry Harpending, an evolutionary biologist at the University of Utah.

According to Cochran, Hardy, and Harpending, the "unique demography and sociology of Ashkenazim in medieval Europe selected for intelligence" (2006, 2). Pointing out that in writing about so-called *Jewish* intelligence they are actually referring to *Ashkenazi* Jewish intelligence—Ashkenazi and not Sephardic or Oriental Jewish communities score above the mean on intelligence tests—Cochran and his colleagues contend that "Jews" are more intelligent due to "occupational selection" (Murray 2007). As summarized by Charles Murray in an article in the Jewish magazine *Commentary*, in the ninth through seventeenth centuries Ashkenazi Jews entered into "occupations involving sales, finance and trade. Economic success in all of these occupations is far more highly selected for intelligence than success in the chief occupation of non-Jews: namely, farming" (2007, 32). In turn, "economic success is related to reproductive success, because higher income means lower infant mortality, better nutrition, and, more generally, reproductive 'fitness'" (32). Confusing the biological and the social in their account of selection, Cochran and his colleagues argue that "Ashkenazi literacy, economic specialization, and closure to inward gene flow led to a social environment in which there was a high fitness payoff to intelligence, specifically verbal and mathematical intelligence but not spatial ability" (2006, 659). Moreover, they contend that the link to Jewish diseases is apparent, first, because a whole cluster of genetic mutations found among Ashkenazi Jews "probably affect early neurological development" (including LSD disorders along with Bloom Syndrome, Fanconi's anemia, and BRCA1 and BRCA2, and they note "Canavan disease looks like a neurological booster that modifies myelin" [684]). Second, according to their quantification, it is unlikely that these "Ashkenazi mutant genes would happen by chance" (684).

That paper was not received as incontrovertible fact. Neither was it systematically dismissed or discredited. One journalist, Jennifer Senior, reported: "Did Jewish intelligence evolve in tandem with Jewish diseases as a result of discrimination in the ghettos of medieval Europe? That's the premise of a controversial new study that has some preening and others plotzing" (Senior 2005, 1). "Most American academics," she wrote, "expected the thing [the article] to drop like a stone. It didn't" (1). A scientific journal agreed to publish it. And "The *New York Times*, the *Economist*, and several Jewish publications risked their own reputations to legitimize it" (1). The American Enterprise Institute sponsored a forum to discuss "Are Jews a race? Is Jewish intelligence genetic?" (Saletan 2007, 1). As William Saletan recounts in a rather tongue-in-cheek article for *Slate* magazine, there was a long back and forth about the idea of the Jewish race, about biology versus "faith and values." "But what if Judaism as a genetic inheritance is compatible with Judaism as a cultural inheritance? And what if the genes that make Jews smart also make them sick? If one kind of superiority comes at the price of another kind of inferiority, and if the transmission of Jewish values drives the transmission of Jewish genes, does that make the genetics and the superiority easier to swallow? Apparently so" (1).

Not for everyone. One prominent geneticist responded by asking, "What are their theories about those on the opposite end of the spectrum? . . . Do they have genetic theories about why Latinos and African Americans perform worse academically" (Kaplan 2009)? Others have argued that the evidence is certainly not in (Goldstein 2008). But the discussion is out there, and it has been held in the press, in synagogues, and before a sold-out audience at the Center for Jewish History in New York City (Pinker 2005). And the reception has been anything but uniformly wary or hostile. At the Center for Jewish History, the audience was certainly intrigued by and perhaps even receptive to the news.

There is no way to police the boundary between cultural or historical claims, on the one hand, and biological claims, on the other. There is no way to ensure that biological data remain *purely* "historical" or "cultural" signs. Family Tree DNA now offers a test for a presumably "biologically meaningful" genetic marker: "Monoamine Oxidase A (Warrior Gene)." "Recent studies have linked the Warrior Gene to increased risk-taking and aggressive behavior in men. The Warrior Gene is a variant of the gene MAO-A on the X chromosome," the website reports. (That is the very same genetic mutation that a researcher working on Jewish genetic origins worried could raise "ethical questions," as I mentioned in chapter 1). Given that "human behavior is complex and influenced by both genes and our environment or circumstances," men who carry the Warrior Gene are not necessarily "more aggressive, but [they] are more likely to be more aggressive than men without the Warrior Gene variant."[26] More broadly, a growing number of geneticists are arguing that, while policing the boundary between culture and nature might be politically correct, it is scientifically unsound (see Goldstein 2008; see also Wade 2008 for an overview of the debate).

The desire to recuperate a traditional Jewish understanding of genealogy might be implausible in general given the past century and a half of work in the biological sciences, the history of anti-Semitism and Nazi genocide, and the emergence of Zionist politics and the Israeli state. It also seems implausible to think that a traditional Jewish understanding of genealogy might be recoverable through the work of a biological science—in this instance, genetic history—that seeks bodily evidence of shared descent for "the Jews," a category so central to race science and racial politics and to the Zionist movement and the establishment of the Israeli state. Neither genetic history nor genetic ancestry testing can escape the legacy of race. History cannot be pried from biology. The search for the biologically meaningless marker cannot be safely cordoned off from the search for culturally, medically, or cognitively meaningful traits. There are no firewalls here.

CHAPTER FIVE

The Right of Return

In contrast to the genetic studies of mainstream Jewish communities, which take for granted the Jewish identity of the people they study, the scientific studies of, and political quests on behalf of, lost tribes and lost Jews seek to answer a different question: Are these people, who believe themselves to be descendants of ancient Jews, truly so? Establishing their phylogenies is a project of identification best described as forensic. It is an attempt to establish a match (however probabilistic) between the population whose origins are being tested and a control group, the known Jewish world as represented by modal haplotypes understood to be traces of ancient Hebrew descent.

This chapter focuses on two communities who believe themselves to be descended from ancient Israelites: the Lemba, a group in southern Africa, and the Bnei Menashe, a group living in the northeastern border states of Manipur and Mizoram in India. In contrast to the commercial market in ancestry testing and genetic genealogical self-fashioning that I explored in the previous chapter, here I consider the explicitly political work that genetic history can do in a context in which recognition is at stake for disempowered or marginalized social groups. I argue that genetic historical studies of the origins of communities who believe themselves to be descendants of one of the ten lost tribes or who have "Jewish roots" are becoming an evidentiary terrain—an archive—around which historical claims are made plausible (or implausible), political practices are made possible, and disputes adjudicated. In making this argument, I focus primarily on the work of a U.S.-based Jewish organization, Kulanu, best described as a missionary group, that is engaged in a multicultural project to recognize nonwhite "would-be" Jews. In so doing, Kulanu seeks to diversify the known Jewish world. As I then illustrate, this effort to support a multicultural Jewish world leads Kulanu activists into a very different political field: efforts to "repatriate" lost tribes (and lost Jews) have emerged as one front in a demographic battle in the Israeli state. A U.S.-based project to support a multicultural diasporic Jewish world converges, in Israel, with a right-wing nationalist agenda that aims to bring members of lost tribes and lost Jewish communities to Israel and settle them in the Occupied Territories.

More broadly, I lay out the contours of a distinctive politics of recognition evident in this work. As Patchen Markell has argued, in the early 1990s prominent theorists of recognition proposed "that many contemporary social and political controversies can be understood . . . as attempts to secure forms of respect and esteem that are grounded in, and expressive of, the accurate knowledge of the particular identities borne by people and social groups" (Markell 2003, 39). According to Charles Taylor, as individuals we are all rooted in a moral space that grounds our ability to evaluate and determine "the good life." That moral space—the good life—is articulated within culture, the "unique, authentic identity of a distinctive 'people' or Volk" (Markell 2003, 154; Taylor 1989; Taylor and Gutmann 1992). In the grammar of multiculturalism, to fail to recognize cultural *difference*—a failure of the liberal (settler) nation-state—is to perpetrate a fundamental injury. It is to deny what it means to be human, that is, that we are all irreducibly always already part of a linguistic community, of a culture (Brown 1995, 2003; Markell 2003; Povinelli 2002; Taylor and Gutmann 1992).

By way of contrast, the grammar of Kulanu's politics of recognition goes like this: On the one hand, it is exciting to find (would-be) Jews who are culturally different from us and thus to diversify the known Jewish world. On the other hand, your cultural/religious difference matters and at the same time doesn't matter. We want to reveal a common genealogical identity on the basis of which you can, following a suitable religious education, be formally reintegrated into the Jewish world, if you so choose. No matter what the rhetoric, in actual practice *Judaism*—and recognizing individuals from lost tribes *as fellow Jews*—requires religious education and formal conversion before a rabbinic court. It involves communities abandoning significant elements of their religious and cultural difference. The common horizon in this politics of recognition is not a humanity characterized by the fact of cultural diversity that must be recognized and preserved. It is a Jewish identity defined on the basis of genealogical descent and, subsequently, normative religious practice. This is a multicultural project configured through the lens of an identity politics that grounds itself in a genealogical self. That genealogical self is produced as "diverse"—as non- or antiracist—by the call to recognize nonwhite Jews as Jewish *kin*.

Through this consideration of the cultural and political practices that have emerged out of research on communities believed to have Jewish roots, I comment more broadly on a key constitutive logic of the genetic historical subject. If, as Elizabeth Povinelli has demonstrated, the politics of recognition presupposes an archive (Povinelli 2002), the genetic historical archive may prove to be (or in Derrida's articulation, may prove *to have been* [1996]) essential to the hope of self-described lost tribes and lost Jews to be recognized by the mainstream of the Jewish world—and by the Israeli state—as (potential) Jews.[1] Not, however, in any simple causal or deterministic sense. The relationship between biology (descent) and culture (Jewishness) requires *work*. As will become clear in the discussion that follows, ambivalence reigns regarding the significance of genetic historical evidence to matters of Jewish identity: it matters, it doesn't matter; it matters, it shouldn't matter; it matters; it doesn't *make* the Lemba or Bnei Menashe (or anyone else) Jews.

In discussions of the Lemba's Jewishness, religion in its distinctly modern configuration (see Asad 1993) mediates the distinction between biology and identity: religion is a matter of *belief*, it involves—for Lemba—the *choice* to learn "proper" Judaism and to convert, to be recognized as Jews by a rabbinic court. Nevertheless, those "choices" are made possible against the background of genetic historical facts that make claims to Judaic origins plausible. The play between biology (as ancestry) and agency (as religious choice) delineates the path for "returning" to Judaism for members of "lost tribes."

Identification

On February 22, 2000 the PBS program *NOVA* aired a documentary, *The Lost Tribes of Israel*. Opening with the image of Jews praying at the Western Wall in Jerusalem, the narrator informs us that "time-honored traditions live on." And those traditions may live on well beyond the boundaries of the known Jewish world: "four thousand miles from Israel, in southern Africa a people called the Lemba also heed the call of the shofar." They believe themselves to be "direct descendants of the patriarchs," the narrator recounts; "however unlikely the Lemba's claims may seem, modern science is finding a way to test them" (*Lost Tribes of Israel* 2000).

The modern scientific tests to which the narrator refers are Y-chromosome studies of Lemba origins. In 1996, the *American Journal of Human Genetics* published a paper by two South African scientists at the University of the Witwatersrand, A. B. Spurdle and T. Jenkins. According to the authors, the use of the Y-chromosome to study Lemba descent is crucial: insofar as

Lemba oral tradition tells of *male* migrants from the ancient Near East who married local women, Y-chromosome polymorphisms are the relevant evidentiary terrain.

In their paper, Spurdle and Jenkins begin with a cultural and historical account of the Lemba. They note a longstanding distinction from "their Bantu-speaking Negroid neighbors" on the basis of several "historical facts": The Lemba "were highly regarded as master metal workers . . . and as skilled potters. . . . The Lemba have also been recognized by some anthropologists as having Caucasoid features" (1996, 1126). Spurdle and Jenkins argue that the Lemba are culturally and racially distinct from the linguistic community of which they are a part, i.e., other "Bantu-speakers."[2] In addition, they "practice a religion that embraces many extraordinary rituals and laws" (1126); they have marriage laws that "encourage strict endogamy"; they practice male circumcision, initiating boys around puberty "in secret closed lodges"; and the Lemba follow food laws that "appear to be essentially Jewish." In the familiar genre of longstanding Christian and Jewish reports of finding lost tribes (Parfitt and Trevisan Semi 2002; Ben-Dor Benite 2009), cultural and religious practices that "resemble" Jewish or Judaic ones are read as providing evidence that this African tribe might have Hebrew origins.[3] Spurdle and Jenkins engage in an evidentiary logic that predates Franz Boas's project of disentangling "culture" from "race": contemporary cultural practices (however degraded or diluted from a possibly original Jewish form) are signs of genealogical descent.[4] The possibility that "cultural traits" could be passed along by "diffusion" rather than by descent does not figure in the paper's analysis.[5]

Spurdle and Jenkins move from an account of those presumably known historical facts to Lemba oral history, which "forms the basis for most theories concerning [Lemba] origins" (1996, 1127). As they explain, according to the Lemba, their ancestors were Jewish traders who "migrated from 'the north' to Yemen in the seventh century B.C.E." and established a community at a place named "Sena." The community migrated to Africa after conflicts between "the Basena [as the Lemba were named] and the Arabs and split in two: one group moved to Ethiopia to become the Ethiopian (or Falasha) Jews and the other migrated to southern Africa to form the Lemba community.[6] Using Y-chromosome evidence, Spurdle and Jenkins test that origin tale.[7]

Based upon a study of forty-nine unrelated Lemba individuals and several control populations ("South African Caucasoids [comprising the Asiatic Indian, Jewish, and European groups], Bantu-speaking Negroids, and a Khoisan sample"), Spurdle and Jenkins conclude that Lemba oral tradition is accurate, broadly speaking (1996, 1127). Via an analysis of four

Y-chromosome markers,[8] they calculate that "greater than or equal to 50% of the Lemba Y chromosomes are Semitic in origin, approximately 40% are Negroid, and the ancestry of the remainder cannot be resolved" (1126). Their conclusions are based upon two arguments: First, Lemba frequencies in each of these categories show a strong "Caucasoid" gene flow into the community that sets the Lemba population apart from African communities. (Lemba frequencies are closest to those of Ashkenazi Jews, Lebanese, and South African Indians.) Second, the Caucasoid gene flow is of Semitic and not of South Asian origin. Let me focus on the second point.[9]

Spurdle and Jenkins analyze several polymorphic sites in order to assess Lemba relatedness to other populations—most specifically to Jews and to Negroids, each of which functions as a normative measure for alternate accounts of male origins. While 36 percent of Lemba Y-chromosomes "appear to be of Negroid origin," 50 percent are attributable to "Caucasoid origin" (1130). More specifically, "Ht7, Ht8, and Ht11 [haplotypes] appear to be typical of Jewish populations, and since they were shown to occur at similar frequencies in a Lebanese sample, they *may be* representative of all Semitic groups. Unfortunately, these haplotypes also occur in the Caucasoid S[outh] A[frican] Asiatic Indian population, and it is impossible to distinguish between Caucasoid gene flow from Semitic and Asiatic Indian sources, by use of frequencies of these three haplotypes alone" (1128–29).

There are additional markers, however, which Spurdle and Jenkins use in order to assess *from which Caucasian population* the Caucasoid gene flow originated. Thirteen percent of South African Asiatic Indians carry the Ht13 haplotype, but it is absent from the Lemba, "suggesting that the Caucasoid male genes present in the Lemba are of Semitic rather than of Indian origin" (1129), a conclusion further supported by the lack among Lemba males of another Y-linked allele characteristic of the South African Asian population. (Spurdle and Jenkins do not consider the fact that theirs is a small sample of a very specific Indian community, South African Asians who migrated to South Africa in the nineteenth century, mainly from Gujarat, and who, due to endogamous marriage practices, founder effects, and genetic drift, could well display different allele frequencies than other South Asian populations both in Africa and on the subcontinent.) "There is no clear evidence for Indian ancestry," Spurdle and Jenkins conclude (1131). The evidence of Semitic ancestry is sustained by the presence among Lemba men of various alleles concluded to be "Jewish in source" (1129).

In the final section of the paper, however, Spurdle and Jenkins display caution regarding what the *genetic data* might actually indicate regarding a specifically Jewish gene flow:

> It is not possible, on the basis of the Y-specific genetic data, to distinguish between Jewish and Arabic Semitic ancestry of the Lemba. This is not surprising, since a common ancient history of Jews and Arabs is reflected both in their language and as similarities in the stories of Judaism and Islam recorded in the Jewish Bible and the Koran, respectively. However, certain features of Lemba culture would seem to suggest that Jewish ancestry is more likely than an Arabic one. These include the practice of separating milk and meat, a dietary law observed in Judaism but not in Islam, and many other food laws that are essentially Jewish. In addition, certain Lemba sacrifices call for the use of liquor, and Muslim law forbids the consumption of alcohol. (1131)

If the genetic evidence, according to Spurdle and Jenkins, can distinguish a (Caucasian) Indian from a (Caucasian) Semitic descent, they turn to cultural evidence to disentangle Jewish from Arab (Semitic) sources.

A few years after the publication of Spurdle and Jenkins's paper, scholars at University College London and the School of Oriental and African Studies revisited the question of Lemba origins on the basis of Y-chromosome evidence. They used new technologies and generated new data sets to try to decipher Lemba origins once again. This study (Thomas et al. 2000) relied upon a set of different techniques for analyzing the Y-chromosomes, using what are considered to be more refined markers that can better distinguish lineages and establish genetic distances.[10] Tudor Parfitt—a scholar at the School of Oriental and African Studies in London who encountered the Lemba some time ago, became interested in their claim to ancient Hebrew descent, and wrote a book exploring their history (Parfitt 1992)—decided that he might be able to use the results of the Cohanim studies (Skorecki et al. 1997; Thomas et al. 1998) in order to solve what he considered the "riddle" of the Lemba. As Parfitt recounts his foray into this research, initially he found Lemba claims to Jewish descent hard to believe (*Lost Tribes of Israel* 2000). The results of the Cohen studies could be used to assess the Lemba's genealogical claim. The Cohen modal haplotype could be used as the measure of ancient Hebrew descent (Parfitt and Trevisan Semi 2002, 47; see also Parfitt and Egorova 2006).

In their study, Mark Thomas and his colleagues (2000) argue that Spurdle and Jenkins were unable to establish whether or not the Lemba have specifically *Jewish* ancestry. They accepted the conclusion that Lemba Y-chromosome haplotypes originated outside of Africa and its indigenous gene pool, and they set out to try and "distinguish a Jewish from a more general Semitic contribution to the Lemba gene pool." Using the Cohen modal haplotype "that is dominant in the Jewish priesthood and that *may be more generally characteristic* of Hebrew ancestry" (ibid., 675, emphasis added), these researchers designed a study to assess whether evidence of a specifically Jewish descent could be found: "The combination of the presence of the CMH at high frequency in the Lemba and its absence in neighboring Bantu populations would be supportive of Lemba claims of a paternal Judaic ancestry, especially if its frequency is relatively low in other Semitic groups" (675). Relying upon various studies that note the CMH's "absence" or "low frequency" in a few populations ("Yakut, Mongolians, Nepalese, Armenians, Greeks, and Cypriots and, interestingly, in Palestinian Arabs"), the CMH is posited as not just Semitic in origin but specifically Judaic. Thomas and his colleagues decided that if the CMH could be found in a similar percentage of the Lemba as it was in the general Jewish population (among "Israelites" at about 9 or 10 percent), the Lemba's assertion of Jewish ancestry might be plausible. This is an apt illustration and extension of Bruno Latour's (1987) argument about the production of scientific facts—that their circulation and citation in scientific papers moves laboratory objects and claims from the domain of controversy into the domain of fact: using the CMH as the measure of Hebrew origins in the design of the Lemba study helped to establish it as evidence of Hebrew ancestry.[11]

Basing their conclusions upon a comparison of Lemba, Bantu, Yemeni, and Ashkenazi and Sephardic "Israelite" (lay Jewish) men, Thomas and his colleagues were more cautious than were Spurdle and Jenkins: "Clearly, there has been a Semitic genetic contribution, including, quite probably one from Arabs, given the Lemba's presence on the eastern coast of Africa, where Arabs have settled for centuries" (Thomas et al. 2000, 685). Given the historical evidence, a Jewish contribution might be less expected: "Both Ashkenazic and Sephardic Israelites are geographically far removed from the Lemba, and, were it not for the Y-chromosome sharing between the Yemeni and Jewish populations, the occurrence of Jewish haplotypes in the Lemba population would be highly suggestive of gene flow between the two groups." In general, whether or not the gene flow came from an Arab source (Yemenis) or a Jewish one is difficult to disentangle. But the presence of the CMH in Lemba samples suggests a possible Jewish source:

> Support for a Jewish contribution to the Lemba gene pool is, nevertheless, found in the presence, at high frequency in the Lemba, of the CMH (.088 of the entire population and .135 of UEP group 1); the CMH is also observed at moderate frequency in Ashkenazic Israelites . . . and Sephardic Israelites

> . . . but it was observed only in a single Yemeni. . . . Furthermore, in an unpublished study of Palestinian Arabs [now published as Nebel et al. 2000] the CMH was present at only very low frequency (<.025). The CMH has been suggested as a signature haplotype for the ancient Hebrew population, *and it may be performing that function in this study*. . . . However, it is possible that the Lemba CMH Y chromosomes are a consequence of a relatively recent event that, in Lemba oral tradition, has acquired a patina of antiquity. (Thomas et al. 2000, 685, emphasis added)[12]

Thomas and his colleagues think they have plausible if inconclusive evidence of a link between the Lemba and ancient Jews. The genetic evidence is consistent with "both a Lemba history involving an origin in a Jewish population outside Africa and male-mediated gene flow from other Semitic immigrants." It is also consistent with a history of "admixture with Bantu neighbors." "There is no need," the authors argue, "to present an Arab versus a Judaic contribution to that gene pool, since contributions from both are likely to have occurred. The CMH present in the Lemba could, however, have an exclusively Judaic origin" (685).[13]

Recognition

Following a lecture given by Vivian Moses, now Director Emeritus of the Centre for Genetic Anthropology at University College London, at the nineteenth annual conference of the International Association of Jewish Genealogical Societies in 1999, a woman in the audience said:

> I would just like to make a comment about the question of are we genetically Jewish. We get half of our genes from each camp. So that means we have potentially one-fourth from each grandparent, one-eighth from each great grandparent, one-sixteenth, you understand that? So once you introduce a non-Jewish ancestor, some of that DNA we inherit. And in order to figure out if we are really Jewish, we are all probably partially Jewish . . . depending on how clear and precise our ancestry is.

Vivian Moses replied: "We are not partially Jewish. Those of us who are Jews are all Jewish, but our ancestors may not have been Jewish." Jewishness, he insisted, is a matter of *self*-designation.[14]

The relationship between ancestry and the question of who is a Jew, however, is far more complex than Moses' reply suggests. The question of whether Jewishness is a racial, a national, and/or a religious identity was rigorously debated in the nineteenth and early twentieth centuries (see chapter 2). And while the early Zionists—and especially, Labor Zionists as they gained prominence in Mandate Palestine and later in Israel—sought to produce a distinctly *national* Jewish identity that would be secular and separated from its religious "roots," genealogical descent, national identity, and religious identity have remained entangled in Israel as well as in the diaspora. Scholars who study the genetic history of modern Jews have argued they may be in a position to resolve the longstanding debate regarding what makes one a Jew—"shared culture or common descent" (Bradman et al. 2004, 89). The public debates over the Lemba's possible Jewishness that emerged in the wake of genetic historical evidence, however, make clear that in actual practice culture (as "religion") and common descent are not alternatives to be decided between.

In presenting their work publicly, researchers insist that genetic history adjudicates neither the question of whether or not one is a Cohanim nor the question of whether or not an individual or a community is Jewish. For example, following the "discovery" of the Cohen modal haplotype, researchers argued that the presence or absence of the haplotype on an individual's Y-chromosome has no bearing on whether or not a particular man is a Cohen, although many a self-identified Cohen has contacted these researchers in order to take the test. In the words of Vivian Moses speaking before the IAJGS convention in 1999:

> There is a significantly greater frequency of this genetic pattern in the Cohanim than there is among the others. . . . People keep asking us, let me give you a bit of my DNA and you tell me whether I am a Cohen. It doesn't work like that. There isn't a Cohen gene. It is a statistical phenomenon that among the Cohanim you find this pattern. Whether it is indicative of Cohenism is another matter. If someone has this pattern but doesn't think he's a Cohen, does it mean he really is? That is a thing one can debate. . . .

Or as a rabbi in Jerusalem who has been involved in the work on the Cohanim and who has a messianic interest in lost tribes said in an interview, vis-à-vis individuals this evidence has no significance. Collectively, "the biology is interesting in a group aspect to show maintenance of a lineage but as far as an individual's stock, *his decisions about himself*, you know is he Jewish or is he not Jewish, it doesn't even reach the table" (emphasis added).

Modal haplotypes are measures of *relative* frequency: the CMH can be modal at 50 percent—its reported frequency among Cohanim—because it is found at a higher rate than among "Israelites," who are the comparison group; the CMH can also be "modal" at 10 percent for Israelites if it is found at a lower frequency in non-Jewish populations. Moreover, the Cohen modal haplotype is found on the Y-chromosomes of Jewish men who do not believe they are Cohanim, and it has been identified on the Y-chromosomes of non-Jewish men as well. As such, the Cohen modal haplotype is not a test for whether or not a particular man is a Cohen any more than it is a test for whether or not a particular Lemba man is a Jew. For that matter, it cannot be used to determine whether or not the Lemba *as a group* are Jews. Genomic facts of generational connection and halakhic traditions of both priestly status and of Jewishness are and must remain distinct, researchers insist.

But let us keep in mind the entangled origins of religion and race in nineteenth century thought, an entanglement especially robust with regard to Semites in general and Jews in particular (Anidjar 2008; Masuzawa 2005). And let us keep in mind the merging of biology (as race, as population), religion, and nation in Zionist thought and the Israeli nation-state (see chapter 2). What, then, in the wake of genetic historical evidence, is the relationship between genealogical and religious answers to the question of who *might be* a Jew?

Assuming for the sake of argument that it has been resolved that the CMH is the (or a) modal haplotype of the ancient Hebrew population,[15] does the presence of the CMH in the Lemba population (at the "right" frequency) make the Lemba Jewish? "Being Jewish is a spiritual, metaphysical state and DNA is a physical characteristic, like nose size," Karl Skorecki has argued, invoking an earlier race science in his choice of phenotypic trait. "But," he continued, "we wouldn't dare go around saying we're going to determine who is Jewish by the length of their nose. Similarly we're not going to determine who is Jewish by the sequence of their DNA" (quoted in Kleiman 2004, 34). The discovery of this genetic trace of Jewish ancestry does not render the Lemba (recognized) Jews. Shaye Cohen, a professor of Jewish Studies now at Harvard University, makes a related point: "As a historian, I find the whole enterprise rather silly. Are the Lemba descendants of the lost tribes who disappeared from the face of the earth? The answer, of course, is no" (quoted in Greenberg 2002). They might well be, however, "a kind of modern lost tribe"—"a group of people unbeknownst to us and to themselves carrying Jewish genetic material." According to Cohen the Lemba will "be accepted as Jews 'if the Jewish people want them to become Jews. And that's the way it's been since Moses and Aaron'" (ibid.).

The work of Thomas and colleagues did not simply transform the

Lemba community into (recognized) Jews—either in the eyes of Lemba (for whom it might have "confirmed" a longstanding belief in their own Jewish origins),[16] or in the eyes of "mainstream" Jewish individuals or groups whose claims to being Jewish are unlikely to be called into question regardless of the results of genetic historical work. Nevertheless, this research has far-reaching implications for the question of recognition. On the basis of what criteria might "the Jewish people want them to become Jews," to return to Shaye Cohen's words? What might the evidentiary terrain of recognition be(come)?

Epistemology and the (Scientific) Grid of Intelligibility

In Kulanu's newsletter, the organization's president, Jack Zeller, wrote an article titled "Proving One's Jewishness":

> It is demeaning for one Jew to have to prove his Jewishness to another Jew. . . . *Kulanu bases identity on oral traditions and practices as if they are one's "notarized signature."* Despite a school of academics who thrive off of "they invented themselves" theories, we think that impoverished communities do not have the communication talent or inclination to formulate these forgeries. Inventiveness on such an intense scale is a product of modern life, especially developed in the imagination of this school of particularly ethnocentric academicians.
>
> Recently, the Lemba, regarded by anthropologists and some Jewish *mavens* of numerous stripes as "non-Jewish," have turned the tables with the active help of Tudor Parfitt, a London-based anthropologist and "honorary Lemba," who was himself helped by a group of eminent geneticists, most of whom are Jewish. (The Lemba have been found to have an anomaly on their Y chromosome that depicts their Middle Eastern heritage and, in fact, marks their priestly clan as closely related to the Cohanim.) Exciting, but also tragic. DNA sequences are interesting, but we have to hide our shame for needing to wait for this scientific evidence to convince us of their Jewishness. (Zeller 2000, 2)

For Zeller, the DNA isn't necessary. "You wouldn't need DNA for proof if you met a Lemba (or a Bnei Menashe from India, for that matter). Meeting a Lemba 'one on one' conveys the palpable Jewishness in his or her soul. Joseph Halevy didn't need DNA 150 years ago when he went to meet the Beta Israel of Ethiopia. It was obvious to him from discussion alone" (ibid.). Or in the words of an Israeli rabbi and activist who works on behalf of lost tribes, "I have many objections to the testing, first of all because they cannot represent a large group. . . . Second, according to the conception of the Cohanim's chromosome, it is connected to men and not to women. . . . In addition, all these issues . . . of testing don't contribute much because a Jew is measured, not by chromosomes nor by the shape of the face, but it is an attribute of a special soul." What needs to be evaluated, he said, are "markers of Judaism that could have only existed among Jews for thousands of years" (interview). Rhetoric as regards its irrelevance aside, DNA evidence has mattered greatly, even for groups such as Kulanu who declare that it does not and should not matter.

For those for whom neither oral histories nor the presence of rituals that "resemble" Jewish ones function as evidence of ancient Hebrew descent, genetic evidence might provide a "background of intelligibility" (Taylor and Gutmann 1992, 37) against which the authenticity of oral traditions can be considered, evaluated, and debated. After all, genetic genealogical testing was central to the *NOVA* video on the Lemba. It was central to a *60 Minutes* segment on the Lemba hosted by Lesley Stahl in which she presumed that the truth of Lemba Jewish origins had been genetically established and asked the daughter of a Lemba leader and prime promoter of the theory of Judaic descent: "In your own mind and in your heart, are you Jewish or are you Christian?" The *New York Times* reported on the finding and suggested that the Lemba may well have ancient Hebrew ancestors (Wade 1999). The BBC covered the topic (Vickers 2010a). And the American Museum of Natural History in New York screened the *NOVA* video and held a public discussion about its findings as part of their special exhibit on genomics (in 2001). Moreover, Kulanu arranged a speaking tour by Tudor Parfitt to Jewish congregations around the U.S. in which he discussed his longstanding research on the Lemba as well as the more recent genetic findings.[17] The belief that the Lemba may be descendants of ancient Jews is no longer as quickly dismissed as the obsession of a "possibly kooky" group, as someone who works at a Jewish cultural institution in New York once described Kulanu to me. (Several years after the interview, that same institution hosted a talk by Tudor Parfitt on his work on the Lemba. Kulanu organized Parfitt's speaking tour.) As proclaimed by the headline of a report by the *HaAretz* news service published in 2010, "DNA tests show the 80,000 strong Lemba people have links to ancient Jews."[18]

The epistemological power of genetic history might best be thought of in terms of the demands of the archive: there are particular "rules of formation" within which certain kinds of statements—historical claims, in this instance—(can) appear as truth statements (Foucault 1972). And those kinds of statements must come in particular forms. The document—collected, classified, and stored in official archives—has been essential to the historical profession since its establishment as a scientific project in the nineteenth century (Steedman 2001). By way of contrast, for (most) practitioners of the historical profession in the nineteenth and twentieth centuries, oral traditions and other kinds of memorial practices or testimonies came to be seen as less reliable, less "objective" sources for building historical knowledge. Genomic databases conform to the demand for historical objectivity. The practices of genetic history generate novel historical archives credible within our "documentary culture" (Carruthers 1990), ones that can be invoked in struggles to constitute a recognizable public (in this instance, the Lemba as Jews), even as they transform what counts as a document and shift the evidentiary terrain to which groups can turn in demanding recognition. Y-chromosome haplotypes can adjudicate the plausibility of Jewish origins and kinship. Given the long history of the centrality of genealogy to Jewish identity, and given the translation of Jewish genealogy into a presumably unproblematic, discoverable, and measurable biological descent principle beginning in the late nineteenth century (see chapter 2), while the empirical terrain of DNA evidence does not produce, in any straightforward manner, the truth of Lemba Jewishness, it does make possible a *reasonable debate* about whether or not they are of ancient Hebrew descent. It makes possible a reasonable debate about whether or not ancestry and identity are distinct (particularly, patrilineal ancestry and Jewishness), and about whether or not the Lemba can or should be recognized as (lost) Jews and brought back into the fold of Judaism.

During the discussion that followed the screening of the *NOVA* program *The Lost Tribes of Israel* at the American Museum of Natural History in the summer of 2001, a woman in the audience, clearly a bit frustrated, asked: "Given the discussion, and the distinction between ancestry and identity being made, I'm still not clear as to whether or not the Lemba really are Jewish, and who gets the final word?" Tudor Parfitt responded:

> That's the point. Who decides? Does Tudor Parfitt decide? In which case, if they say they are Jewish, and *they are practicing some kind of religion that looks like Judaism*, and everyone around them says, ah, yes, these are the black Jews, in that sense yes, they are Jewish. Whether the DNA evidence changes their status, I don't think it does. Particularly, all that it shows is that probably at some time in the past, Jews went to central Africa. And we didn't know that and that is a remarkable finding. . . . But if you were to take this film or our discussion or any number of papers that have been written on the subject

> to the Chief Rabbinate in Israel and ask them to decide on the basis of this whether the Lemba are Jewish or not . . . the answer would be no. Because none of these features has got anything to do with the halakhic definition of what a Jew is. (emphasis added)[19]

A second member of the audience objected to the entire discussion about the Lemba's Jewishness and to Parfitt's willingness to accept them as Jews, and he did so on the grounds of religious practice. "Another feature of what determines Jewishness is belief in the Bible and the use of certain prayers." The Lemba demonstrate neither, he said. In other words, Jewishness requires the demonstration of a normative set of practices and beliefs. Parfitt responded: "The fact is that the Lemba do have a whole range of practices and prayers which, while not being very similar to those used by mainstream Judaism, nevertheless do appear to be Jewish in some way. They look rather like Old Testament practice, and they certainly seem to predate the coming of the Christian missionaries."

But in the wake of genetic evidence, the question of recognizable—or normative—religious belief and practice has become more convoluted than this interchange suggests. Since this work became publicly known in 1999, various Jewish groups—especially in the U.S. and to a lesser extent in South Africa—have begun to visit Lemba communities and attempt to integrate them into the mainstream of Judaism. Kulanu (which means "all of us" in Hebrew) declared that, in light of "discoveries indicating that the Lemba tradition may be exactly right," "the naysayers are being silenced" (Heppner 1999, 1). The organization has dispatched various visitors, including more than a few rabbis, and it has funded Jewish education programs in Lemba communities. For example, as reported by Rabbi Yaacov Levi in the spring of 2000, "we have established a Jewish presence in the center of the Lemba area, and will soon be reaching out to those in Zimbabwe" (Levi 2000, 13). That Jewish presence is largely devoted to an educational program designed to prepare Lemba for formal conversion. Rabbi Bernard of Johannesburg agreed to facilitate Lemba conversion. And as Levi's report implies, there is much to be done before such conversions are possible: "I will be speaking for a lengthy time on our education and service plans [before the Lemba Cultural Association] meeting and also providing Jewish prayer to open and close the meeting. Last meeting, much to my surprise, the 'chaplain' called on a Lemba who opened his New Testament and made a prayer from it" (ibid.). He had explained the problem at greater length in a dispatch dated December 27, 1999:

> There is little knowledge existent today among the Lemba in South Africa on Judaism. There is a strong Jewish identity among them and many seriously want to be educated. From what little I have seen so far, they have the strongest Jewish identity here in S[outh] A[frica].
>
> We are going to be doing an on site education program here at the farm while the main site is prepared about 10 kilometers from here. . . . We will recruit 6 men, one from Zimbabwe, five from distant communities to come and train for six months, and then as lay leaders, to form congregations in their home areas. This will be a major INTENSIVE 6 months. They will be in contact with me weekly after returning home. Hopefully one year from now we will have seven new congregations here of the Lemba.[20]

Levi ends one of his letters back to Kulanu on an upbeat note: "We are preparing for Passover already in the Northern province. The Christian influence resulted in the local Lemba ignoring the Jewish Festivals. This year the Lemba will be eating *matzah*" (Levi 2000, 15).

Levi's visit to the Lemba was not the only one. Rabbi Leo Abrami also visited the Lemba from January to April 2002 with the support of rabbis in Johannesburg (Parfitt 2003, 117). He brought a number of books with him to be included in the Lemba cultural center and synagogue then under construction. Kulanu was involved in fundraising for these buildings. Abrami reported that he brought "copies of *What is a Jew* by Kerzer, *This is my God* by Herman Wouk, a copy of the new JPS Hebrew-English Bible, and dozens of other books on Jewish Theology, several manuals of Jewish history, Hebrew text-books, over a hundred small Shabbat manuals, *talilot* and other educational material" (Abrami 2001). In addition, Kulanu representatives orchestrated "contact" between Lemba and South African Jewish youth; Karen Primack reported in Kulanu's newsletter in the summer of 2000 (Primack 2000, 1) that a "mini-camp" for ten Lemba youth and "their counterparts in the white Betar youth movement" had been held and that there were plans to have Lemba youth participate in a Betar camp. According to Rufina Mausenbaum, a South African involved in work with Lemba communities, "we sincerely hope that the acceptance and socializing will help to heal the pain of centuries of rejection—it is a beginning, and I hope within a few years we will have a totally integrated group of Lemba Betar!" (quoted in Primack 2000, 7). She explained:

> We are accepting the Lemba as Lemba who have Jewish ancestry and not because we expect them to return to Judaism. . . . By going away together the

> [Lemba] youth [who have "had little or no exposure to Judaism"] will have the opportunity to experience first hand a little about Judaism and a Jewish way of life. Once they have an understanding of what Judaism is all about an informed choice can be made. This is where socializing and exposing the Lemba to a Jewish environment becomes an important factor. (Quoted in Primack 2000, 8)

Whatever the caveats or cautions of the genetic studies of the Lemba, at the popular level their conclusions have been translated into a near certainty. Tudor Parfitt argues, the "Jewishness of the Lemba" is now "more or less standard in popular academic discourse" (Parfitt 2003, 116). And that "fact" has had particular consequences: Jewish groups and communities, however limited they may be, have initiated sustained relationships with Lemba communities—*as fellow Jews*. Culture and descent are inextricably intertwined (Michaels 1992): suddenly Judaism becomes *their* religion and Jewishness becomes *their* culture by virtue of their genealogical descent, even if they do not at present engage in the doing of "Jewish things." In turn, though now recognizable Jews by virtue of their descent, *their* religion is something that they need to be taught.

In published papers as well as public forums and newspaper accounts, researchers involved in this work refer to Judaism as a "religion," to Jewish identity as a matter of religious belief and practice and as a matter of choice. Insisting on religion as the appropriate category for understanding what makes a Jew a Jew enables them to separate genealogy from identity, science from politics, and genetic history from race science. Religion, in their understanding, is about choice, about the faith that we—as modern subjects—voluntarily embrace (see Asad 1993). But as was true of nineteenth-century philology that articulated the concept of "language families," genetic history cannot escape its own genealogical logic. The cultural and political implications of genetic history are not dissimilar from those of the science of philology that rose to prominence more than a century ago:

> [One] implication of the language family seems to be the capacity to isolate any instances of historical—meaning, "accidental"—transmission of language and keep such cases separate from the question of genealogy or descent. By this reckoning, learning another's language, or being born into a language of another, would not alter one's inherent identity. (Masuzawa 2005, 169)

Conversion to Judaism, of course, has been possible for many centuries. And becoming a Jew (*giyur)* involves taking on a new genealogical identity as the son or daughter of Abraham. But that does not mean that converts are necessarily seen as equal Jews—or as equally Jewish (see Cohen 1999). As one Kulanu activist wrote:

> An acute sense of loss compels me to spend my time on genealogy. . . . I feel a personal sense of victory when I hear about someone returning—yes, we've gotten one back!—one who had been tragically lost to us as a people. We are reclaiming our own. With one caveat I do not feel the same sense of victory when I hear about someone converting.

The caveat? "A mystical belief of mine: I firmly believe that many Jews by choice are descendants of Jews where the knowledge of the historical linkage has been lost. What draws them back is some kind of subconscious historical memory" (Primack 1998, 273).

Researchers in genetic history present their work as nonpolitical: the Cohen modal haplotype does not adjudicate the Jewishness of the Lemba. It simply reveals a potentially interesting and heretofore unknown historical fact. But perhaps phylogenetic evidence is better named "prepolitical"—that is, in terms of how it is represented by journalists who report it, activists who invoke it, audiences who seem fascinated by it, and scientists who engage in double-talk on the matter (it matters, it doesn't matter) (see chapter 3). This kinship "is . . . the sphere that conditions the possibility of politics without entering into it," to borrow Judith Butler's words (2000, 2). As genealogical facts, Y-chromosome haplotypes reveal a history that, if true, makes possible the granting of Jewish "citizenship" to the Lemba—but only following the choice to convert to Judaism, after which Lemba individuals can achieve recognized membership in the Jewish world (which brings financial benefits as Jewish groups in the U.S. support projects in "developing" Jewish communities) or more literally, citizenship in the Israeli state. That political rationale for bestowing recognition discloses the dynamic relationship between "biology" and "choice"—between genetic haplotypes, religious faith, and group membership. In addition, it demonstrates how *in practice* the relationship between individual and group identification is impossible to disentangle: as a statistical fact, the results of the Lemba study open up the possibility that this group really might be descendants of ancient Jews. In turn, Judaism can be encouraged and taught and, after a suitable education, individuals who choose to learn and practice normative Judaism will be able to be recognized, in accordance with (orthodox) rabbinical legal definitions, as ("returning") Jews.

Liberalism, Colonization, and the Politics of Jewish Inclusion

What commitments motivate Kulanu's work? Why might Jewish activists want to bring groups such as the Lemba into the Jewish fold? In the introduction to an edited volume, *Jews in Places You Never Thought Of*—a compilation of accounts of their work by Kulanu members—Karen Primack writes:

> This book is about an unusual subject, presented in an unusual way. It is *not* an encyclopedic directory of information about the little-known communities it describes, and it is not meant to include all such communities in existence. Rather, it comprises personal accounts of interactions with these communities by the authors of the articles. With the exception of a few historical overviews, and views expressed by members of the communities themselves, the pieces were written by people who actually visited these far-off (and often inaccessible) groups—talking, singing, eating, dancing, and praying with them. It is a book about these communities, but it is also a book about personal involvement. (Primack 1998, xi)

"Personal involvement," a "Jewish handshake," "a practice of *doing*" (ibid., xvi)—such are the things that Kulanu strives for. A largely web-based organization run by volunteers, its members are engaged in a search for meaning, for connection—for authenticity—that saturates the discourse of Kulanu members; theirs is a desire to visit and to "help" a group of Jews who are somehow simpler, more authentic, more devout, and also a folkloric fascination that merges tourism with a missionary zeal. The religious desire to convert lost Jews back to (proper) Judaism is expressed throughout Kulanu's publications: as correspondence sent back by visitors to Lemba communities points out over and over again, the Lemba are at a loss for Jewish religious leaders, and the consequences are dire.[21] Yaacov Levi, for example, reports:

> After meeting with elders throughout the area, I can see a need for teachers and lay leaders much sooner than envisioned. To this end I am going to be meeting with Professor Mathivha this week and other leaders of the Lemba to begin an immediate program to give instruction to a group of lay leaders

> in the daily and Shabbat services and to receive the weekly parasha to teach to congregations. It has been the absolute absence of this that has led to the assimilation of many of the Lemba and conversion to Christianity and Islam. (Levi 1999a, 12)

In a more urgent tone, "The Lemba in Mozambique, Soweto and Zimbabwe have been and are being 'evangelized' by Muslims. This is a real and continuing situation" (Levi 2000, 13).

The fear of "losing Jews" to other faiths might best be understood in the broader context of anxiety about the decline of the U.S. Jewish population (sometimes referred to as a "second holocaust")—a phenomenon documented, reported on, and discussed by American Jewish organizations and the Jewish press throughout the 1990s. Skyrocketing rates of intermarriage between Jews and non-Jews have raised fears that the children of "mixed marriages" will not consider themselves Jewish (Wertheimer 1994). To quote one Kulanu board member:

> Not a week goes by that I do not receive passionate e-mails from emerging Jewish communities all over the developing world. In Africa alone, new native communities in Nigeria, Ghana, South Africa, Zimbabwe and other places are literally pleading with us to send them Jewish teachers, books, and religious articles.
>
> We Jews do not have a demographic problem, we have an attitude problem. That problem is called racism, and we need to deal with it directly and forthrightly. (Cotel 2003, 5)

Within the context of that demographic anxiety, it is easy to understand why Levi reports, with much optimism: "When back in the States I had only a vague idea of the numbers of those who wished to return to Judaism. This is 'firming up' to an expectation of thousands, scattered over several hundred kilometers in numerous communities" (Levi 1999a, 12). Or as a second Kulanu activist reported, "over the last decade . . . the great fortune of visiting over 100 Jewish communities in 20 countries on five continents, I can honestly say that I have never encountered a more dedicated Jewish community than the Bnei Menashe [in north-east India]. Unlike many of our Jewish communities around the world, which are losing members to aliyah and assimilation, the Bnei Menashe are gaining new converts by the hundreds, grown men undergoing the painful circumcision process to find spiritual peace in Judaism" (Schwartz 2001, 2). It is worth highlighting that he mentions "losing members to aliyah," i.e., emigration to Israel, as a problem. This activist is most committed to the vitality of Jewish life outside of Israel.

Kulanu frames its missionary project of encouraging a return to Judaism in terms of an antiracist, multicultural commitment. According to Jack Zeller, "one purpose of this work is to preserve and enhance the enthusiasm of one Jew meeting another, even if it occurs in the most unlikely place, and the other Jew is of a darker skin color or different appearance, or practices a Judaism that is non-rabbinic in origin, or is a newly arrived Jew by choice" (Primack 1998, xv). As explained by Rufina Mausenbaum, "all South Africans 'come with a huge amount of baggage—very recent and painful—still a part of our lives as we try to overcome the years of apartheid'" (quoted in Primack 2000, 8; see also Mausenbaum 1999).

Kulanu has its roots in American Jewish efforts in the 1980s on behalf of Ethiopian Jews—the campaign to get the State of Israel to accept Ethiopian Jews as new Jewish immigrants *(olim)* and to airlift them out of Ethiopia (see Parfitt 1985).[22] After 1991, a Kulanu member explained in an interview, "the American Association of Ethiopian Jews closed . . . and I asked myself, what did I want to do with the rest of my life?" The 1980s campaign on behalf of Ethiopian Jewry is key to understanding Kulanu's political commitments and projects. In that struggle, major Jewish American figures and organizations positioned themselves as fighting for the rights of nonwhite Jews: as a post-civil rights, U.S.-based movement on behalf of "black Jews," Jewish activists produced Israel's racism—a racism recognized along the black-white spectrum of American race relations—as a problem and positioned themselves as Israel's antiracist conscience. That legacy is evident in Kulanu's rhetoric as well as its ongoing work. In his editorial that declared "it is demeaning for one Jew to have to prove his Jewishness to another Jew" (quoted above), Jack Zeller prefaced his remarks on the Lemba with remarks about Ethiopian Jews: even though Ethiopian Jews were recognized as "authentic" and thus given the right to immigrate to Israel, it would be hard "to find an Ethiopian Jew who was not (unnecessarily) asked at some point to prove his Jewishness in one way or another." And Zeller is not alone in drawing parallels with the treatment of Ethiopian Jewry. Another Kulanu activist reported: "I had Shabbat dinner along with Erev Shabbat at Rabbi Barnard's [the Johannesburg rabbi who agreed to oversee the formal conversion of Lemba]. He is with Chabad, and he is 'on our side'; he knows full well the situation with the Ethiopians in Eretz [Israel] and is supportive of our step-by-step plans" (Levi 1999a, 12).

Kulanu positions its work in terms of a commitment to social justice, tolerance, multiculturalism, and antiracism. The newsletter displays a sustained commitment to promoting "diversity"—whether by announcing the publication of its books (*Jews in Places You Have Never Thought Of*, or, *Under One Canopy: Readings in Jewish Diversity*) or by advertising lectures organized by Kulanu activists: for example, a seminar at the 92nd Street Y in Manhattan to discuss "exotic" Jews (*Kulanu* 7, no. 3 [Autumn 2000]: 9). Kulanu sees itself as part of the Jewish Multiracial Network. Its newsletter announces that the network needs new members and advertises the various events that the Jewish Multicultural Network plans (*Kulanu* 8, nos. 3 [2001]; 6; 11, no. 4 [Winter 2004–5]: 5). Kulanu's newsletter once declared that "lessons on Diversity [are] needed," while reporting on the work of the Coalition for the Advancement of Jewish Education to "publish a Jewish diversity curricular resource for educators that could be utilized across various Jewish disciplines and subjects to explore the topics of Jewish cultural diversity; intra-group relations; and the subject of inclusion and welcome of 'difference' in the home, synagogue and community." Do Kulanu activists have "any lessons that could be incorporated into this publication" (*Kulanu* 11, no. 2 [Summer 2004]: 4)? More generally, as evidenced in the Kulanu Briefs and Publications of Note sections of each newsletter, the organization devotes itself to promoting knowledge and activism vis-à-vis a racially and culturally diverse Jewish world.

From Kulanu's perspective, the results of genetic historical work on the Lemba can be harnessed to a political project that seeks the *diversification* of the Jewish people. Irwin M. Berg, a Kulanu member, pointed out in an article that not all groups have been as lucky as the Lemba:

> There is a priestly class of Lemba whose DNA shows a connection with Jewish *Kohanim* (Priests). Other than the Lemba, these peoples [Jews from places as diverse as sub-Saharan Africa, Afghanistan, India, and China] have no 'proof' that they were once Jews. They have their customs and traditions, which they compare to the customs and traditions of the children of Israel as reported in the Torah. . . . Even though Jewish tradition does not report on these peoples, early Arab historians have identified some of them as Jews. Speaking as a Jew as well as a dilettante historian, I think we should be respectful of the claims of peoples whose customs and traditions bear the indices of ancient Jewish practices. As I was told in a remote village in Mali near Timbuktu, Africans remember forever the oral traditions of their ancestors, particularly those relating to their origins. (Berg 2006, 14)

But whether or not people *should be* respectful of indigenous claims, Berg makes clear he does not think they have "proof," that is, evidence that would be widely recognized as credible by the Jewish world.

The "facts" of genetic history—however "tragic" the necessity of having to use them—are powerful in Kulanu's struggle to diversify the Jewish world. And Kulanu is not alone in recognizing that reality. Consider the following rather extensive exchange between a member of the audience and Tudor Parfitt at the American Museum of Natural History session on "The Lost Tribes of Israel":

> AUDIENCE MEMBER: I personally feel that the whole concept of the lost tribes is, to me, utterly ridiculous in the sense that, I heard a very famous rabbi say once, that these tribes are not lost to us, we are lost to them. I kind of like that particular concept instead. And myself, being a Jew of color historically with roots in Morocco, I personally do not consider myself to be lost to anyone. My question is, therefore, I personally feel that the genetic research can also be divisively used in a harmful way in some respects. I would like to know what benefit can these findings give to the Jewish community, but to the non-Jewish community as well. How can it benefit our understanding of the idea of diversity in the Jewish community, which especially needs to be taught today in Western societies?
>
> PARFITT: Let's take the Lemba as one example. . . . I take what you are saying about the lost tribes of Israel, it is a kind of shorthand for saying all kinds of other things. But I think it has been very useful for the Lemba, and also in a way for their neighbors. After all, for the last hundred years the Lemba have made attempts to get themselves accepted in the context of South Africa by the white Jews in Johannesburg and elsewhere, and indeed, one or two committees of inquiry were sent to the Lemba villages by the white Jewish community, who always declared that they were not Jews. And that was the end of it, and there was no contact between them. The only contact between them was that sometimes Lembas were servants in white Jewish households, and that was the beginning and the end of it. And now, since the genetic work was done I know that there has been a proper symbiosis between at least some aspects of the white Jewish community and the Lemba. I know that the Betar youth group, for instance, has had joint events with Lemba groups. And the Lemba have taken a certain amount of pride in the fact that they have got similar antecedents and ancestors as do the white Jews. So in terms of Jewish diversity, I think it has proved to be extremely useful. The very fact that in North America now, as I said, the Lemba are considered by very many, particularly liberal Jews, to be part of the family of Israel, very

> largely because of this genetic work, is precisely along the lines of increasing Jewish diversity.[23]

From the perspective of those who champion Lemba integration—as well as the integration of other, nonwhite communities with "Jewish roots"—genetic history has enabled the recognition of nonwhite Jews. It has legitimated previously dubious claims to ancient Hebrew descent. The Jewish world, thereby, can be made (more) multiracial, more diverse. In short, a multicultural discourse with regard to race and inclusion structures the work of groups such as Kulanu. In addition, it structures the press coverage of and documentaries made about the discovery of the Cohen modal haplotype among the Lemba and public discussions of the consequences or effects of the scientific work. The results of that work have been translated, in a Latourian sense,[24] as they travel through social networks that have appropriated them for their own purposes: to expand membership in the (recognized) Jewish world by integrating "lost"–generally, nonwhite—Jews. And in the struggle to expand the known Jewish world, the critique of those who deny the claims of groups such as the Lemba to Jewishness often involves the accusation of racism or Eurocentrism. For Kulanu activists, it is the fact of recognizing Jewish kinship *despite phenotypic difference*, despite the fact that the Lemba are Africans and the activists white, that makes evident the ways in which Kulanu's political project works against the grain of race.

Kulanu's liberalism belongs to a very particular historical moment. It belongs to a "'post-civil rights' context in which 'most Americans believe themselves and the nation to be opposed to racism and in favor of a multiracial, multiethnic pluralism'" (Povinelli 2002, 26–27). More specifically, it operates within the grammar of a multicultural political imagination, part of "a global adjustment of the constitution of public and legal national imaginaries as state institutions and public sympathy attempt to address the multiplicity of social identities and traditions constituting and circulating through the contemporary nation" (ibid., 26). In this instance, however, that nation is a diasporic nation of Jews. From the perspective of the desire to expand diversity *within* the Jewish world, the commitment to diversity—to antiracism—takes a very particularistic form: it involves helping other (would-be) Jews.

Multiculturalism has a distinctive bedfellow in this social field. The story of the lost tribe carries messianic overtones. As explained by Zvi Ben-Dor

Benite, a "tension between history and prophecy" stands at the heart of this story: as history, "the tribes existed, they disappeared, they exist somewhere right now"; as prophecy, "they shall return." "This underlying framework of loss and redemption," Ben-Dor Benite argues, has grounded all accounts of and sources on the lost tribes over the centuries—be they Jewish or Christian, secular or religious (2006, 7). In its work in search of lost tribes, then, Kulanu joins together the multicultural gesture of Jewish inclusion not just with a missionary fantasy but also with a messianic impulse. Moreover, as I demonstrate below, Kulanu cooperates with Israelis of the religious-nationalist camp, which has its own investment in lost tribes and would-be Jews. Out of that articulation, what is perhaps best described as a postcolonial colonial politics is being forged.[25]

In writing about whether or not Kulanu has become overextended—"overly ambitious, understaffed and passionately concerned about every person who wants to participate in Jewish life" (Zeller 2001, 6)—Zeller describes Kulanu's work as a Zionist cause that *should be picked up* by others:

> How did Kulanu get so over-extended? We underestimated numbers [of those who want to "return" to Judaism]. In addition, some of us thought that one of the major Jewish organizations would take over our cause much the way virtually all Jewish organizations are now Zionist and involved with Soviet Jewry. Will that yet occur? What Jewish civilization would abandon thousands of Jewish descent from the Iberian Peninsula, India and Africa? Thousands who want to be Jewish for an authentic reason? (ibid.)

A second Kulanu activist articulates the organization's role at the forefront of a new Zionism: "The similarities between modern Zionism in its early stages and Kulanu are fascinating." As is true of Kulanu's work today, the early Zionists were faced with a Jewish community not yet committed to their project. Reading those early Zionists as devout (the Zionist movement was "a cultural-religious-national Jewish movement"), Joshua Fox understands Kulanu as "just beginning to plant the seeds of a Jewish revival, by reaching out to lost and scattered Jews worldwide," be those "communities lost to us for centuries, or *Gerim* [converts to Judaism] joining our people—from the four corners of the earth." Fox notes that Kulanu faces much opposition in its work both from people who see their work as destroying "local customs" and from Jews "who may be uncomfortable with the great fervor that new returnees or converts show for the Jewish religion." But he has no doubt that Kulanu can overcome the opposition. "Kulanu is gaining the recognition and support of wide sectors of the Jewish people. Religious

Jews appreciate the addition of deeply believing and devoted Jews. Humanists see that bringing in Jews from different races and ethnic groups helps us actualize the principles of ethics and humanism that we have always worked for" (Fox 2003, 7). But Fox's Zionism, it is worth emphasizing, is a diasporic one: his primary commitment is to add deeply believing and devoted Jews to the Jewish *people* and not necessarily to the Jewish state.

Fox's emphasis on the religious "fervor" of returnees and converts, and the anxiety of secular Jews towards that fervor, is essential to understanding the complex international political field within which Kulanu's mission operates. Even though not all Kulanu activists are motivated by religious commitments, the importance of faith is evident in its work and in its newsletters: members report on having eaten in a Kosher restaurant (in Johannesburg); they write of the desire to teach returnees proper Judaism. And a few want to learn from the very different rituals of others, to recognize those rituals as Jewish, and to become more tolerant Jews.[26]

Those religious commitments take on a particular political significance in the context of the Israeli political scene, as distinct from the American one. Kulanu's Israeli organizational equivalents, Amishav (My People Return) and, more recently, Shavei Israel (Israel Returns) are Orthodox, whereas Kulanu is not. And those Israeli groups fight the Israeli government bureaucracy in their efforts to "return" lost tribes and lost Jews (*Anusim*, that is, Jews who were "lost" to the Spanish inquisition) not just to the Jewish *people*, which is the rhetorical and, by and large, practical emphasis of Kulanu's work, but to the Jewish *state*. Amishav and Shavei Israel work to bring these "old-new" Jews, to borrow Theodore Herzl's (1997) turn of phrase, "home." And when they have succeeded, those "returnees" have been integrated into the settler movement. Kulanu works closely with Amishav and Shavei Israel. On behalf of both Amishav and Shavei Israel, Kulanu members protest Israeli government policies, raise money, and fund aliyah.

Rabbi Eliyahu Avichail founded Amishav in the 1970s, during the "messianic fervor" that followed the 1967 war. Rabbi Avichail is said to have given a lecture on lost tribes at Yeshivat Mercaz ha-Rav in Jerusalem. Rabbi Zvi Yehuda Kook, the inspiration for the Gush Emunim (Block of the Faithful) settler movement, attended the lecture. (Mercaz ha-Rav Yeshiva was Kook's yeshiva.) Following that lecture, reportedly, Rabbi Kook encouraged Avichail to pursue his work on lost tribes in a more systematic manner.[27]

The story of the ten tribes articulated with a Zionist imagination in the early twentieth century through the work of figures such as Jacques Faitlovitch, Zvi Kasdio, and Yitzhaq Ben-Zvi, who was to become Israel's second president. As Ben-Dor Benite has argued, that very conjuncture reconfigured the meaning of the originally biblical story in important ways: it now operated in the context of a "very real possibility of Jewish political sovereignty," thereby politicizing the possibility of the tribes' return (2006, 221). And it was within a theological vision of modern Zionism that the story of the lost tribes took far stronger hold: Abraham Isaac Kook (1865–1935), the father of a messianic Zionist politics, which sees the modern Jewish state as a sign of the beginning of redemption, "forged a direct link between the ten tribes and Zionism. Just as redemption involved ending Jewish exile, it also meant ending the exile of the ten tribes" (ibid., 222).

Avichail followed the advice of Zvi Yehuda Kook (Abraham Isaac Kook's son), and since the late 1970s he has been working with one group in particular: the Bnei Menashe, a subgroup of the Mizo/Kuki (also referred to as Shinlung) in India. In the words of the Kulanu activist quoted above, when it comes to the Bnei Menashe, "Avichail gets all the credit. I'm essentially marketing the [Bnei] Menashe. We're an amplifying network."

Speaking before the Knesset's Committee on Immigration, Absorption, and Diaspora Affairs during a hearing on the Bnei Menashe (October 5, 2000), Avichail described how he initially became involved in working with this group. In 1979, when at Hebrew University (as rabbi), a friend of his at the National Library—an Indian named Shimon Shimon—showed him a letter from a group who "called itself the 'Jews of [North]-East India' who had turned to him [Shimon] and asked for help." "They lived as Jews," Avichail testified. He made contact with them, and over time he came to "the conclusion that there is no doubt that these people are tied to the Jewish people," even though it took him more time to ascertain that their connection was specifically to the tribe of Manassas (Committee on Immigration, Absorption, and Diaspora Affairs/1490, October 5, 2000, Hebrew). As reported by an anthropologist, Shalva Weil, who testified before the same committee meeting, the tribe had a tradition of having descended from one of the ten lost tribes of Israel. It was Rabbi Avichail, however, who identified them as the Bnei Menashe (Parfitt and Trevisan Semi 2002, 34). According to Avichail, in testimony before the Knesset committee, among the Shinlung only "those who live a Jewish life" are called Bnei Menashe.

Avichail has been traveling to India for the past few decades, meeting and working with the community and, in the words of one letter of appreciation for his work, "guiding them to develop as Orthodox Jews, and helping the first members of this community settle in Israel" (Samra 2005, 7). Advocating for the Bnei Menashe in Israel has involved working with the

Ministry of Interior as well as with Israel's Chief Rabbinate. An article in *HaAretz* from 2005 explained the situation:

> Since 1992 Avichail has had an "understanding" with a number of interior ministers regarding permission to bring a few [Bnei Menashe] each year, at most, to Israel. The procedure is that those who are granted a permit to immigrate arrive in Israel as tourists, without immigrant rights. Here they undergo a conversion process and only afterward officially receive immigrant status and citizenship. To date [2005], about 800 of the Bnei Menashe have come to Israel, a number lower than the quota they are allowed. (Sheleg 2005b)

As the hearing before the Knesset committee indicated, by 2000 Bnei Menashe immigration—however small in absolute numbers—was something the government decided to learn more about. Naomi Blumenthal, the committee's chair, noted that their purpose was neither to decide on the Jewishness of the Bnei Menashe nor to legislate on their right of return; it was "information" that the committee sought.[28]

Those Bnei Menashe who had managed to immigrate to Israel—initially under tourist visas until converted by the Rabbinate and given citizen status as Jews—were settled in the territories.[29] According to a Peace Now activist knowledgeable about Amishav's work on behalf of the Bnei Menashe, "the settlements are actually the pivot. As soon as they arrive in Israel they arrive in settlements and that is where they have their first encounter with Israeli society" (interview). He elaborated: "The people who are behind this are mainly settlers or people who support settlements—Amishav and Shavei Israel, who have an office in Kiryat Moshe in Jerusalem, which is an Orthodox-nationalist neighborhood with the Mercaz Ha-Rav Yeshiva within it, a grow house for the more Orthodox-nationalist and conservative conceptions among the religious-nationalist public." Following their arrival in Israel, Bnei Menashe were settled in Gush Katif (a settlement block in Gaza evacuated in August 2004; see Hason 2004) and in various settlements on the West Bank—Ofra, Beit El, and Kiryat Arba, for example (Sheleg 2005b). Although the number of Bnei Menashe immigrants is not large, nevertheless, the Peace Now activist pointed out, "if five to six hundred people live in [the settlement of] Eina—in fact, 569 people lived there according to the Peace Now 2007 report—bringing thirty to forty people is highly significant for the community. The demographic field, the growth of Jews, inflames all of the claims that are based on 'settler's natural growth'—more homes, it produces more jobs since every family that arrives has to be guided and taught Hebrew, and this is the money" (interview).

Education is key to the process of Bnei Menashe absorption, and what it means to "educate" incoming Bnei Menashe is broad. Those Bnei Menashe arriving in Israel undergo a process of training and Orthodox conversion. Some of that goes on "ad hoc before they arrive," the Peace Now activist explained. And it continues in the settlements. According to one Bnei Menashe immigrant, "Since the summer of 1995, I have managed to explore a wide range of activities. I recall clearly the evening spent together with a group of youngsters at Rav Avichail's *shiur*, and the *brit mila* the next day in Kiryat Arba" (Mordechai 1998, 9). (Kiryat Arba is a settlement near Hebron on the West Bank, one of the earliest settlements and one of the more virulently ideological ones.) Writing in 2000, Rabbi Avichail reported that the young men arriving that year would study "in a new education program in Efrat [a West Bank settlement] under the auspices of Rabbi Shlomo Riskin" (Avichail 2000, 1).[30] In addition, Avichail reports, in order "to ensure that the Bnei Menashe have the proper tools for integrating fully into Israel's high-tech society, an arrangement has been made with the College of Judea in Kiryat Arba to accept 13 young men and women from the community who have already completed the conversion process." Avichail then calls upon Kulanu activists to donate money for the education program (ibid.). In the summer of 2001 Avichail put together a three-month summer program in Kfar Etzion (in Gaza) for about twenty to twenty-five Bnei Menashe students. They were taught Hebrew and Judaic studies with a view towards training them for "teaching roles in their own community." With the aim of raising $100,000, Avichail called upon Kulanu activists once again to donate money.[31] And in 2005, with the support of Shavei Israel and the local council of Kiryat Arba, the Bnei Menashe opened their first community center "in Israel."[32]

In 2003 the Minister of Interior, Avraham Poraz, temporarily put a stop to Bnei Menashe immigration—a dispute, in short, over their Jewishness, over settlements, and over secularism. One cannot understand the argument over the Bnei Menashe without understanding the debate over non-Jewish immigrants to the Israeli state in the late 1990s/early 2000s, which was precipitated by the immigration of thousands of apparently non-Jews from the former Soviet Union a decade or more before. As recounted by Orit Shohat in "Who Decides Who's Jewish?" (*HaAretz*, November 26, 1999, English), when Rabbi David Beniziri declared that "the Russian immigration of recent years is for the most part not Jewish, which is having a negative effect on the Jewish character of this country," a whole public argument ensued that focused on the Who is a Jew provision in the Law of Return. The Law of Return's "secular provision," Shohat argues, enabled this apparently non-Jewish immigration to occur: "Even a Christian whose grandfather was Jewish is entitled to immigrate to Israel with all his family." There were various logics for establishing the law's genealogical criteria: to encourage massive immigration, certainly. But, Shohat argues, "it was also legislation designed to demonstrate to the religious community that the religious criterion as to who is a Jew, i.e., someone with a Jewish mother or who has undergone an Orthodox conversion, cannot dictate secular Israel's immigration policy" (ibid.).

What is clear throughout the Absorption Committee's hearing is that Knesset members were dubious—as were a few witnesses before the committee—about the Jewishness, or Jewish ancestry, of the Bnei Menashe. In their testimony Avichail and Hillel Halkin repeatedly emphasized the "sincerity" of the Bnei Menashe: They are committed to Judaism. They are not economic migrants. Others raised questions about the plausibility of their Jewish ancestry. Naomi Blumenthal, the committee's chairperson, asked Rabbi Avichail: "When you arrived there, how did you find the people? Did they themselves come and say to you: We have distinctive customs. We are different from the others? Or did you say to them: Perhaps you have something that reminds one of the tribe of Menashe?" Blumenthal also questioned a member of the Bnei Menashe living in Israel:

> KNESSET MEMBER NAOMI BLUMENTHAL: When did you know that you are a member of the tribe of Menashe?
>
> RIVKA MORDECHAI: . . . Already in childhood, we knew about Shabbat and the holidays. When I was in primary school my parents and a group of people created a connection with Rabbi Avichail . . . and that connection, little by little, grew stronger.
>
> BLUMENTHAL: Were you different from the population among which you lived?
>
> MORDECHAI: In Mizoram, we speak the same language, and our traditional customs are the same as they were before the arrival of Christianity. Today most of the population is Christian, but they have a strong connection with the people of Israel and with the people of the State of Israel. All of them love Israel and the people of Israel and all of them believe that we are the descendants of the tribe of Menashe, that we are a part of the Jewish people, even though the majority still do not want to make aliyah to Israel.

The former head of the Jewish Agency's Office of Aliyah made it quite clear he thought the whole thing was nonsense. And others asked over and over again for proof. As one Knesset member asked of Avichail: You said you have "clear evidence" of their descent. What is it?

Hillel Halkin helped get a genetic historical study of the Bnei Menashe off the ground in 2003. In his testimony before the committee Halkin had explained that there is "a possibility that there is an ancient connection, I don't know if with the Bnei Menashe, . . . but there is certainly a possibility of an ancient connection with the People of Israel." He then argued that there needs to be "serious research, and here we are talking about linguistic, ethnographic, anthropological, and historical." But what kind of a genealogical connection could be proved on the basis of those evidentiary sources? Genetics could provide the proof.

While initiated in part by Hillel Halkin, it was Karl Skorecki (of Haifa's Technion) and Professor Laldena (of Manipur University, Department of History) who began a genetic study of the Bnei Menashe. As of the summer of 2009, Karl Skorecki said that his study was not yet final: it had not yet been submitted to a journal or peer-reviewed. As such, he did not want to publicly pronounce on the results, but he did note: "I do know that there are nonscientific interests in this area that I really don't like, and it's a shame. This is possibly the most striking case of research I have been involved with, the only one I think, that I feel has included more than purely scientific interests" (interview). He elaborated, there are interests of several kinds: "Let's say groups or people or individuals, who for political or other reasons want to prevent the immigration of these people and their acceptance as citizens of Israel . . . or the opposite, people who want to find a link to the tradition . . . to show that they are descendants of the ancient Hebrews." As Yair Sheleg noted in April 2005, given the recent change of the Israeli Minister of Interior (Poraz stepped down in December 2004), "the two genetic studies are taking place during a period that is . . . crucial for the Jewish future of the Bnei Menashe" (2005b).

Two scientists at the Central Forensic Laboratory in Kolkata had also designed a genetic study of the Bnei Menashe, part of a program for investigating minority group genetics commissioned by the Indian federal government, and they published it online in *Genome Biology*—without peer review—in December 2004. While Mizo-Kuki Y-haplotypes showed no evidence of Jewish (or Middle Eastern) descent, the authors argued that the maternal line did (Maity et al. 2004).

Karl Skorecki, among others, questioned the results (Sheleg 2005b). But as Skorecki has argued, a lack of evidence is not an evidence of lack.

As Skorecki said in an interview, "if we haven't sampled each and every one of the Bnei Menashe or any other community, then a representative sample [may] represent the group [statistically speaking.] But we may have missed the four people who carry the markers that testify to a source of a shared ancestry and we haven't sampled them by chance." In other words, as Skorecki explained to Sheleg in 2005, "the absence of a genetic match still does not say that the Kuki do not have origins in the Jewish people, as it is possible that after thousands of years it is difficult to identify the traces of the common genetic origin. However, a positive answer can give a significant indication" (quoted in Sheleg 2005b). According to Yair Sheleg, Hillel Halkin notes that "while the research is interesting, it will not tip the balance for him." Rather, it is the "many Jewish traditions among the tribe . . . as well as texts and prayers that are very reminiscent of the Jewish liturgy" that Halkin found convincing (ibid.).

Genetic evidence, however, may tip the balance for others. The Peace Now activist doubts a genetic connection will be found with the Bnei Menashe: the story of the lost tribes is a "myth," and the "aspiration" to find them goes back a long way in Jewish history. But if such proof were found, "I have no doubt it would . . . be an influential factor, out of my familiarity with Israeli society and how things work here. . . . The state will collaborate with it—elements in the state" (interview). And that statement is not far fetched. According to Karl Skorecki, there are ways in which DNA is beginning to enter into questions of who is a Jew. Rabbis have approached him for advice about the plausibility of using DNA in order to resolve halakhic questions. For example, if a given woman has "the sequence of mitochondrial DNA [that] is the same as those sequences we've identified . . . is there a scientific logic that can say that this testifies to a maternal chain? Can you draw this conclusion and with it rule for example that this person is evidently from a maternal chain of people who are identified as Jews, and so is also [Jewish]" (interview). And it is not rabbis alone who are beginning to explore questions about the possible relevance of DNA. As Skorecki noted, while he did not participate, under Prime Minister Olmert "there was some kind of discussion on the topic of genetics and the conclusions for [would-be-Jewish] communities" (interview). More generally, if one keeps in mind the Knesset Absorption Committee's recurring questions regarding evidence and proof—and if one simultaneously keeps in mind that the *secular* provision of Israel's Law of Return specifies biological descent relationships not restricted to religious definitions of Jewish identity as criteria for citizenship in the Jewish state—it is plausible that positive proof, genetic style, could make inroads into the terms of the debate, helping to establish the historical credibility of ancestral claims.

Following Avraham Poraz's decision to close the door to Bnei Menashe immigration, Michael Freund—the former executive director of Amishav (who had taken over from Avichail for a few years) and founder of Shavei Israel—took his fight to Israel's Chief Rabbinate. As explained in an article in Shavei Israel's newsletter, "After Poraz's decision was announced [and the author identifies Poraz as "Poraz of the Shinui party," a not so subtle reminder of the secularism and "left-wing" leanings that lead to this decision], Freund . . . began lobbying to receive official rabbinical recognition of the Bnei Menashe as a means of circumventing the Interior Minister's decision" (Freund 2005a). In March 2005 Israel's Chief Sephardic Rabbi Shlomo Amar recognized the Bnei Menashe as "descendants of Israel."[33] Amar, Freund argues, thereby "paved the way for their conversion and eventual aliyah" (ibid., 3). The article then notes: "all those converted plan to move to Israel." Despite the fact that conversion is allowed in Judaism, for these would-be-Jews, recognition of their Jewish ancestry seems to be a requirement that must precede both conversion and the possibility of immigrating *as fellow Jews* to the Israeli state. And facilitating aliyah—a legal and not just symbolic recognition—is the goal of Amishav and Shavei Israel as organizations.

As of January 2011, 1,700 members of the Bnei Menashe community in India had immigrated to the Jewish state, and the Knesset's Committee for Immigration, Absorption, and Diaspora Affairs decided that the Israeli government should "bring the rest of them [7, 232 individuals] home" (Zuroff 2011). According to Michael Freund, testifying before the committee in December 2010, "despite being cut off for more than 2,700 years, the Bnei Menashe never forgot who they are or where they come from, and never gave up on the dream of returning to Israel" (quoted in ibid.).[34]

While they coordinate their efforts, especially on behalf of the Bnei Menashe, Kulanu and its Israeli equivalents operate with very different political rhetorics. Kulanu responds to Israeli government decisions with which they disagree with the charge of racism. For example, the following is Jack Zeller's response to Poraz's decision to stop the immigration of Bnei Menashe:

> It is fairly obvious by now from the example that Avraham Poraz set with Beta Israel [a second group of Jews] from Ethiopia, that he wants to keep out *olim* who are not of European descent. . . . Every Rabbi in every synagogue in the United States needs to speak up. Nothing would upset Poraz more than

> knowing that on his next trip to the United States he would be banned from entering every synagogue or Jewish communal building, no matter what! This is a leverage not often used. But if not here and now, when?[35]

That rhetorical critique of racism is largely absent from the political language of the Israeli groups. Amishav emphasizes the religious sincerity of Bnei Menashe immigrants. They are devout. They lead a *more Jewish* life than do most Israeli Jews. And in his testimony before the Knesset Special Committee in 2000, Avichail emphasized Amishav's commitment to making sure the process of *giyur* is properly managed. Therefore, even if Israel keeps its doors open to the Bnei Menashe, "thousands if not millions" of Shinlung immigrants *are not* going to arrive on its doorstep. Amishav cannot handle the conversion of more than fifty to one hundred individuals per year. And their capacity to properly facilitate conversion limits the number of entry permits that Amishav asks for.

Under the leadership of Michael Freund—a *Jerusalem Post* syndicated columnist, former Deputy Director of Communications and Policy Planning in the Prime Minister's office (1996–99, under the government of Benjamin Netanyahu), and a hawkish blogger on Arab and Palestinian matters—Shavei Israel articulates the significance of its work in terms summed up in the title of an editorial Freund wrote in the *Jerusalem Post*: "Our Communities Can Help Israel's Demographic Crisis" (Freund 2001). Israeli policymakers would be "ill advised to overlook [the] dire warnings" of a recent demographic study presented to the Knesset: that Arabs will outnumber Jews within Israel's 1967 borders by 2035. Given that "world Jewry is simply not rushing to Israel," Israel needs to think "more creatively about how to address the ongoing erosion in the country's demographic profile." There are lots of possible "sincere" immigrants among the Bnei Menashe, the Lemba, the Abayudaya (a group of Ugandan Jews with which Kulanu does extensive work) and Crypto-Jews (Anusim). As Freund notes in a second *Jerusalem Post* article (Freund 2005b), if one totals the numbers of Falash Mura, Subbotniks (Jews in Siberia), and Bnei Menashe, there are over forty thousand immigrants "all of whom will tie their fate with the people of Israel and make *aliya*." While Israel should not "become a missionary state, seeking to convert the entire world to Judaism . . . if there are various groups who have already taken the first step and demonstrated a genuine and heartfelt commitment to Judaism and Israel, then why should they be overlooked or ignored?" Israel needs to start reaching out to "lost Jews," "assessing their claims to Jewish ancestry" (Freund 2001). More in tune with an Israeli political landscape—even if a rather recent immigrant from North America—Freund frames the importance of his work within a very different language than that of multiculturalism. His rhetoric is that of the (demographic) security of the Jewish state. As Tudor Parfitt has put it, in 1993 residents of Gush Katif asked Avichail "to supply Bnei Menashe laborers to replace the Palestinian workers, who were increasingly considered to be a security risk." For settlers the Bnei Menashe "were a god-send—front line troops for Israel's demographic war with the Palestinians" (Parfitt and Trevisan Semi 2002, 160).

Kulanu's (postcolonial) multicultural politics has come full circle here with the colonial project of settlement. When asked about their cooperation with the settler movement, an activist said in an interview, "I have a realistic view of what the Rabbi [Avichail] did. [The West Bank] is the easiest place to settle because it costs much less money. For the Jewish agency to resettle one Jew, it costs $100,000. In the settlements, it costs pennies." Moreover, he argued, the "only place you can put people to work the next day is the territories. People are very authentic. It's a natural place. I don't necessarily support the territories, but I would have to be insane to criticize his selection." At any rate, you "can't allow anyone to mouth off—either supporters or critics of the territories. We would lose focus. It's petty squabbling, one of the big problems with American Jewry. The big responsibility is to be supportive in general and not to micromanage decision-making processes." In sum, the politics of settlement is sidelined (as "petty squabbling") and nonwhite Jews become the site for discussions of Jewish racism, which is viewed as an entirely internal Jewish problem. The question of Palestine, the realities of a colonial present, and its very violent forms of racism in a state structured around the distinction between Jew and non-Jew, subject and citizen, and movement and enclosure are displaced.[36] As it is articulated within the terms of a Jewish identity politics, a self-declared anti-racist, multicultural, and humanist political stance takes a very particularistic form: helping one's own at the direct expense of those not within one's expanding—and no longer exclusively "white"—Jewish world.

The Return to Jewishness

In an incisive article, Steven Kaplan considers the persistent racial underpinnings of modern Jewish identity. Examining the case of Ethiopian Jewry, he argues that arguments about whether or not they are Jews point to the ways in which Jews do indeed consider themselves a racial group. "The physical, genetic, and historical characteristics which appear to separate Ethiopian Jews from other Jews, and are the subject of so much attention, are only a problem if one operates under the assumption that these (traditional racial) markers are shared by other Jewish groups." In sum, "Ethiopian Jews are . . . the topic of so much interest and discussion precisely because they challenge existing 'racial' ideas of Who is a Jew" (Kaplan 2003, 80).

It was not just in the late nineteenth and early twentieth centuries that scholars arguing about the Jewishness of Ethiopian Jews identified phenotypic characteristics said to distinguish them from other Ethiopians. "Indeed, the massive Ethiopian aliyah of the 1980s produced in its wake a wealth of material claiming to distinguish Jewish Ethiopians from other Ethiopians on the basis of their physical appearance" (ibid., 80). They are not as black as other Ethiopians, for example (83). Descent, another constitutive component of racial thought, Kaplan argues, also dominates discussions of the Jewishness of Ethiopian Jews. In 1973, Rabbi Ovadia Yosef, then Israel's chief Sephardic rabbi, declared Ethiopian Jews to be "Children of the Tribe of Dan" (Ben-Dor Benite 2009, 224). Scholars who have documented that Ethiopian Jews are descendants of fourteenth- to sixteenth-century converts and not of the biblical patriarchs have been accused of undermining Ethiopian Jewish claims to Jewishness. Moreover, recent genetic studies which reported that Ethiopian-Jewish Y-chromosome modal haplotypes fall outside of the general Jewish Y-chromosome pool have raised political fears: Israeli scientists were reluctant to publish the results, a "recognition of the reality that membership in the Jewish people is commonly . . . believed to be based upon racial-genetic ties" (87).

Kaplan makes an important point. Contrary to a rhetoric that writes and speaks of the consequences of genetic historical studies as "antiracist," this genealogical science does not unmake race or racial thought. Ancestry—that which was believed to be visible in particular phenotypic properties—was central to (Jewish) racial thought. And ancestry as a scientifically knowable fact is getting a new lease on life here: biological evidence is being used to evaluate the credibility of oral traditions and beliefs about descent from Jewish ancestors. At the same time, however, given the evidentiary terrain of anthropological genetics, this science allows for the separation of phenotype from descent in the biological imagination, thus allowing activist groups (and some scholars) to imagine that this cannot be race all over again. Racial groups, after all, are getting all mixed up here. Phylogenetics is allowing us to establish kinship across what were previously considered impassable racial divides. By going "beneath" racial phenotype, Y-chromosome research has revealed a kinship between (known) Jews and (particular) Africans, between whites and blacks. I once asked a researcher in an anthropological genetics lab working on African American origins whether it really was possible for the police to use Y-chromosome or mtDNA testing to determine the race of a suspect. He answered, of course. Certain Y-chromosomes have a distinctly African origin, for example. He then paused and said: "But that doesn't mean the suspect will actually look black."

In this scientific epistemology, the *phenotypic* differences of Ethiopian Jews do not render them non-Jewish, not according to a biology based upon the search for shared origins and continuous descent that uses noncoding markers, i.e., genetic markers with no (known) phenotypic effects. At the same time, however, in this scientific epistemology—and not just in contemporary social and political imaginations—shared descent is not enough to make one a Jew. There is no straightforward causal or constitutive relationship here. Scientists merely reveal mathematical distributions of noncoding markers. Being Jewish, as a community or as an individual, scientists and journalists argue, is a matter of "religion"—that which they represent as a matter of belief or choice. For these (nonwhite) groups or individuals to be recognized as Jews, they need to demonstrate their willingness to convert, of their "sincere" desire to practice Judaism. But conversion, choosing Jewishness, becomes possible—it becomes socially and politically felicitous—only in light of the facts that genetic history provides or not. For the Lemba, in the aftermath of genetic historical studies, converting emerges as a choice *to return*. And the question of return matters not just for activists with some mystical or religious desire to find lost tribes and lost Jews—to "get one back," to return to the Kulanu member's words quoted above. It matters for the possibility of recognition by the Israeli state. As Kaplan points out with respect to the Falash Mura, a group of Ethiopian self-defined Jews who want to immigrate to Israel but whose ancestors had converted to Christianity, descent is a key legal criterion determining their "right of return." In the words of one Israeli government decision, only "members of the Ethiopian community who are defined as the 'seed of Israel' and have *returned* to a Jewish way of life—are entitled to make aliyah" (Kaplan 2003, 89, emphasis added). Individual Falash Mura must prove either that they had familial ties to Ethiopian Jews already in Israel or that their family was historically Jewish. In the words of one informant, "Peoplehood trumps theology" (interview).[37] As reported in *HaAretz* in 2005, Lemba individuals are converting to Judaism and making aliyah.[38]

If looking at groups who challenge the taken-for-granted assumptions about who is a Jew helps to reveal the persistent and yet complex power or meaning of genealogical-biological definitions of Jewishness first articulated in the late nineteenth century, so too does such a view from the margins lend insights into the centrality of the State of Israel for the Jewish diaspora's self-definition. Kulanu's primary commitment is to Jewish life *in the diaspora*—to discovering it, recognizing it, and facilitating it. As stated in Kulanu's mission statement, Kulanu will support "aliyah" when and if communities ask Kulanu to help them (generally financially) to immigrate to Israel. But Jack Zeller notes, to date most of Jewish life has happened in the diaspora and that is likely to be the case for many generations to come (Primack 1998).

While committed to life in the diaspora, however, the State of Israel looms large on the horizon. For the kinds of political projects in which Kulanu activists engage, a complete "return" to Judaism for individuals and communities with presumably "Jewish roots" presupposes the imprimatur of the Israeli state—as something that *could be* attained if one so desired. From Kulanu's perspective, once someone *can make* aliyah—even if he or she chooses not to—then he or she has been recognized by "the Jewish world."

Kulanu's commitment to the diaspora is perhaps best described as a diasporic Zionism. It is a politics that seeks to sustain and nourish the diaspora that is caught in the long shadow of the Israeli state, the only authority that can finally determine who is a Jew, even as that resolution is infinitely deferred. As Kaplan points out, the establishment of the Jewish state "transformed the 'Who is a Jew?' question from a theoretical or existential question to a practical issue with broad political implications" (2003, 89). Today, the theoretical or existential question of "Who is a Jew?" can never be severed fully from the practical, political concerns regarding citizenship in the Jewish state, not even for those living in and committed to sustaining the diaspora. It is no accident that the various streams of American Judaism—Orthodox, Conservative, Reform, and Reconstructionist—take their battles over "Who is a Jew?" to the Israeli state, lobbying the Ministry of Interior or the Ministry of Religion to recognize one criterion or another (Brackman 1999; Rabinowitz 1997). As summed up by one Jewish genealogist not interested in genetic ancestry testing, "Why should I care about it? I wouldn't have any trouble getting a rabbi in Israel to recognize me as a Jew" (interview). The diasporic Jewish self has been fundamentally reconfigured by Israel—by its nation-statist logic, and by its political, cultural, and religious authority (and authorities). And the genetic historical archive may well emerge as a powerful evidentiary (back)-ground for both existential *and* practical resolutions to the question of membership for those who wish to "return."

CHAPTER SIX

The Things We Carry[1]

History through the Molecular Optic

"What is DNA, this molecule that allows us to travel so far back into the past—this history book we carry around like a gift from a long line of ancestors," asks Spencer Wells, director of National Geographic and IBM's Genographic Project (2006, 15)? The belief that DNA is a historical document is not new. As I demonstrated in chapter 1, that premise drove the work of Emile Zuckerkandl and Linus Pauling several decades ago. Nevertheless, there is a renewed emphasis on—and a novel prominence, both scientific and public, to—the claim that history is to be found in our genes. As David Goldstein puts it, "genetics is slowly earning a place in the historical sciences" (2008, 3).

With the birth of molecular anthropology in the 1960s, information became "a metaphor for [the] 'historical record'" (Diaz 2007, 650). By the mid-1970s many of the assumptions and commitments held by the initial architects of molecular anthropology had been integrated into the study of human origins and prehistory. More recently, anthropological genetics has ridden upon the coattails of the ongoing work and technological innovations of the Human Genome Project and post-genomics to make those assumptions ever more true: the idea that "the anthropological gene and genome . . . [are] documents of human history" (Sommer 2008, 474) is taken for granted by scientific practitioners and increasingly by a variety of publics.

That genes and genomes are widely taken to be documents of human history, however, tells us only so much. What is "history" understood to be? How do scientists find cultural and historical significance in DNA? Moreover, why and in what ways might a distinctly genetic understanding of history matter? What difference does it make to how individuals and collectives understand who they are?

I have argued that noncoding regions of the Y-chromosome and the control regions of the mitochondrial genome presuppose different understandings of biology and of biological causality than did race science. I have argued as well that claims about population origins and phylogenies generated on their evidentiary ground presuppose different understandings of the relationships among human agency, culture, and nature than did the phenotypic signs upon which race science was built. In this concluding chapter, I extend my arguments about the epistemological politics of contemporary studies in genetic history by considering the implications of treating DNA as "a history book" for our understandings of both "history" and of its relationship to the self. More specifically, I consider the kinds of data upon which genetic history relies, the organization of this data into narrative accounts, and the genre of history that is made. Based upon my readings of textbooks, scientific papers, and popular accounts of this rapidly growing scientific field, I ask, what are the "styles of reasoning" (Hacking 2002, 178) through which genetic data is transposed into a historical narrative? And what are the evidentiary and interpretive problems that arise?

I begin with a discussion of the epistemological assumptions that drive anthropological genetics, that is, deep historical claims about human geographic origins and migrations, often referred to as human prehistory. I analyze the work of anthropological genetics in relation to the conceptual models it draws upon in the broader field of evolutionary biology, which examines the origins and development of biological species, human and nonhuman. That discussion sets the stage for analyzing the evidentiary problems faced by more temporally recent and population-specific projects in genetic history. Reconstructing the "histories" of a contemporary sociocultural group on the basis of conceptual models drawn from evolutionary biology has epistemological consequences, not the least of which is that it generates a gap, I argue, between what genetic evidence might actually be able to demonstrate and the historical claims and narratives presumably built upon its basis.

If built upon the evidentiary assumptions and methods of evolutionary biology and, of course, of contemporary genomics, genetic history derives its cultural and political significance from a broader discursive field: a distinctly modern sensibility in which we have come to understand who we are, as individuals and as collectives, in terms of our pasts. The disciplines of biology and history emerged in the late eighteenth and early nineteenth centuries as fields that reflected and produced a whole constellation of ideas about origins, continuity, and truth. Moreover, both of these disciplines have made assumptions and confirmed a set of ideas about interiority, authenticity, identity, and selfhood that endure today. I sketch that constellation of ideas and commitments by giving a brief account of the emergence of a discursive field in which the historical profession, theories of childhood, and the importance of the past and of (repressed) memory to knowing the self converged in the mid-to-late nineteenth century. Genetic history is the latest instantiation of a perduring belief in both the importance and knowability of the past: that fundamental aspects of who one is are determined by one's past and that the past can be reconstructed and known on the basis of the remainders it has left behind—documents, artifacts, psychic memories, and most recently, genetic mutations. At the same time, the historical sensibilities and commitments that came into being in the nineteenth century have shifted in important ways. In the discourse of genetic history the question of historical determination is constantly qualified and conditioned by invoking notions of culture, human agency, and individual choice.

Contemporary projects in genetic history draw their authority as much from the norms of the historical profession and the assumptions of a modern historical sensibility, first articulated well over a century ago, as from the power and epistemological credibility of contemporary genomics. At the same time, however, the specific evidentiary nature of genetic history fashions a distinctive kind of history and historicity for the contemporary self: there is a particular immediacy, intimacy even, built into a historical practice in which each of us is believed to carry within a scientifically legible archive of our past. To quote David Goldstein, *Jacob's Legacy* "describes a few specific aspects of Jewish history that have been studied using the genetic legacies carried *by living people* who call themselves Jews" (2008, 3, emphasis added).

In emphasizing and specifying the nature of anthropological genetics as a historical science, I conclude this book by highlighting a fundamental continuity between race science and anthropological genetics often sidelined by all the talk of biology that dominates meta-critiques of contemporary studies of human biological diversity. Where race science and anthropological genetics converge most powerfully is in their shared presumption regarding the importance of "history": who we are collectively and individually is given by and legible in biological data which testifies to our origins, to our pasts, to who we really are. Given the proliferating scholarship on biological citizenship and the shift that medical genomics is bringing about in understandings and practices of the self (e.g. Rose 2007; Petryna 2009; Heath et al. 2004; Rabinow 1996a), it is worth keeping in mind that, *from the perspective of a different genomic science*, what might easily be glossed as yet another instance of "biological citizenship" (Petryna 2002; Rose 2007) is perhaps more aptly described as a rearticulation of the historical assumptions that underwrote race science. For race science, history was driven by a teleological logic, that is, by the biological character and destiny of different and unequal races. For anthropological genetics—and more specifically, for genetic history—biology is a sign of human "history," that is, of the origins and phylogenies of human *cultures*, understood to be distinct and (at least at one time) endogamous population groups, and of their enduring legacy and significance for selfhood today. The genetic historical self is an individual who is the beneficiary of prior choices made by ancestors of whom she carries the trace in her body, of whom she is made up. If that authentic self is to endure, she must repeat the choices that her ancestors made over and over again. There is no destiny, no teleology in this historical imagination. There is only the risk of loss.

History in Molecules

With a flourish characteristic of his bestselling books, Bryan Sykes, Professor of Genetics at Oxford University and founder of Oxford Genetics, one of the first genetic genealogical-testing companies to enter the market, explains the way that neutral mutations work:

> [They] just get passed from one generation to the next, their fate entirely out of their hands. They risk elimination if they end up in someone who has no children or can do well if they find themselves in a large family. They might lead less dramatic lives than the mutations that bring success or devastation. But it is these, the silent passengers of evolution, that are its most articulate chroniclers. (2006, 96)

But what is it that those silent passengers of evolution are chronicling? According to Spencer Wells, those silent passengers, that is, junk DNA, are "our text, and [they] provide us with the story of our ancestors" (Wells 2006, 15).

Our past, our ancestors: anthropological genetics is a "discourse of the continuous" (Foucault 1972, 12). It presupposes and generates an understanding of the past as continuous with and a precursor to *our* present, to *our* existence. "When did *our* ancestors live? When did humans first inhabit different parts of the world?" (Relethford 2001, ix, emphasis added).

While other forms of evidence (fossils, for example) have allowed evolutionary biologists and biological anthropologists to study proto-human lineages and other species now extinct, for a long time that was not a possibility in anthropological genetics. In the 1960s, Emile Zuckerkandl, one of molecular anthropology's initial architects, understood the inability to study or reconstruct extinct species to be a serious drawback for the molecular approach to evolution (Zuckerkandl 1963, 258).[2] By the late 1980s, however, the ability of molecular evidence to lend insight into *ourselves* was argued to be one of anthropological genetics' primary virtues: genetics provides us with evidence of "real ancestors," in Rebecca Cann's words (1988, 127). As Bryan Sykes has written, the discovery of a shin bone from a quarry on the Sussex coast or of a tooth from a cave in north Wales—both "over a quarter of a million years old and both the remains . . . of much sturdier, large boned humans, more like Neanderthals than our own species"—is interesting. But those finds cannot compete with the knowledge we gain from studying our own genomes: "Fascinating though these finds are, they are merely glimpses into the world of long-extinct humans who came and went but left no lasting impression on the Isles, small bands of roving hunters whose luck finally ran out. These were not our ancestors" (2006, 13).

Evolutionary theory began with the question of origins and speciation. As the prominent evolutionary biologist Ernst Mayr has explained, "The main concern of evolutionists in the period 1859 to about 1895 was the proof of evolution and the establishment of the various lines of common descent. Phylogenetic research (the study of the evolution of a given species) was a major preoccupation of evolutionists" (1982, 572).[3] In his *On the Origins of Species* Darwin proposed "the theory of common descent" and sought to explain the mechanisms of evolutionary development and change (ibid., 582), which he based upon a tripartite process: individual variation, selective advantage, and the shifting genetic composition of any given population over time (Marks 2011, 42). Darwin described the evolutionary process as one of descent with modification. And if descent with modification is the still enduring theory, phyletic trees have long been evolutionary biology's representational form: "Ever since Darwin," writes Luigi Luca Cavalli-Sforza, "we have thought of evolution in terms of trees that trace the relationships among species and their ancestors" (2000, 36; see Barnes and Dupré 2008, chap. 4). Since humans are a single species, our descent from a single human origin to multiple and diverging populations is not a "dendritic" process: populations do not diverge once and for all.

They "sometimes branch and sometimes reunite" (Marks 2011, 50). Nevertheless, the story of human origins (unity) and migrations (splitting) is represented on phylogenetic trees.

In evolutionary biology, the process of analyzing origins and mapping phylogenies is a heuristic used to gain insight into the dynamics of speciation, genetic mutation, and natural selection. Evolutionary biology is a history of *life*. It is an effort to understand the biological processes through which organisms and species emerge, reproduce, adapt, survive, die, and die out.

For its part, anthropological genetics—studies of human prehistory and history—generates evidence of the geographic origins of human beings (an original "unity") and tracks divergence and descent. It is not focused on the dynamics of evolution per se. To borrow Ernst Mayr's characterization of the nineteenth-century writings of George Cuvier and Charles Lyell: "Origins rather than evolution [is] the explanatory concept" (1982, 320). For example, relying on a comparison of archaeological and genetic data, and interested in the relationship between what he identifies as "biological evolution" and "cultural evolution" (the invention and "selection" of certain ideas and practices by human populations over time [Stone and Lurquin 2005, 99–101]), Cavalli-Sforza has sought to understand how farming spread to the European continent during the Neolithic period, approximately 11,000 to 5,000 years ago. Was it a result of "cultural" or of "demic" diffusion? In other words, was it the result of the spread of "ideas and practices . . . without major population movement," or was it "spread through migrations of farmers themselves" (ibid., 87)? For their part, genetic historians ask temporally and geographically more circumscribed questions. They are interested in *population specific* origins and descent. They inquire into the geographic origins and phylogenies of human groups defined along contemporary "cultural" or "ethnic" lines. In the studies I examined in this book, what is the level of kinship and diversity within the contemporary Jewish gene pool? What does that tell us about Jewish origins and about Jewish life in the diaspora? Might the Lemba be descendants of ancient Jews?

In *Strange Dislocations*, Carolyn Steedman argues that over the past century "a change took place in the way that people understood themselves—indeed, came to new understandings of what a self was, and how a self came into being" (1995, 4). Weaving an explication of cell theory in nineteenth-century biology together with an account of the emergence of the concept of childhood, the founding of history as a scientific discipline, and Freud's development of the theory of the unconscious, Steedman presents a nuanced reading of the appearance of "the past" as the locus of identity and subjectivity, both individual and collective. Victorian society, she argues, came to believe that "the core of an individual's psyche and identity was in his or her lost past, her childhood." The metaphor of childhood was "commonly used to express the depth of historicity within individuals." And that historicity came to be understood as fundamental to the self. The modern idea of history and "modern conventions of historical practice" emerged within the same discursive terrain. The historian's desire for the archive, Steedman argues, is "emblematic of a modern way of being in the world . . . expressive of the more general fervor to know and to have the past" (75).

In her rich account of the entanglement of discourses of biology, psychoanalysis, and history, Steedman provides a genealogy of the desire to know and to have a past, a widespread cultural and political sensibility in the contemporary world (76). As indicated most forcefully by the emergence of identity politics in the late twentieth century, there is a widespread presumption in the Euro-Atlantic world that our true selves are to be found in our past—as individuals, as groups. Moreover, as Ian Hacking has argued, still heirs to a discourse that came into being in the mid-to-late nineteenth century, we take for granted that that there is a past—a depth—to be had (Hacking 1995). And that "depth" is configured as an interiority of the self from which "a wholeness . . . will figure itself forth, from inside to outside" (Steedman 1995, 15).

Many historians, philosophers, and cultural critics have laid out the relationship between interiority and the modern subject. According to Charles Taylor, one of the three major facets of modern identity is its "inwardness, a sense of ourselves as beings with inner depths, and the connected notion that we are 'selves'" (1989, x). In those inner depths, Taylor argues, is lodged our "authenticity": a moral voice that directs us not towards some external "source"—God or the Idea of the Good—but to an "inner moral source" (x). With the birth of genetic history and genetic ancestry testing, we are witnessing the emergence of a different kind of talk about the self, albeit one still true to key premises of the modern subject that Taylor has described. We are witnessing the emergence of a new kind of "source within": the genome as an empirical and legible record of our authentic, cultural, and historical selves.

What most interests me in Steedman's argument is not the general story of a trajectory towards inwardness or even towards history and memory that so many others have told (see Schorske 1998; Bann 1984; Hacking 1995; Sherman 1999). I am interested in her account of the ways in which physiological theories of interiority and essence have shaped understandings of the modern subject. And I am interested in that account because there are strong resonances between nineteenth-century physiological theories of the cell and contemporary anthropological genetic theories of and rhetoric about the genome as a historical record.

Nineteenth-century theories of the cell, Steedman argues, gave "figurative shape to the body's 'minute components'" (1995, 55). Between 1840 and 1870, physiologists transmitted their theories to a more general audience and "provided an image, of the smallest place within: the fundamental unit of life" (59). In the face of Darwin's theory of evolution, cell theory provided an alternative vision of the nature of life (93). Much historical scholarship has assumed that in the aftermath of the publication of the *On the Origin of Species*, "the public needed to understand and assimilate . . . the great order of consanguinity"(82). But even more radical and transformative, Gillian Beer argues, was that evolutionary theory "emphasized extinction and annihilation equally with transformation—and this was one of its most disturbing elements" (2000, 12). Darwin had suggested "irretrievable loss" (Steedman 1995, 93). By way of contrast, cell theory envisioned "continuance, survival, the essential self transposed but not obliterated by transformation." Physiological cell theory operated with a temporality in which the fundamental unit of life, the essential self—embodied in the cell—changed but could not become extinct (ibid., 92). "The cell was the final place, the thing that simply could not be dispersed." It was "the entity and the place that was 'not capable of loss of existence'" (60).[4]

In the discourse of early twenty-first century anthropological genetics, we hear powerful echoes of the language and logic of nineteenth-century cell theory. In this turn of the millennium version, the genetic polymorphism is the "tiny thing" that has not been dispersed. And that tiny thing is a witness to events long past. To quote from the introduction to a textbook on genetics and the search for human origins: "This book is about the search for human history not in documents or records, or even in ancient archaeological remains, but in the genetic material that we all carry" (Relethford 2003, x). Genetic traces are passed down from one generation to the next. We each carry "pieces of history" within (ibid.). If, as Steedman argues, "'history' in its conventional modern meaning suggests that by a painstaking dredging through the detritus left behind it might be possible to conjure the past before our eyes" (1995, 12), in the practices of genetic history, it is biological detritus—junk DNA and the noncoding regions of the mitochondrial genome—that are believed to be the (most transparent and powerful) historical traces through which we can conjure the past before our eyes. And insofar as that biological detritus, as I have demonstrated in chapter 4, can reveal knowledge about our pasts that we did not already know, it can have the power to adjudicate between and rearticulate the truth claims made by other forms of historical knowledge (oral traditions, family lore). And it can displace other ways of being. For example, a Lemba who believes him- or herself to have Judaic roots but who is, nevertheless, a practicing Christian, may, in the aftermath of genetic studies that brought the Lemba community to the attention of the Jewish world, follow the encouragement of Kulanu activists to convert to Judaism and become a practicing Jew.

Even as it functions as a historical archive, however, the genome is not like the institutions—national, municipal, and church archives, for example—to which historians go. The genome does not house "selected and consciously chosen documentation from the past." Nor does it house "mad fragmentations [notes, lists, letters] that no one intended to preserve and just ended up there" (Steedman 2001, 68). How then are we to understand the genome as a historical archive? What practical and conceptual differences might it make to our understandings of and identifications with history—with "the past"—that this is genetic and not documentary or artifactual data?

Many scholars have argued that, as sciences that seek to "read" genetic information, genetics, genomics, and post-genomics are best understood as information sciences. Since the 1960s, genetics has been configured as a practice that involves "reading" the "Book of Life." And genomics and post-genomics are increasingly understood as involving its rewriting (Kay 2000). Given the language of "code" and of the "Book of Life," alongside the computational technologies involved in processing and reading genomic information, it is easy to see how scholars have come to argue that the computer database is the genome's "natural habitat" (Haraway 1997, 83). The natural habitat of the anthropological genome, however, remains the body, the *individual* human body that is aggregated with other presumably similar individual bodies to produce a population. And the power of that literalized and empiricized interiority—of those genetic polymorphisms that never go away—is evident in the range of practices I have discussed in this book, from scientific studies of population origins and relatedness (often framed as studies of the self) to genetic ancestry-testing companies and the practices of self-fashioning proliferating on the basis of test results. Those little things within confirm or reveal truths about our past(s) and about ourselves that no other evidence has been able to either demonstrate as robustly or to reveal at all—such is genetic history's epistemological conceit.

If genetic history has generated the possibility of "an identity through the processes of historical identification" (Steedman 2001, 77), its practices of identification involve relating to a history *within*—to "those fragments of history that have been passed down [in the case of mtDNA] in the *bodies* of women" (Sykes 2006, 117, emphasis added).

Traces of the past are literally, empirically, a part of us. They are found in those little, embodied things that have become legible to the scientific gaze. And as I illustrated in chapter 4, that fact is fashioning a novel immediacy and an intimacy for practices of historical identification occurring on anthropological genetic terrain. Consumers of genetic ancestry tests are not identifying with texts, artifacts, or images, for example, objects that their ancestors may have made but which remain external to them. The past—our past—is to be found within (each of) us: "Every one of us is carrying his or her personal history book around inside us—we simply need to learn how to read it" (Wells 2002, xvi). The power of that personal history book—the power of *self*-discovery—is evident in the ways in which individuals search for and come to embrace that which they find out they or others have always already been (see chapters 4 and 5).

This self who carries his or her history book around inside is a self born of the genomic age: In the genetic historical imagination, we do not find evidence of ourselves in psychic memories, long repressed in our unconscious, as we did for Freud and psychoanalysis (see Hacking 1995). Nor do we discover our authentic selves by listening to an "inner voice" that orients us in a moral space, as Charles Taylor characterizes the distinctively "modern" self (1989, 13). We discover ourselves in genetic polymorphisms read by genome sequencing machines and interpreted by scientific experts as historical documents. And each of us carries presumably distinctive and distinguishing genetic polymorphisms within. Given the increasing technological prowess of genomics, the scientific and commercial promise is that each of our genomes will become increasingly knowable and both the risks that we face and our most fundamental selves will become ever more legible. From the perspective of medical genomics, consumers are being lured into purchasing genetic tests that provide information on disease risk and drug response in the promise of better medical knowledge to come (Fortun 2008; see Rose 2007; Sunder Rajan 2005).[5] For their part, genetic genealogists believe it essential to collect genetic material now before relatives die on the promise that techniques will improve and become less expensive, genetic ancestry databases will expand, and they or their descendants will be able to learn something more about themselves (see chapter 4). To quote Nicholas Wade: "Geneticists may in future be able to trace back human lineages or pedigrees to all times and places, providing a genetic framework for exploring almost every historical period. Meanwhile a promising start has been made" (2006, 234).

But what has this promising start told us thus far? What kinds of substantive historical claims, what kinds of facts, and what kinds of narratives are being built on genomic evidence? And how?

From Genealogy to Genetic History

John Relethford sums up the search for human origins by asking, "What is our place in nature? What can genetics tell us about the pattern and degree of this relationship? Who are our closest living relatives, and how long have we been on a separate evolutionary path from them" (2003, 17)? Genetic historians ask an analogous set of questions about population-specific origins: What is the "pattern and degree" of the "relationship" of one human population to another? What other populations are our "closest living relatives"? How long ago did populations migrate and thereby "split"? "History" in this imagination is a biological-descent relationship tracked along particular (and until recently, almost exclusively) lineal molecular lines. History is understood to be genealogy, a historically deeper version of efforts to reconstruct family trees. As I show, however, the move from family genealogy to genetic history is not so straightforward. The nature of the evidence renders what it is we can know about familial versus population histories and how it is that we produce narrative accounts of those "pasts" radically different.

Spencer Wells explicates the work of anthropological genetics in the following terms:

> In a sense, I met the genealogists halfway. People constructing a family tree are typically investigating events from the past few centuries, while population genetics starts there and pushes further into the past. Most of us have a sense of our family history, but eventually we all hit a brick wall. Our DNA breaks through that wall, providing a unifying path that leads from the present into the realm of deep ancestry. (2006, 13)

In a similar vein, John Relethford tells the story of his journey into anthropological genetics. He remembers reading Alex Haley's *Roots*:

> I found something very appealing about the idea of finding one's 'roots' and extending family history into the past. . . . Many of you may have similar

> interests in family genealogy. It is natural to be curious about where you come from. I also find it fascinating to consider that every person we know or meet also has a past and that we all have connections at points in the past, either recent or ancient. . . . My specific interest in my family history grew into a broader interest in history and ancestry in general. (2003, 1–2)

The move into "deep ancestry" (Wells 2006) is not as seamless as such narratives suggest, however. The kinds of questions researchers ask and answer change significantly the more remote the past becomes. "Think about your own roots," Relethford has written. "How much information do you have about your ancestors?" Your parents:

> In many cases, you will have a fair amount of information on them, perhaps where and when they were born, where they went to school, and the names of members in their immediate family. Most likely, you also have a lot of other information based on your life with them, including knowledge of their hobbies, favorite foods, and their dreams and fears. How about your grandparents?

What happens when Relethford's account moves back further in time?

> Although I am fortunate to have inherited family genealogical data, I don't know the names of most of my ancestors beyond a few generations. I am luckier than most since my mother's family did not move around much over the past few centuries. Even given this information, the percentage of my known ancestors decreases significantly with each generation. After seven generations, I know less than 5 percent of my ancestor's names, let alone anything else about them. . . . By eleven generations in the past, I have very little information; I know the names of only 2 out of a maximum of 2,048 ancestors, which is less than 1.0 percent. (2003, 5)

Relethford then points out: "There is a definite limit to our written records. No one knows the name of even one ancestor living 10,000 years ago. If we consider the limits on our direct genealogical information in terms of the long span of human history (in the broad sense), it becomes apparent that we know very, very little about our ancestors using a traditional genealogical approach" (7).

But note the shift in the kinds of questions asked and the kinds things we might come to know as Relethford moves from documentary and oral evidence to genetic evidence, in other words, as he moves back in time. He asks his readers to consider the following:

> *Where* do your ancestors come from? Again, we seldom have specific information for more than a few generations. . . . Questions of ancestry, both for individuals and for populations can be answered using historical data, but obviously for only limited periods of time. . . . Over long periods of time . . . this information gets lost, for one reason or the other. Genetics provides a way of uncovering some of this information. Past events—*from migration of people from one group to another to changes in population size*—may have left a record behind in our genes. Our written and oral histories are incomplete and lack much time depth, but we carry a genetic signature of past events. In this sense, the study of genetics in living people can provide clues to human history. (2003, 7, emphasis added)

There is an important difference between the kinds of questions Relethford asks about his family genealogy and the kinds of questions he asks regarding genetic history. Familial genealogies are not just matters of where someone came from. (And "changes in population size" figure nowhere in such quests.) Family genealogists seek to know information about hobbies, dreams, and fears, and at the very least, names. But we cannot learn anything so specific from genetic historical data.

In *Saxons, Vikings, and Celts: The Genetic Roots of Britain and Ireland*, Bryan Sykes takes the radical position that genetic evidence is "entirely independent of these other sources [material artifacts, written documents, human remains]. It does not rely on them" (2006, 2). He insists genetic evidence can generate knowledge heretofore unknown. But what kind of knowledge or historical claims can genetic evidence—*on its own*—actually produce? "Since every ancestor was an individual," Sykes explains, "I was determined to treat the DNA sequences as individuals. Each one had, at some time, set off from some distant land and stepped ashore on the Isles, soaked with salt spray and red-faced from the cold" (112). Building his historical narrative on the basis of one such "individual," Sykes writes "Through this girl, or her descendants, this new sequence left the island of Rona and found a home on nearby Skye, where it still remains. From there, perhaps one of the daughters in the next generation went to live on Lewis while another traveled down to Glasgow. I cannot tell exactly *when* this happened, but the journeys have been recorded by the genes of the descendants" (114–15, emphasis added). What is evident from reading his book, however, is that

Sykes cannot actually tell exactly *what* happened either. It is not just a problem of "when." It is a problem of what and, moreover, it is a problem of why. Providing our "ancestors" with names and treating genetic haplotypes as the story of individuals, Sykes weaves a fascinating tale. But his narrative does not derive from the genetic data (see Sykes 2006, 2001). We do not know if it was one individual or a group of individuals who left the Island of Rona. We do not know if these presumed individuals moved from Rona to Skye, or whether they stopped somewhere en route and their descendants in those other locations did not survive to pass the lineage on. We do not know who traveled to either Lewis or Glasgow, a daughter, a granddaughter, a great-granddaughter, a group of daughters. And we do not know why. For that matter, we cannot be certain that anyone traveled at all: perhaps it was a matter of convergence, a mutation that happened more than once in different locales and at different times.

Genetics may help us discover patterns of human and population geographic origins and migrations (although one has to keep in mind that since individual genetic loci may evolve differently than has the population as a whole, a gene tree may not correspond to a population tree). But out of those statistical patterns, specific and detailed historical narratives are being built. As Nicholas Wade recounts, the *historical* as distinct from the evolutionary questions to be answered by genetic historians have to do with "the fragmentation of the once united human family into different races and warring cultures" (2006, 11). At the broadest level, genetic historians are asking, where did specific contemporary population groups originate? To where did they migrate? From whom do contemporary populations, say, the contemporary inhabitants of the British Isles, descend? According to Wade, genetically exploring "the historical past"—as distinct from human prehistory—can yield "greater returns because it can be related to known people or events. DNA can be used to analyze populations, saying who came from where, which helps understand mixtures of people. . . . It faithfully records who slept with whom throughout the ages, a matter of historical interest in cases like the secret family of Thomas Jefferson. And with populations that have married within themselves for centuries, like those of Jews, DNA can reach back to the time of the patriarchs" (234).

History writing has been about much more than origins and migration events, however. It certainly has been about much more than who slept with whom, who was endogamous and who was not. As Foucault has written of the historical profession, ever since the founding of history as a scientific discipline in the nineteenth century, "documents have been used, questioned, and have given rise to questions; scholars have asked not only what these documents meant, but also whether they were telling the truth, and by what right they could claim to be doing so, whether they were sincere or deliberately misleading." In all of this, "the document was always treated as the language of a voice" (1972, 7). Foucault, as is well known, developed his archaeological method as an alternative to and a critique of the epistemological assumptions and methodological practices of the historical profession. He emphasized the importance of identifying discontinuities, ruptures, thresholds, and transformation rather than continuities, and he read archival documents in terms of their "rules of formation," that is, in terms of what kinds of statements could count as candidates for truth or falsity, rather than in search of hidden meanings or in order to unearth a lost voice. Nevertheless, as Foucault among many others has noted, retrieving voice from historical documents emerged as a key fantasy of the historical profession in the nineteenth century, and it has not disappeared today: as Jules Michelet explained in an 1869 preface to his *L'histoire de France*, "he had wandered in 'those solitary galleries' of the Archives 'in . . . deep silence,' until with Vico there had come to him 'the whispers of the souls who had suffered so long ago and who were smothered now in the past'" (quoted in Steedman 2001, 70). If making "ink and parchment" speak is the traditional historian's dream (70), the dream of anthropological genetics is to make the genome speak.

In recounting the potential contributions of "a field now often called genetic history," David Goldstein argues that it will never be able to "replace, or even really compete with, the painstaking work of archaeology, philology, linguistics, paleobotany, and the many other disciplines that have helped resurrect some of the lost stories of human history." Nevertheless, "our narratives describing the histories of peoples and events, from Aryan invaders of India to the Viking attacks on the British Isles, are all being augmented and refined by genetic analysis." Goldstein continues: "whereas the real grandeur and detail of human history can only be seen in the context of our archaeological and written legacies, and of course, our memories . . . at times genetic history stretches the boundaries of its scientific formalism and hints at answers to bigger questions: What makes a people a people? What binds them together through time? What alienated them from some and aligns them to others" (2008, 3–4)?

But on what basis can one answer such questions? How does one derive meaning, truth, intention, voice, or consciousness—what makes a people a people, what binds them together through time—from DNA? More generally, how does one extract narratives from the anthropological genome?

Narrating DNA

Marianne Sommer and Edna Suarez Diaz have argued that advocates for a molecular approach to evolution had more than disciplinary visions and interests in mind. They also made explicit political claims. Molecular evidence was argued to be less vulnerable to subjective judgment. That was a powerful claim in a field long mired in debates about the relationship of human beings to primates and about intrahuman phylogeny. As I discussed in chapter 1, that claim to objectivity rested in part on the discontinuous nature of protein sequences. It also rested on the mathematical and technical character of molecular analysis. Advocates for a molecular approach represented it as relying upon "rigorous mathematical logics and technologically-driven quantitative approaches" (Sommer 2008, 479). Central to molecular anthropology's mathematical logic were the eventually twinned notions of the molecular clock (the rate at which random mutations occur) and neutral evolution (selectively neutral evolutionary change): "proteins, like organisms," it was assumed, "evolve, and [therefore] two groups of organisms that at one point shared a common ancestor would at that point also have shared a molecular makeup. After separation, an ancestral molecule would have evolved independently in the two lines of descent while maintaining structural homologies" (ibid., 480). As Zuckerkandl and Pauling first proposed, by tracking *protein evolution* instead of shifts in morphological traits evident in fossil records, one should be able to "see" ancestral relationships between two distinct species, even if those ancestral relationships were not visible in phenotypic forms. One should be able to see descent with modification (in this instance, *neutral* modification), to return to Darwin's phrase.

Ever since Motoo Kimura articulated the relationship between neutral evolution and the molecular clock in 1968 (Kimura 1968), there have been various problems that have troubled the "rate" question: At what rates do proteins/genes mutate? Do different proteins/genes mutate regularly? Do different proteins/genes mutate at different rates? The *rate* of mutations (the molecular *clock*) is a stochastic process and thus difficult, perhaps even impossible, to calculate accurately (see Barnes and Dupré 2008, chap. 4).

In order to generate historical accounts, genetic history—in other words, that subfield of anthropological genetics that is concerned with the *"histories"* of particular contemporary sociocultural groups and specific (known) historical events—must translate mathematical calculations of stochastic genetic occurrences into meaningful and substantive historical narratives. Researchers must read DNA in such a way as to generate a historical interpretation and a story. And they do so primarily by reading stochastic genetic events via already existing historical narratives (oral, documentary, or pseudo-documentary as in the case of the early biblical tales).

In his book *Mapping Human History*, Steve Olson, a science writer who has published in the *Atlantic Monthly*, the *Washington Post Magazine*, *Slate*, and *Science* among other places, explains: "Geneticists are just beginning to read the story written in our DNA. . . . They are learning *how* groups mixed and diverged over time" (2002, 7, emphasis added). Geneticists might be able to learn *that* groups mixed and diverged over time, which is the focus of much anthropological genetic work on early human migrations, "pre-history," that is. However, the *stories*, the *how* and for that matter the why, must come from elsewhere. And as I have demonstrated in earlier chapters with respect to the histories of specific culturally and politically recognizable population groups, even *what* is discovered, what scientists and publics learn about origins, events, populations, genealogies, and practices, does not come from the biological data itself (chapters 2, 3, and 5). Take the example of Genghis Khan and his apparently newly discovered genetic legacy. Under the heading "The Secret Strategy of Genghis Khan," Nicholas Wade explains what a team of geneticists led by Chris Tyler-Smith (at Oxford University) discovered. They analyzed the Y-chromosomes of approximately two thousand men from populations across "the Eurasian land mass." They "noticed" a single cluster for many of the Y-chromosomes. "The striking feature of the cluster was that the owners of its Y chromosomes did not all come from a single population, as would have been expected, but from regions all over Eurasia." Noting that the "master sequence" of this Y-chromosome type was common in Inner Mongolia,

> Tyler-Smith and his colleagues believe that the master sequence chromosome must be that of the Mongol Royal House. As members of the Mongol Royal House, Genghis Khan and his male relatives, who he sent to administer the regions of his vast empire, would have carried that Y-chromosome sequence. Dating methods suggest the cluster started to form around 1,000 years ago, the time that Genghis's dynasty began its ascent to power. (Wade 2006, 235)

According to Wade, Mongol soldiers doubtless raped many women during their cruel and murderous campaigns. But there may be a more significant reason for the existence of so many men carrying the specific chromosome of the Mongol royal house: "Genghis accumulated a large harem in which he seems to have labored with surprising industry. The fourteenth century

Persian historian Rashid ad-Din, who served as chief minister of the Mongol government of Persia, wrote that Genghis Khan had nearly 500 wives and concubines, and that it was his practice to take women into his harem as booty whenever he conquered a new tribe" (236–37). As calculated by Tyler-Smith and his colleagues, Genghis Khan succeeded in fathering "an astonishing 8% of [contemporary] males throughout the former lands of the Mongol Empire" (ibid.).

The Y-chromosome evidence does not generate those historical details. It is not the evidentiary source for the historical narrative that either Tyler-Smith and his colleagues or Nicholas Wade tell. Already available historical facts, known historical figures, and stories are needed if any narrative is to be made from the Y-chromosome evidence. In turn, in a circular logic of discovery, evidence, and proof, the genetic data seems to prove the story of Genghis Khan and his maniacal "interest in procreation."[6] But it is not possible to date the Y-chromosome evidence with any precision to Genghis Khan's time. As far as we know from the historical record, Genghis Khan was born in 1162 and died in 1227. According to the original scientific paper, the 95 percent confidence interval of the time to most recent common ancestor (TMRCA) of the Y-chromosome lineage is 700–1,300 years ago, with other programs for estimating origin going farther afield, 590–1,300 years ago. Given uncertainties regarding the molecular clock and generation time, moreover, even the dating of 590–1,300 years may be way off. Date spreads of six to seven hundred years are not significant for evolutionary questions nor even for questions regarding human prehistory—say, the spread of agriculture across the European continent—but they are significant for historical ones. By their own estimate of thirty years per generation, there are about twenty-three generations in a seven-hundred-year time period. The Y-chromosome sequence could have originated at any moment during that time period. It could have originated before Genghis Khan's time. And it could just as well have originated after his death.

There is no way to tie this Y-chromosome cluster to *Genghis Khan*, a known historical figure who *might* be its "father" but who might just as well not be. It is *possible* that this is Genghis Khan's lineage. But that link remains circumstantial and speculative. Nevertheless, on the basis of historical sources that suggest that Genghis Khan and his male relatives "had many children," and by arguing that the genetic evidence for the origins of this Y-chromosome haplotype corresponds (roughly) to that of the Mongol Empire, researchers have concluded that they have found Genghis Khan's lineage. In addition, they have argued, "Our findings . . . demonstrate a novel form of selection in human populations on the basis of social prestige" (Zerjal et al. 2003, 720). In other words, the social power of Genghis Khan and his relatives allowed them to reproduce and spread their genetic legacy, a fact that scholars in history and anthropology would well have been able to argue without recourse to the biological notion of selection.

That DNA does not tell such stories on its own is not, of course, limited to the case of Genghis Khan. There is no way to tie the Cohen modal haplotype (which now turns out to be four unrelated lineages; Hammer et al. 2009) to the biblical periods or to the biblical figure of Aaron without an already existing historical narrative that there was a figure called Aaron who fathered the priestly line. And that story, along with the whole story of Exodus it is worth noting, is widely believed to be mythical rather than historical in biblical studies and biblical archaeology circles today (see Herzog 1999; Thompson 1992, 1999). The time to most recent common ancestor of the Cohen modal haplotype is estimated to be "2,650 (3,180) years before present [84–130 generations], dating the coalescence of the Cohanim chromosomes to between the Exodus and the destruction of the first Temple in 586 B.C." (Thomas et al. 1998, 139). Even that time span, however, could be way off, given the problems in accurately calculating the molecular clock. As the authors explain, "uncertainty in the mutation rate significantly broadens these intervals (conservatively taking 95 percent confidence intervals on both the distance and the mutation rate leads to an interval of 34–455 generations)" (ibid., 139); that is, when calculated at their estimate of twenty-five years per generation, it leads to an interval of up to 11,375 years. Moreover, even if the initial time estimate were correct, it can hardly be used as independent evidence of the existence of a particular biblical figure.

Similarly, as I argued in chapter 5, it is not possible to discover that the Lemba have ancient Israelite ancestors without a Lemba oral tradition which, first, generated the historical questions that framed the Y-chromosome studies of Lemba men and, second, provided the narrative structure to interpret the genetic evidence collected. Bracketing recent reassessments of the Y-chromosome haplotypes common in the Jewish priesthood (Hammer et al. 2009), let's assume a man—a trader—came to southern Africa and "fathered" approximately 10 percent of the contemporary Lemba male population. What if this man was neither a practicing Jew nor of Hebrew descent but simply carried what has come to be known as the Cohen modal haplotype?

I make the above arguments not to prove that this is bad science. And researchers themselves vary in their own interpretations of the extent to which genetic history adds anything to longstanding forms of historical evidence and modes of historical inquiry, as attested to by the different accounts of genetic evidence and of its knowledge-making capacities given by Sykes and Goldstein that I quoted above. Moreover, there are other sorts of questions that anthropological geneticists ask: in what ways did the size and makeup of a given population shift over time, for example? Was a population's genetic makeup more affected by genetic drift, natural selection, or interbreeding? And there are other forms of evidence such as skeletal and archaeological remains to which anthropological geneticists (mostly, biological anthropologists) turn in order to make sense of genetic data and to ask broad anthropological questions about biology and culture, health and diet, or about power, labor, and violence and its effects on life-span, for example, as has been true of research at the African Burial Ground in New York City. What I am arguing is that even as good science—even as science practiced according to the norms of evidence, experiment, technology, and the ethics of its time—when genetic historians try to contribute to the story of a specific "people's" history or a particular historical figure or event, they are engaged in a category mistake: if the aim is to uncover evidence for particular historical traditions or for specific already "known" historical persons and events, then genetic data is the wrong kind of data. It is out of its historical depth. And that category mistake extends to what seems to be a widespread practice in genetic history: extending the natural scientific principle of parsimony to history itself.

"The principle of parsimony states that the simplest explanation consistent with a data set should be chosen over more complex explanations, and is a guiding tenet in scientific study," Caro-Beth Stewart explains (2000, 48). In the practices of evolutionary biology and anthropological genetics, the principle of parsimony was long relied upon to construct phylogenetic trees: if the goal is to "discover the genealogical relationships between 'taxa' (biological entities such as genes, proteins, individuals, populations, species or higher taxonomic units)," the epistemic logic went, then "by the criterion of parsimony, the 'best' or most parsimonious tree is one that requires the fewest total events" (51). The most parsimonious tree, however, may not always be either accurate or immediately evident, Stewart warns. Taxa that appear to be similar may not share a common ancestor: "who among us would assume that an Elvis Presley look-alike is actually his twin brother" (49).[7] Given the phenomenon of homoplasy—genetic mutations that are the same but do not derive from the same ancestral types—identical haplotypes may have different origins. Many biologists and geneticists have more recently turned to other models because they no longer believe that the principle of parsimony can adequately take into account the more complex processes through which DNA sequences evolve over time. Likelihood or Bayesian analyses, for example, seek to take account of such complexities, incorporating more fully the fact of homoplasy and genetic convergence into the production of phylogenetic trees.

If as these biologists and geneticists maintain, the principle of parsimony is not the best route for understanding and reconstructing the evolution of DNA sequences, it poses far greater problems still for reconstructing population histories. For Spencer Wells, the transition from biological explanation to historical explanation presents no problem. Just extend the principle of parsimony. "Nature usually does favour simplicity over complexity," he asserts. "So, if we accept that when nature changes, it tends to do so via the shortest path from point A to point B, then we have a theory for inferring things about the past" (2002, 23). But that assumption, itself a theoretical commitment regarding how nature works, does not provide us with a reliable or plausible theory for inferring historical events. Even if the most parsimonious tree is the evolutionarily correct tree for that particular DNA sequence or locus, it may not actually generate the right *history* of a given group. There is the technical matter: phylogenetic trees are trees of *haplotype* origins and descent lines. They do not represent the origin of the species, as Spencer Wells points out. Nor do they represent the origin of any given population. Phylogenetic trees "represent the time [and the marker], peering back into the past, when we stop seeing genetic diversity in our mtDNA and Y-chromosome lineages" (ibid., 55), when we "simply loose the signal." And we loose that signal because "all the genetic diversity present today coalesces to a single ancestor" (59). In other words, mtDNA and Y-chromosome trees only tell us about those lineages that *have survived*: older populations could well have been made up of lineages now extinct and thereby invisible to scientists who reconstruct "the past" on the basis of genetic material "carried by living people," to return to Goldstein's words (2008, 3). Those lineages would or could tell a different history. And there are more analytic problems still, and especially for studies that rely on modal haplotypes: there are many individual members of today's "populations"—in the case of this book, of contemporary Jewish communities—who do not carry the haplotypes that, mathematically speaking, characterize (are "modal" for) the population of which those individuals are a part.

And that is not all. Is it reasonable to assume that history is best understood through a minimalist, parsimonious logic?[8] What if it takes many more "events" to explain the presence of the CMH amongst the Lemba: First, there was a trader, perhaps of Middle Eastern but not Judaic/Jewish origins, who migrated to southern Africa and through an extreme founder effect, fathered 10 percent of Lemba men and 50 percent of Lemba men of the Buba clan. Second, perhaps many centuries later, Christian missionaries came to southern Africa, and as they did so many times and in so many places, they identified a lost tribe, converted that tribe to Christianity, and began to impart to the Lemba a belief in their ancestral Judaic origins. Third, presume there was a Lemba figure, a Christian minister who promoted that belief in being a lost tribe and rallied increasing numbers of individuals in Lemba communities to his call? (We know, for example, in the case of the Bnei Menashe, just such an indigenous Christian leader emerged in the mid-twentieth century and became a charismatic leader of the community). Moreover, to project this story into the future, presume that, in the last decades of the twentieth century, along came geneticists, cheek swabs in hand, who collected genetic material from a certain percentage of Lemba men and informed them that "the Lemba" do indeed seem to have ancient Judaic genetic material (based upon the presence of the Cohen modal haplotype that was discovered, about a decade later, to not be a single haplotype at all). In turn, as I elaborated in chapter 5, assume that rabbis funded by an American Jewish missionary group visited Lemba communities and taught them Jewish ways and that as a consequence a growing number of Lemba individuals chose to convert formally to Judaism. A generation or two hence, the Lemba may well turn out to have been one of the most longstanding Jewish communities in the world. That may not be the most parsimonious explanation we can produce. It may not be the story that most simply and seamlessly squares the Lemba oral tradition with the genetic data. But given what we know about missionary fantasies and activities in the eighteenth and nineteenth centuries, it is certainly a plausible historical account. And given what we know of the work of genetic historians and Jewish activists at the turn of the millennium, it is also certainly a plausible future.

DNA evidence does not generate historical stories. As detailed narrative accounts, genetic histories are forged at the intersection of genomic evidence read against documentary sources and oral traditions that provide the interpretive frame. Given the available historical narratives—given narratives that define a priori what will count as a significant "event," an important "discovery" to be made, or "truth" to be verified (or not)—what is the most parsimonious and therefore the most plausible historical account that squares with the genetic evidence? There are times when the genetic data does not square at all: as I discussed in chapter 3, Levites do not seem to originate in the ancient Near East, and, according to one prominent study, neither do Judaism's "founding mothers" (Behar et al. 2003; Thomas et al. 2002). In those instances alternative historical accounts are proposed. But when genetic facts *can be squared* with the evidence, the most parsimonious explanation seems to carry the day. One (known) event is better than two. The messiness and density of the social world is reduced to clean lines on phylogenetic trees. Moreover, working within longstanding assumptions about the objectivity and transparency of documents that underwrote the work of professional historians (and that indeed has not disappeared today), historical documents, biblical texts, and even oral traditions (retroactively, that is, once "proved" by genetic data) are treated as mere accounts of past events (for critiques of those assumptions, see Novick 1988; Steedman 2001; White 1973). But archives and documents are enactments of power and desire, they project and create not just pasts but possible futures, and they need to be read as much for what they don't or can't say or know as for what they do (Foucault 1972; Stoler 2009; Povinelli 2002; Derrida 1996). Instead, "historical sources" (oral, biblical, archival) simultaneously frame the historical questions to be investigated by genetic historians and, in turn, provide explanations of the genetic data and generate the details on the basis of which narratives are produced. That is how genetic history is made.

The Historical A Priori

In the "Paradoxes of Universality," Etienne Balibar argues that conflicts over (human) universality and difference have historically unfolded within "the space of humanism" (1990, 288–89). Since the mid-nineteenth century, those conflicts have been waged in significant part on the terrain of the biological sciences. For race science, the question of origins was always simultaneously a question about human unity: are we one species or more? Though race science struggled with this question, in the long shadow of eugenics, Nazism, and genocide (to simplify a rather complex history of science and society), population genetics has been clear: we are one species. Nevertheless, as Jenny Reardon has argued, "human population geneticists' claims that we are all one human species depend upon first dividing us into racial and ethnic groups and studying our differences" (Reardon 2004, 41).

Studying those differences, however, has involved producing not just inter-group distinctions. It has generated intra-group distinctions as well, and not just in terms of identifying and sometimes celebrating the "multicultural" nature of the Jewish world, as I have shown in this book. Anthropological genetics, like population genetics before it, produces intragroup distinctions every time it writes the history of "a population"—a mathematical concept translated into a substantive thing. Those intrapopulation distinctions are most consequential politically and culturally when we are dealing with the histories—the genealogies—of contemporary communities.

Genetic histories such as those I have discussed in this book are built upon Y-chromosome and mtDNA evidence. In so doing, the history of the "group"—the Jews, in this instance—is built upon evidence of two lineal descent lines, maternal and paternal. And as I mentioned above, it is built on the basis of those lineages that have survived. Just as significant is the fact that the history of "the Jews" is built upon a mathematical distribution of data. In other words, on the basis of Y-chromosome evidence what we know at best is that a certain percentage of Jewish men have paternal lines that likely originated in the Near East (possibly ancient Palestine). Unless we narrate Jewish history as the descent of men, we certainly do not know that "the Jews" originated in the ancient Near East, as I argued in chapter 3. More pertinent to the argument I want to make here, we only know this origin story to be true of "the Jews" if we assume that the descent of Jewish men is the story of the descent of *those men* that carry these presumed traces of a presumably ancient Hebrew patriline. The same is true for the origin of Ashkenazi women: if 40 percent descend from just four maternal founders, what about the other 60 percent (see chapter 3)?

Lineal descent trees produce "clean" histories of "populations." (According to one genetic anthropologist, those clean trees are an artifact of how researchers interpret mtDNA and Y-chromosome evidence rather than a property of genetic systems themselves.) In this instance, they produce clean histories of "the Jewish people." (For that matter, lineal descent trees produce clean genealogies of individuals [see chapter 4]). They privilege the statistical norm—those with shared descent—and ignore those lineages or mutations that are outliers. Lineal descent trees privilege the 40 percent of Ashkenazi women who share four ancestors over the 60 percent who do not and whose phylogenies scatter and form no discernible shared tree.

There are newer statistical techniques that make use of all the sampled haplotypes or SNPs. Principal components analysis, for example, one biological anthropologist explained to me, aims not to privilege certain mutations or lineages over others, and that may solve some of the problems I raise above. But such statistical techniques do not resolve all of the obstacles involved in translating genetic evidence into narratives about the origins and history of a particular human collectivity. If, on the one hand, a very strong majority clusters around particular lineages, then one would still need to privilege that majority in order to tell a history cast in a narrative of origins and descent. If, on the other hand, the population scatters into numerous lineages, if it forms a phylogenetic "bush" rather than a phylogenetic tree,[9] then historical narration becomes impossible, at least in the terms of a historical sensibility that wants to know from where we originate and from whom we descend. One could interpret the data in other ways, asking questions about population structure, for example, but one would need to leave aside the central historical questions and collective and personal attachments, desires, and identifications that have thus far driven this still emerging and proliferating scientific and social field.

There are, likewise, newer genomic techniques that attempt to take into account far more genetic loci than did Y-chromosome- and mitochondrial-based work. There have been two recent studies of Jewish origins based upon data from genome-wide sequences (Atzmon et al. 2010; Behar et al. 2010). In the abstract to their article, Atzmon and his colleagues reiterate a statement that has been made over and over again as novel biological techniques and epistemic objects have been brought to bear on the question of Jewish origins: neither "previous genetic studies of blood group and serum markers" nor "successor studies of monoallelic Y chromosomal and mitochondrial genetic markers" have resolved "the issues of within and between-group Jewish genetic identity" (2010, 850), something the current study hopes to do. "This study touches upon an issue that was raised over a century ago by Maurice Fishberg, Joseph Jacobs, and others about whether the Jews constitute a race, a religious group, or something else" (ibid., 857). Using the latest genomic techniques, Atzmon and his colleagues aim "to improve the understanding about the relatedness of contemporary Jewish groups" (850). And what they report, as have many before them, is that "Jewish populations from the major Jewish Diaspora groups—Ashkenazi, Sephardic, and Mizrahi—formed a distinctive population cluster . . . albeit one that is closely related to European and Middle Eastern, non-Jewish populations" (857).

In using non-sex-specific genetic data, these studies escape the problem of narrating Jewish history as the descent of men. They are building conclusions on a much wider genetic data set. But they do not necessarily avoid the problem of privileging certain "kinds" of contemporary Jews over others. The researchers collected samples from individuals whose four grandparents were all Jewish. (More specifically, each of their four grandparents had to come from the same Jewish community, say Ashkenazi or Syrian). What if scientists sampled a random set of contemporary Jews, many of whom may not have two Jewish parents let alone four Jewish grandparents? What would "Abraham's Children in the Genome Era" (Atzmon et al. 2010) look like then?[10]

If one reads against the grain of standard evidentiary assumptions and practices in genetic history, there is ample evidence in statistical distributions for producing counter-narratives. In the case of Jewish history, not only did the maternal lines most likely not originate in the Middle East, neither did *many* of the fathers (chapters 1 and 3). And as someone once asked me: what about the *other* 50 percent of the Cohanim (see chapter 1)? The other 50 percent emerges as the residual. Individuals who are statistical (or sampling) outliers become irrelevant to reconstructing "the history" of "the group"—the Cohanim, the Jews.

More broadly, genetic history recognizes only certain kinds of human collectivities. Only certain kinds of human identifications are legible through its molecular optic. A "population" such as the Mizrahim—a group born of a political struggle in the modern State of Israel (chapter 3)—emerges as the "Other" in a history told through genetic and genomic evidence and techniques. In fact, all sorts of identifications are rendered outliers in this scientific practice: I can be an authentic Cherokee or Navaho. And insofar as Jewishness is not "merely" a religion (see chapter 2), I can be an authentic Jew. But I cannot be an authentic—a legible—Christian or Muslim, conservative or liberal. Our truest selves are defined according to descent lines from ancestors understood to represent particular ethnic lineages. As articulated most explicitly in the technology developed to measure "admixture": from which ancestral—from which *original*—populations do we descend (see Fullwiley 2008a)? Vertical inheritance (that is, the tree instead of the bush) may not be the best optic through which to reconstruct and tell the history of human communities. But that is how histories have long been told—histories of the nation and histories of the race, for which origins or moments of birth define identities and for which continuous descent from a common ancestor substantiates the fact of the group as an enduring, substantive thing.

In identifying the shift from a "classical" to a "modern" episteme, Foucault identified history as "the very medium in which beings develop and acquire characteristics which in turn cannot be considered or explained independently of this development" (Han-Pile 2005, 589). According to Foucault, by the early nineteenth century history had become both "the fundamental mode of being of empiricities" (1970, 219)—that is, of "things and men" (276)—and it became "a certain *modern* manner of knowing empiricities" (250). As Béatrice Hans-Pile points out, perhaps most exemplary of that historical a priori was Darwin's theory of evolution (2005, 589).

Genetic history is firmly lodged within this manner of being and knowing: a population develops and acquires genetic "characteristics" through time. More specifically, a population develops and acquires genetic characteristics through a succession of evolutionary events together with the kinship practices the population sustains and the choices its human actors make. In turn, a population is known—and is *knowable, it exists*—via those now scientifically legible traces of a population's historical existence. If in the genetic historical studies I have discussed in this book, what makes Jews a meaningful population at the outset is self-designation, what makes Jews a meaningful population as a consequence of the genetic data is the evidence of shared Y-chromosome and mtDNA haplotypes (and more recently, of shared SNPs from across the genome).[11] Those haplotypes substantiate a Jewish collectivity as "a population" in genetic history's style of reasoning.

Genetic history is not a straightforward heir to Darwin's theory of evolution, however. It does not generate a history of life. It is not interested in the ways in which "the living organism manifests . . . the functions that keep it alive," as was true of nineteenth-century biology (Foucault 1970, 290). Nor in a contemporary genomic or post-genomic vision of life does it seek to understand the genetic "code" or the processes of transcription, expression, and mutation through which organisms reproduce, adapt, survive, and die. In short, we need to take seriously that this is a genomic discipline that, while looking at "biological" data, is looking for "historical" signs. Genetic historians presume a priori that "profoundly historical mode of being of . . . men" of which Foucault wrote (276). But in so doing, they are writing not of man as an organism or of man as a species. They are writing about man as a Subject (see Foucault 1970, chapter 9). More specific yet, for genetic history man's—and woman's—historical mode of being is found in signs of stability and evidence of endurance, of having *carried on* as a group, even if, as in the case of the search for African American origins, it is "ancestral populations"—those "ethnic groups" that genetic studies identify in Africa and use as the basis for determining the "cultural" or "ethnic" origins of African American individuals—that have carried on.

Race science also proffered a theory of history, of course. In racial thought, history was believed to be "the consequence of a 'secret' hidden and revealed to men about their own nature and birth." It presumed an "*invisible cause* of the destiny of societies and peoples, ignorance of which account[ed] . . . for degeneration or for the historical power of evil" (Balibar, quoted in Goldberg 1990, 287). In racism's "historio-biological" discourse history was cast within the "theme of racial confrontations" understood as "struggles for existence" (Foucault 2003, 60). And that struggle for existence was understood in distinctly *biological*-qua-political terms. As Foucault has written,

> evolutionism . . . not so much Darwin's theory itself as a set, a bundle, of notions (such as: the hierarchy of species that grow from a common evolutionary tree, the struggle for existence among species, the selection that eliminates the less fit) . . . became . . . not simply a way of transcribing a political discourse into biological terms, and not simply a way of dressing up a political discourse in scientific clothing, but a real way of thinking about the relations between colonization, the necessity for wars, criminality, the phenomena of madness and mental illness, the history of societies with their different classes, and so on. (ibid., 257)

In the historical imagination of nineteenth and early twentieth century racial thought, the struggle for political existence was inseparable from the struggle for biological existence. As I discussed in chapter 2, for leading Jewish scientists and political figures who sought to understand the racial character of the Jews, *degeneration*—physical and psychological—was understood to be a result of a history in which Jews had long lived in "exile" from their "homeland." Zionism would revive the Jewish people culturally *and physically* by returning them home. In other words, there were eugenic dimensions to the Zionist belief that the national project would succeed only if the new Hebrew—a revived and regenerated Jewish biological and cultural self—could be born.

As was true of race science, so too does the genetic historical imagination posit the significance of biological origins: that knowledge of one's ancestors tells one something significant about who one is. And that enduring historical-biological sensibility indexes a crucial continuity between the underlying assumptions of contemporary practices of anthropological genetics and those that drove race science in the late nineteenth and early twentieth centuries. Nevertheless, there are also important ways in which the similarity between the historical-biological sensibilities of these two disciplines and cultural and political fields ends there: for genetic historians, history is not characterized by a biological struggle for survival. It is not characterized by the need to fight off "threats"—biological threats—to one's purity and existence. Even in the case of religious nationalists in Israel importing members of so-called lost tribes, the threat to Israel's Jewish character is seen in demographic and political, not strictly biological (pathology, disease, degeneration) terms.

In contrast to the "historico-biological" discourse that Foucault described, genetic history is perhaps better described as a biologico-historical one. Genetic history is built upon a genomic "record" of shared origins out of which genetic kinship and distance are measured. But *history itself* is attributed to the religious and cultural (here, understood as "kinship") practices of women and men in the past and through time. In engaging in such practices—in *choosing* to do so, as cultural practices are often articulated vis-à-vis the history of the Jews—so too do men and women end up "making-up" themselves as members of a meaningful and scientifically legible human collectivity (Hacking 1986, 2006). "Survival" in a genetic historical imagination refers to *cultural* survival: to the fact that Jews have endured *as a self-identified and therefore also as a scientifically legible group* despite a long history of exile and dispersion. The biological is an evidentiary window into that historical fact. It is an evidentiary window into those facts of human culture, agency, and choice.

In contrast to race science and racial thought, genetic history embraces a liberal commitment to human agency and choice in its epistemological and commercial presumptions, practices, and norms. But as I have argued in earlier chapters, that does not mean that it has no determining authority. As most evident in practices of genetic ancestry tests in which individuals discover "facts" about themselves they never knew and vis-à-vis groups of possibly ancient Jews (the Lemba of southern Africa or the Bnei Menashe of India), genetic history fashions a post-facto determinism. Genetic history reveals to us *in actual fact* our true (biologico-historical) selves. The choice to learn about myself, to remain who I am or to realign my sense of self vis-à-vis newly revealed bodily facts about who I have always already been, remains mine to make. To choose otherwise, however, is to abandon a historically authentic self that I carry around within. It is to abandon a self that is now scientifically legible, epistemologically robust, ever more felicitous culturally and politically—and that is very much at home—in this genomic age.

REFERENCES

Abrami, Leo. 2001. "Visit to the Lemba." *Kulanu*, January. http://www.kulanu.org/lemba/visitlemba.php (accessed 9/27/2005).

Abu El-Haj, Nadia. 2001. *Facts on the Ground: Archaeological Practice and Territorial Self-Fashioning in Israeli Society*. Chicago: University of Chicago Press.

———. 2007a. "Rethinking Genetic Genealogy: A Response to Stephan Palmié." *American Ethnologist* 34 (2): 223–26, doi: 10.1525/ae.2007.34.2.223.

———. 2007b. "The Genetic Reinscription of Race." *Annual Review of Anthropology* 36 (1): 283–300.

———. 2010. "Racial Palestinianization and the Janus-Faced Nature of the Israeli State." *Patterns of Prejudice* 44 (1): 27–41.

Almog, Oz. 2000. *The Sabra: The Creation of the New Jew*. Translated by Haim Walzman. Berkeley and Los Angeles: University of California Press.

Anderson, Amanda. 2006. *The Way We Argue Now: A Study in the Cultures of Theory*. Princeton: Princeton University Press.

Anderson, Warwick. 2006. *The Cultivation of Whiteness: Science, Health, and Racial Destiny in Australia*. Durham: Duke University Press.

Andolfatto, P. 2005. "Adaptive Evolution of Non-Coding DNA in Drosophila." *Nature* 437 (7062): 1149–52.

Anidjar, Gil. 2008. *Semites: Race, Religion, Literature, Cultural Memory in the Present*. Stanford: Stanford University Press.

Armelagos, George J. and Alan H. Goodman. 1998. "Race, Racism and Anthropology." In *Building a New Biocultural Synthesis: Political-Economic Perspectives on Human Biology*, edited by Alan H. Goodman and Thomas Leland Leatherman, 359–78. Ann Arbor: University of Michigan Press.

Asad, Talal. 1993. *Genealogies of Religion: Discipline and Reasons of Power in Christianity and Islam*. Baltimore: Johns Hopkins University Press.

Aschheim, Shlomo. 1982. *Brothers and Strangers: The East European Jew in German and German-Jewish Consciousness*. Madison: University of Wisconsin Press.

Atzmon, Gil, Li Hao, Itsik Pe'er, Christopher Velez, Alexander Pearlman, Pier Franceso Palarma, and Bernice Morrow. 2010. "Abraham's Children in the Genome Era: Major Jewish Diaspora Populations Comprise Distinct Genetic Clusters with Shared Middle Eastern Ancestry." *The American Journal of Human Genetics* 86 (6): 850–59.

Avichail, Eliyahu. 2000. "Bnei Menashe Update: 100 New Immigrants in 2000!" *Kulanu* 7 (3): 1, 4.

Baker, Lee. 1998. *From Savage to Negro: Anthropology and the Construction of Race, 1896–1954*. Berkeley and Los Angeles: University of California Press.

Balibar, Etienne. 1990. "Paradoxes of Universality." In *Anatomy of Racism*, edited by David Theo Goldberg, 283–94. Minneapolis: University of Minnesota Press.

Bamshad, Michael, Toomas Kivisild, W. Scott Watkins, Mary E. Dixon, Chris E. Ricker, Baskara B. Rao, and J. Mastan Naidu. 2001. "Genetic Evidence on the Origins of Indian Caste Populations." *Genome Research* 11 (1): 994–1004.

Bann, Stephen. 1984. *The Clothing of Clio: A Study of the Representation of History in Nineteenth-Century Britain and France*. Cambridge: Cambridge University Press.

Banton, Michael. 1978. *The Idea of Race*. Boulder: Westview Press.

Barkan, Elazar. 1992. *Retreat of Scientific Racism: Changing Concepts of Race in Britain and the United States between the World Wars*. Cambridge: Cambridge University Press.

Barnes, Barry, and John Dupré, 2008. *Genomes and What to Make of Them*. Chicago: University of Chicago Press.

Baudrillard, Jean. 1981. *For a Critique of the Political Economy of the Sign*. Translated by Charles Levin. St. Louis: Telos Press.

———. 2005. *The System of Objects*. Translated by James Benedict. New York: Verso.

Beck, Ulrich. 1992. *Risk Society: Towards a New Modernity*. London: Sage Publications.

Beer, Gillian. 2000. *Darwin's Plots: Evolutionary Narrative in Darwin, George Eliot, and Nineteenth-Century Fiction*. 2nd ed. Cambridge: Cambridge University Press.

Behar, Doron M., Michael F. Hammer, Daniel Garrigan, Richard Villems, Batsheva Bonné-Tamir, Martin Richards, and David Gurwitz. 2004. "Mtdna Evidence for a Genetic Bottleneck in the Early History of the Ashkenazi Jewish Population." *European Journal of Human Genetics* 12 (5): 355–64.

Behar, Doron M., Ene Metspalu, Toomas Kivisild, Alessandro Achilli, Yarin Hadid, Shay Tzur, and Luisa Pereira. 2006. "The Matrilineal Ancestry of Ashkenazi Jewry: Portrait of a Recent Founder Event." *The American Journal of Human Genetics* 78 (3): 487–97.

Behar, Doron M., Ene Metspalu, Toomas Kivisild, Saharon Rosset, Shay Tzur, Yarin Hadid, and Guennady Yudkovsky, Dror Rosengarten, Luisa Pereira, Antonio Amorim, Ildus Kutuev, David Gurwitz, Batsheva Bonné-Tamir, Richard Villems, and Karl Skorecki. 2008. "Counting the Founders: The Matrilineal Genetic Ancestry of the Jewish Diaspora." *PLoS ONE* 3 (4): e2062.

Behar, Doron M., Mark G. Thomas, Karl Skorecki, Michael F. Hammer, Ekaterina Bulygina, Dror Rosengarten, and Abigail L. Jones. 2003. "Multiple Origins of Ashkenazi Levites: Y Chromosome Evidence for Both near Eastern and European Ancestries." *The American Journal of Human Genetics* 73 (4): 768–79.

Ben-Dor Benite, Zvi. 2009. *The Ten Lost Tribes: A World History*. New York: Oxford University Press.

Berg, Irwin. 2006. "Historical Reflections on 'Who Is a Jew?'" *Kulanu* 13 (1): 14, 16.

Bloom, Etan. 2007 "What 'the Father' Had in Mind? Arthur Ruppin (1876–1943), Cultural Identity, *Weltanschauung* and Action." *History of European Ideas* 33: 330–49.

———. 2008. "Arthur Ruppin and the Production of the Modern Hebrew Culture." Ph.D. diss., Tel Aviv University.

Bolnick, Deborah, Duana Fulwilley, Troy Duster, Richard S. Cooper, Joan H. Fujiura, Jonathan Kahn, and J. S. Kaufman. 2007. "The Science and Business of Genetic Ancestry Testing." *Science* 31: 399–400.

Bondi, Ruth. 1981. *Sheba, a Doctor for Everyone* [Hebrew]. Tel Aviv: Zmora-Bitan-Modan.

Bonné-Tamir, Batsheva. 1980, "A New Look at Jewish Genetics." [Hebrew] *Mada* 24, no. 4–5: 181–86.

Boyarin, Daniel, and Jonathan Boyarin. 1993. "Diaspora, Generation and the Ground of Jewish Identity." *Critical Inquiry* 19 (4): 693–725.

Brackman, N. 1999. "Who Is a Jew? The American Jewish Community in Conflict with Israel." *Journal of Church & State* 41 (4): 795–824.

Bradford, Nicole. 2008. "Riding the 'Genetic Revolution': Family Tree DNA Applies Science to Ancestry Research." *Houston Business Journal*, February 22.

Bradman, Neil, Mark G. Thomas, Michael E. Weale, and David B. Goldstein. 2004. "Threads to Antiquity: A Genetic Record of Sex-Specific Demographic Histories of Jewish Populations." In *Traces of Ancestry; Studies in Honour of Colin Renfrew*, edited by M. Jones, 89–98. Cambridge: McDonald Institute

Brodkin, Karen. 1998. *How Jews Became White Folk and What That Says About Race in America*. New Brunswick: Rutgers University Press.

Brodwin, Paul. 2002. "Genetics, Identity, and the Anthropology of Essentialism." *Anthropological Quarterly* 75 (2): 323–30.

Brown, Wendy. 1995. *States of Injury: Power and Freedom in Late Modernity*. Princeton: Princeton University Press.

———. 2003. "Neo-Liberalism and the End of Liberal Democracy." *Theory & Event* 7 (1): 1–28.

Burchard, Esteban Gonázlez, and Elad Ziv. 2003. "Human Population Structure and Genetic Association Studies." *Pharmacogenomics* 4 (4): 431–41.

Butler, Judith. 2000. *Antigone's Claim: Kinship between Life and Death*. New York: Columbia University Press.

Calhoun, Craig. 2006. "The Privatization of Risk." *Public Culture* 18 (2): 257–63, doi: 10.1215/08992363-2006-001.

Callaway, Ewen. 2010. "Neanderthal Genome Reveals Interbreeding with Humans." *New Scientist*, May 6.

Canguilhem, Georges. 1989. *The Normal and the Pathological*. New York: Zone Books.

Canguilhem, Georges, and François Delaporte. 1994. *A Vital Rationalist: Selected Writings from Georges Canguilhem*. New York: Zone Books.

Cann, Rebecca L. 1998. "DNA and Human Origins." *Annual Review of Anthropology* 17 (1): 127–43, doi:10.1146/annurev.an.17.100188.001015.

Cann, R. L., M. Stoneking, and A. C. Wilson. 1987. "Mitochondrial-DNA and Human-Evolution." *Nature* 325 (6099): 31–36.

Capelli, Cristian, Nicola Redhead, Julia K. Abernethy, Fiona Gratrix, James F. Wilson, Torolf Moen, and Tor Hervig. 2003. "A Y Chromosome Census of the British Isles." *Current Biology* 13 (11): 979–84.

Carruthers, Mary J. 1990. *The Book of Memory: A Study of Memory in Medieval Culture*. Cambridge: Cambridge University Press.

Cavalli-Sforza, Luigi Luca. 2000. *Genes, Peoples, and Languages*. New York: North Point Press.

Chatterjee, Partha. 1986. *Nationalist Thought and the Colonial World: A Derivative Discourse?* London: Zed Books.

Chetrit, Sami Shalom. 2010. *Intra-Jewish Conflict in Israel: White Jews, Black Jews*. London: Routledge.

Cochran, Gregory, Jason Hardy, and Henry Harpending. 2006. "Natural History of Ashkenazi Intelligence." *Journal of Biosocial Science* 38 (5): 659–93, doi: 10.1017/S0021932005027069.

Cohen, Debra Nussbaum. 1997. "Kohen Gene Pioneers Fear Misuse." *The Jewish News Weekly of Northern California*, January 7, 1997.

Cohen, Shaye J. D. 1999. *The Beginnings of Jewishness: Boundaries, Varieties, Uncertainties*. Berkeley and Los Angeles: University of California Press.

Cohen, Steven Martin, and Arnold M. Eisen. 2000. *The Jew Within: Self, Family, and Community in America*. Bloomington: Indiana University Press.

Collins, F. S. 2004. "What We Do and Don't Know About 'Race,' 'Ethnicity,' Genetics and Health at the Dawn of the Genome Era." *Nature Genetics* 36 (11): S13–S15.

Comaroff, John L., and Jean Comaroff. 2009. *Ethnicity, Inc.* Chicago: University of Chicago Press.

Cooper R. S., J. F. Kaufman, and R. Ward. 2003. "Race and Genomics." *New England Journal of Medicine* 348 (12): 1166–70.

Cotel, Moshe. 2003. "Letters: Racism Charged." *Kulanu* 10 (1): 16.

Daston, Lorraine, and Peter Galison. 2007. *Objectivity*. New York: Zone Books.

Davidovitch, Nadav, and Shifra Shvarts. 2004. "Health and Hegemony: Preventive Medicine, Immigrants and the Israeli Melting Pot." *Israel Studies* 9 (2): 150–79.

Davison, Charlie. 1996. "Predictive Genetics: The Cultural Implications of Supplying Probable Futures." In *The Troubled Helix: Social and Psychological Implications of the New Human Genetics*, edited by Theresa Marteau and Martin Richards, 317–30. Cambridge: Cambridge University Press.

Derrida, Jacques. 1996. *Archive Fever: A Freudian Impression*. Chicago: University of Chicago Press.

Diaz, E. S. 2007. "The Rhetoric of Informational Molecules: Authority and Promises in the Early Study of Molecular Evolution." *Science in Context* 20 (4): 649–77.

Dietrich, M. R. 1994. "The Origins of the Neutral Theory of Molecular Evolution." *Journal of the History of Biology* 27 (1): 21–59.

Du Bois, W. E. B. 1897. *The Conservation of Races*. The American Negro Academy Occasional Papers, no. 2. Washington, D.C.: Published by the Academy.

Dumit, Joseph. 2004. *Picturing Personhood: Brain Scans and Biomedical Identity*. Princeton: Princeton University Press.

Duster, Troy. 1998. "Persistence and Continuity in Human Genetics and Social Stratification." In *Genetics: Issues of Social Justice*, edited by Ted Peters, 218–38. Cleveland: Pilgrim Press.

———. 2003. *Backdoor to Eugenics*. 2nd ed. New York: Routledge.

Efron, John M. 1994. *Defenders of the Race: Jewish Doctors and Race Science in Fin-de-Siècle Europe*. New Haven: Yale University Press.

Endelman, Todd M. 2004. "Anglo-Jewish Scientists and the Science of Race." *Jewish Social Studies* 11 (1): 52–92.

Entine, Jon. 2007. *Abraham's Children: Race, Identity, and the DNA of the Chosen People*. New York: Grand Central Publishing.

Epstein, S. 1995. "The Construction of Lay Expertise: Aids Activism and the Forging of Credibility in the Reform of Clinical-Trials." *Science, Technology & Human Values* 20 (4): 408–37.

Epstein, Steven. 2007. *Inclusion: The Politics of Difference in Medical Research*. Chicago: University of Chicago Press.

Eyal, Gil. 2006. *The Disenchantment of the Orient: Expertise in Arab Affairs and the Israeli State*. Stanford: Stanford University Press.

Falk, Raphael. 1998. "Zionism and the Biology of the Jews." *Science in Context* 11 (3–4): 587–607, doi: 10.1017/S0269889700003239.

———. 2006a. "Zionism, Race and Eugenics." In *Jewish Tradition and the Challenge of Darwinism*, edited by Geoffrey Cantor and Mark Swetlitz, 137–62. Chicago: University of Chicago Press.

———. 2006b. *Zionism and the Biology of the Jews* [Hebrew]. Tel Aviv: Ressler.

Farber, Paul Lawrence. 2000. *Finding Order in Nature: The Naturalist Tradition from Linnaeus to E.O. Wilson*. Baltimore: Johns Hopkins University Press.

Fish, Stanley. 1997. "Boutique Multiculturalism, or Why Liberals Are Incapable of Thinking About Hate Speech." *Critical Inquiry* 23 (2): 378–95.

Fishberg, Maurice. 1911. *The Jews: A Study of Race and Environment*. Contemporary Science Series. London: The Walter Scott Publishing Co.

———. 1923. "Intermarriage between Jews and Christians." In *Scientific Papers of the Second International Congress of Eugenics: Held at American Museum of Natural History, New York, September 22–28, 1921*, edited by Charles Benedict Davenport, 132–33. Baltimore: Williams & Williams.

Fischer, Michael M. J. 2003, *Emergent Forms of Life and the Anthropological Voice*. Durham, N.C.: Duke University Press.

———. 2009. *Anthropological Futures (Experimental Futures)*. Durham, N.C.: Duke University Press.

Fortun, Michael. 2008. *Promising Genomics: Iceland and Decode Genetics in a World of Speculation*. Berkeley and Los Angeles: University of California Press.

Foster, Eugene A., M. A. Jobling, P. G. Taylor, P. Donnelly, P. de Knijff, T. Zerjal Rene Mieremet, and C. Tyler-Smith. 1998. "Jefferson Fathered Slave's Last Child." *Nature* 396 (6706): 27–28.

Foucault, Michel. 1970. *The Order of Things: An Archaeology of the Human Sciences*. London: Tavistock Publications.

———. 1972. *The Archaeology of Knowledge*. Translated by A. M. Sheridan Smith. New York: Pantheon Books.

———. 1973. *The Birth of the Clinic: An Archaeology of Medical Perception*. New York: Pantheon Books.

———. 1990. *The History of Sexuality*. Vintage Books ed. 3 vols. New York: Random House.

———. 2003. *Society Must Be Defended: Lectures at the Collège de France, 1975–76*. New York: Picador.

———. 2007. *Security, Territory, Population: Lectures at the Collège de France, 1977–1978*. Translated by Graham Burchell. Edited by François Ewald and Alessandro Fontana. Basingstoke: Palgrave Macmillan.

Fox, Joshua. 2003. "Kulanu and Zionism: Reviving the Jewish People." *Kulanu* 10 (2): 6–7.

Freedman, Samuel G. 2000. *Jew vs. Jew: The Struggle for the Soul of American Jewry*. New York: Simon & Schuster.

Freund, Michael. 2001. "Our Communities Can Help Israel's Demographic Crisis." *Jerusalem Post*, September 12.

———. 2005a. "Hundreds of Bnei Menashe in India Return to Judaism." *Shavei Israel Roots* 6:3, 7.

———. 2005b. "Let My People In!" *Jerusalem Post*, November 2.

Frudakis, Tony. 2008. "The Legitimacy of Genetic Ancestry Tests." *Science* 319:1039.

Fullwiley, Duana. 2008a. "The Biologistical Construction of Race: 'Admixture' Technology and the New Genetic Medicine." *Social Studies of Science* 38 (5): 695–735, doi: 10.1177/0306312708090796.

———. 2008b. "Out from under the Skin: Disease Etiology, Biology and Society: A Commentary on Aronowitz." *Social Science & Medicine* 67 (1): 14–17, doi: 10.1016/J. Socscimed.2008.02.020.

Galison, Peter. 1987. *How Experiments End*. Chicago: University of Chicago Press.

———. 1997. *Image and Logic: A Material Culture of Microphysics*. Chicago: University of Chicago Press.

Galison, Peter, and David J. Stump. 1996. *The Disunity of the Sciences: Boundaries, Contexts, and Power*. Stanford: Stanford University Press.

Gannett, Lisa. 2001. "Racism and Human Genome Diversity Research: The Ethical Limits of 'Population Thinking.'" *Philosophy of Science* 68 (3): S479–S492.

———. 2004. "The Biological Reification of Race." *British Journal for the Philosophy of Science* 55 (2): 323–45.

Gannet, Lisa, and James B. Greisemer, 2004. "The ABO Blood Groups: Mapping the History and Geography of Genes in *Homo sapiens*." In *Classical Genetic Research and its Legacy: The Mapping Cultures of Twentieth Century Genetics*, edited by Hans-Jörg Rheinberger and Jean-Paul Gaudillière, 119–72. New York: Routledge.

Garifo, Chris. 2000. "U of A Researcher Heads Breakthrough Genetic Study." *Jewish News of Greater Phoenix*, May 19.

Gates, Henry Louis. 2006 and 2008. *African American Lives 1 and 2*. PBS miniseries first broadcast in February 2006 and February 2008. Available on PBS Home Video.

———. 2007. *Finding Oprah's Roots: Finding Your Own*. New York: Crown Publishers.

———. 2010. *Faces of America: How 12 Extraordinary People Discovered Their Pasts*. New York: New York University Press.

Giddens, Anthony. 1999. "Risk and Responsibility." *The Modern Law Review* 62 (1): 1–10.

Gilman, Sander L. 1985. *Difference and Pathology: Stereotypes of Sexuality, Race, and Madness*. Ithaca: Cornell University Press.

———. 1991. *The Jew's Body*. New York: Routledge.

Goldberg, David Theo. 1990. *Anatomy of Racism*. Minneapolis: University of Minnesota Press.

———. 1993. *Racist Culture: Philosophy and the Politics of Meaning*. Cambridge, Mass.: Blackwell.

———. 2009. *The Threat of Race: Reflections on Racial Neoliberalism*. Malden, Mass.: Wiley-Blackwell.

Goldschmidt, Elisabeth. 1963. *The Genetics of Migrant and Isolate Populations: Proceedings of a Conference on Human Population Genetics in Israel, Held at the Hebrew University, Jerusalem*. Baltimore: Published for the Association for the Aid of Crippled Children by the Williams & Wilkins Co.

Goldstein, David B. 2008. *Jacob's Legacy: A Genetic View of Jewish History*. New Haven: Yale University Press.

Goldstein, Eric L. 1997. "Different Blood Flows in Our Veins: Race and Jewish Self-Definition in Late Nineteenth Century America." *American Jewish History* 85:29–55.

———. 2006. *The Price of Whiteness: Jews, Race, and American Identity*. Princeton: Princeton University Press.

Goodman, Alan H. 2007. "Toward Genetics in an Era of Anthropology." *American Ethnologist* 34 (2): 227–29.

Goodman, Alan H., Deborah Heath, and M. Susan Lindee. 2003. *Genetic Nature/Culture: Anthropology and Science Beyond the Two-Culture Divide*. Berkeley and Los Angeles: University of California Press.

Gould, Stephen Jay. 1981. *The Mismeasure of Man*. New York: Norton.

Grady, Denise. 1997a. "Finding Genetic Traces of Jewish Priesthood." *New York Times*, January 7.

———. 1997b. "Father Doesn't Always Know Best." *New York Times*, January 19, 1997.

Greely, Henry T. 2008. "Genetic Genealogy: Genetics Meets the Marketplace." In *Revisiting Race in a Genomic Age*, edited by Barbara A. Koenig, Sandra Soo-Jin Lee, and Sarah Richardson, 215–34. Piscataway, N.J.: Rutgers University Press.

Greenberg, Eric. 2002. "Kohens in Unlikely Places." *The Jewish Week*, January 9.

Gudding, Gabriel. 1996. "The Phenotype/Genotype Distinction and the Disappearance of the Body." *Journal of the History of Ideas* 57 (3): 525–45.

Gurevitch, J., D. Hermoni, and E. Margolis. 1953. "Blood Groups in Kurdistani Jews." *Annals of Human Genetics* 17 (1): 94–95.

Gurevitch, J., E. Hasson, and E. Margolis. 1956. "Blood Groups in Persian Jews." *Annals of Human Genetics* 21 (2):135–38.

Gurevitch, J., and E. Margolis. 1955. "Blood Groups in Jews of Iraq." *Annals of Human Genetics* 19 (4): 257–59.

Gurevitch, J., E. Hasson, E. Margolis, and E. Poliakoff. 1955a. "Blood Groups in Jews from Cochin." *Annals of Human Genetics* 19 (4): 254–56.

———. 1955b. "Blood Groups in Jews from Tripolitania." *Annals of Human Genetics* 19 (4): 260–61.

Habermas, Jürgen. 2003. *The Future of Human Nature*. Cambridge, U.K.: Polity Press.

Hacking, Ian. 1983. *Representing and Intervening: Introductory Topics in the Philosophy of Natural Science*. Cambridge: Cambridge University Press.

———. 1986. "Making up People." In *Reconstructing Individualism: Autonomy, Individuality, and the Self in Western Thought*, edited by Thomas C. Heller, Morton Sosna, and David E Wellbery, 222–36. Stanford: Stanford University Press.

———. 1990. *The Taming of Chance*. New York: Cambridge University Press.

———. 1995. *Rewriting the Soul: Multiple Personality and the Sciences of Memory*. Princeton: Princeton University Press.

———. 2002. "'Style' for Historians and Philosophers." In *Historical Ontology*, 159–77. Cambridge, Mass.: Harvard University Press.

———. 2005. "Why Race Still Matters." *Daedalus* 134 (1): 102–16.

———. 2006. "Genetics, Biosocial Groups and the Future of Identity." *Daedalus* 135 (4): 81–95.

Hagen, J. B. 1999. "Naturalists, Molecular Biologists, and the Challenges of Molecular Evolution." *Journal of the History of Biology* 32 (2): 321–41.

Haley, Alex. 1976. *Roots*. Garden City: Doubleday.

Halkin, Hillel. 2002. *Across the Sabbath River: In Search of a Lost Tribe of Israel*. New York: Houghton Mifflin Harcourt.

Hammer, M. F., A. J. Redd, E. T. Wood, M. R. Bonner, H. Jarjanazi, T. Karafet, and S. Santachiara-Benerecetti. 2000. "Jewish and Middle Eastern Non-Jewish Populations Share a Common Pool of Y-Chromosome Biallelic Haplotypes." *Proceedings of the National Academy of Sciences of the United States of America* 97 (12): 6769–74.

Hammer, M. F., D. M. Behar, T. M. Karafet, F. L. Mendez, B. Hallmark, T. Erez, L. A. Zhivotovsky, S. Rosset, and K. Skorecki. 2009. "Extended Y Chromosome Haplotypes Resolve Multiple and Unique Lineages of the Jewish Priesthood." *Human Genetics* 126 (5): 719–24.

Handler, Richard. 1988. *Nationalism and the Politics of Culture in Quebec*. Madison: University of Wisconsin Press.

Han-Pile, Béatrice. 2005. "Is Early Foucault a Historian? History, History and the Analytic of Finitude." *Philosophy Social Criticism* 31 (5–6): 585–608, doi: 10.1177/0191453705055491.

Haraway, Donna J. 1989. *Primate Visions: Gender, Race and Nature in the World of Modern Science.* New York: Routledge.

———. 1991. *Simians, Cyborgs, and Women: The Re-Invention of Nature.* London: Free Association.

———. 1997. *Modest_ Witness@ Second_ Millennium. Female-Man©_Mouse Meets_ Oncomouse™*: Feminism and Technoscience. New York: Routledge.

Hart, Mitchell Bryan. 2000. *Social Science and the Politics of Modern Jewish Identity.* Stanford: Stanford University Press.

———. 2005. "Jews, Race, and Capitalism in the German-Jewish Context." *Jewish History* 19 (1): 49–63.

———. 2007. *The Healthy Jew: The Symbiosis of Judaism and Modern Medicine.* New York: Cambridge University Press.

Hason, Nir. 2004. "A Lost Tribe Fears Being Lost Again" [Hebrew]. *HaAretz,* August 25.

Hauskeller, Christine. 2004. "Genes, Genomes and Identity: Projections on Matter." *New Genetics and Society* 23 (3): 285–99.

Heath, D., R. Rapp, and K. S. Taussig. 2004. "Genetic Citizenship." In *Companion to the Handbook of Political Anthropology,* edited by D. Nugent and J. Vincent, 152–67. London: Blackwell.

Helmreich, Stefan. 2000. *Silicon Second Nature: Culturing Artificial Life in a Digital World.* Updated ed. Berkeley and Los Angeles: University of California Press.

Heppner, Max Amichai. 1999. "Science Ties the Lemba Closer to Mainstream Jews." *Kulanu* 6 (2): 1, 13.

Herzberg, Arthur. 1968. *The French Enlightenment and the Jews.* New York: Columbia University Press.

Herzl, Theodor. 1988. *The Jewish State.* New York: Dover Publications.

———. 1997. *Old New Land.* Translated by Lotte Levensohn. Princeton, N.J.: M. Weiner.

Herzog, Zeev. 1999. "The Bible: There Are No Finds in the Territory" [Hebrew]. *HaAretz, Weekend Supplement,* October 29.

Hess, Jonathan. 2002. *Germans, Jews and the Claims of Modernity.* New Haven: Yale University Press.

Hey, Jody. 2001. *Genes, Categories, and Species: The Evolutionary and Cognitive Causes of the Species Problem.* Oxford: Oxford University Press.

Hirsch, Dafna. 2008. "Interpreters of the Occident to the Awakening Orient." *Comparative Studies in Society and History* 50 (1): 227–55.

———. 2009. "Zionist Eugenics, Mixed Marriage, and the Creation of a 'New Jewish Type.'" *Journal of the Royal Anthropological Institute* 15:592–609.

Hubbard, Ruth, and Elijah Wald. 1993. *Exploding the Gene Myth: How Genetic Information Is Produced and Manipulated by Scientists, Physicians, Employers, Insurance Companies, Educators, and Law Enforcers.* Boston: Beacon Press.

Hudson, Kathy, Gail Javitt, Wylie Burke, and Peter Byers. 2007. "ASHG Statement on Direct-to-Consumer Genetic Testing in the United States." *The American Journal of Human Genetics* 81 (3): 635–37.

Hutton, Christopher. 2005. *Race and the Third Reich: Linguistics, Racial Anthropology and Genetics in the Dialectic of Volk.* Cambridge: Polity Press.

Jacobs, Joseph. 1891. *Studies in Jewish Statistics, Social, Vital and Anthropometric.* London: D. Nutt.

Jacobson, Matthew Frye. 1998. *Whiteness of a Different Color: European Immigrants and the Alchemy of Race*. Cambridge, Mass.: Harvard University Press.

———. 2006. *Roots Too: White Ethnic Revival in Post-Civil Rights America*. Cambridge, Mass.: Harvard University Press.

Jameson, Fredric. 1981. *The Political Unconscious: Narrative as a Socially Symbolic Act*. Ithaca: Cornell University Press.

Jobling, M. A. 1994. "A Survey of Long-Range DNA Polymorphisms on the Human Y-Chromosome." *Human Molecular Genetics* 3 (1): 107–14.

Jobling, M. A., and C. Tyler Smith. 1995. "Fathers and Sons—the Y-Chromosome and Human-Evolution." *Trends in Genetics* 11 (11): 449–56.

Judt, Tony. 2009. "Israel Must Unpick Its Ethnic Myth." *Financial Times*, December 7.

Kahn, Jonathan D. 2003. "What's the Use? Law and Authority in Patenting Human Genetic Material." *Stanford Law & Policy Review* 14 (2): 417–44. doi: 10.2139/ssrn.409220.

———. 2004. "How a Drug Becomes 'Ethnic': Law, Commerce, and the Production of Racial Categories in Medicine." *Yale Journal of Health Policy, Law, and Ethics* 4 (Winter): 1–46.

Kaiser, J. 2003. "Genomic Medicine—African-American Population Biobank Proposed." *Science* 300 (5625): 1485.

Kalmus, H., A. Amir, Ona Levine, Elishiva Barak, and Elizabeth Goldschmidt. 1961. "The Frequency of Inherited Defects of Colour Vision in Some Israeli Populations." *Annals of Human Genetics* 25:51–55.

Kaplan, Karen. 2009. "Jewish Legacy Inscribed on Genes?" *The Los Angeles Times*, April 18.

Kaplan, Steven. 2003. "If There Are No Races, How Can Jews Be a 'Race'?" *Journal of Modern Jewish Studies* 2 (1): 79–96.

Kay, Lily E. 1993. *The Molecular Vision of Life: Caltech, the Rockefeller Foundation, and the Rise of the New Biology*. New York: Oxford University Press.

———. 2000. *Who Wrote the Book of Life? A History of the Genetic Code*. Stanford: Stanford University Press.

Keller, Evelyn Fox. 2000. *The Century of the Gene*. Cambridge, Mass.: Harvard University Press.

Kevles, Daniel J. 1985. *In the Name of Eugenics: Genetics and the Uses of Human Heredity*. New York: Knopf.

Kimura, Motoo. 1968. "Evolutionary Rate at the Molecular Level." *Nature* 217 (5129): 624–26.

———. 1983. *The Neutral Theory of Molecular Evolution*. Cambridge: Cambridge University Press.

Kirsh, Nurit. 2003. "Population Genetics in Israel in the 1950s—the Unconscious Internalization of Ideology." *Isis* 94:631–55.

———. 2007. "Genetic Studies of Ethnic Communities in Israel: A Case of Values-Motivated Research Work." In *Jew and Sciences in German Contexts: Studies from the 19th and 20th Centuries*, edited by Ulrich Charpa and Ute Deichmann, 181–94. Tübingen: Leo Beck Institute.

Kleiman, Yaakov. 2004. *DNA & Tradition: The Genetic Link to the Ancient Hebrews*. New York: Devora.

Knorr-Cetina, Karin. 1999. *Epistemic Cultures: How the Sciences Make Knowledge*. Cambridge, Mass.: Harvard University Press.

Knorr-Cetina, Karin, and Alex Preda, eds. 2005. *The Sociology of Financial Markets*. New York: Oxford University Press.

Koenig, Barbara A., Sandra Soo-Jin Lee, and Sarah S. Richardson. 2008. *Revisiting Race in a Genomic Age*. New Brunswick: Rutgers University Press.

Koestler, Arthur. 1976. *The Thirteenth Tribe: The Khazar Empire and Its Heritage*. New York: Random House.

Lakoff, Andrew. 2005. *Pharmaceutical Reason: Knowledge and Value in Global Psychiatry*. Cambridge: Cambridge University Press.

Lander, E. S., L. M. Linton, B. Birren, C. Nusbaum, M. C. Zody, J. Baldwin, and K. Devon. 2001. "Initial Sequencing and Analysis of the Human Genome." *Nature* 409 (6822): 860–921, doi: 10.1038/35057062.

Latour, Bruno. 1987. *Science in Action: How to Follow Scientists and Engineers through Society*. Cambridge, Mass.: Harvard University Press.

Lederer, Susan E. 1997. *Subjected to Science: Human Experimentation in America before the Second World War*. Baltimore: Johns Hopkins University Press.

Lenoir, Timothy. 1997. *Instituting Science: The Cultural Production of Scientific Disciplines*. Stanford: Stanford University Press.

Levi, Yaacov. 1999a. "Journal from Lemba Lands." *Kulanu* 6 (4): 1, 12–13.

———. 1999b. "Lemba Dispatch #1." *Kulanu*, December 27. http://www.kulanu.org/lemba/lemba1.php (accessed April 15, 2011).

———. 2000. "Journal from Lemba Lands." *Kulanu* 7 (1): 13, 15.

Lewontin, Richard C. 2000. *The Triple Helix: Gene, Organism, and Environment*. Cambridge, Mass.: Harvard University Press.

Lifton, Robert Jay. 1986. *The Nazi Doctors: Medical Killing and the Psychology of Genocide*. New York: Basic Books.

Lipmann, Abby. 1991. "The Geneticization of Health and Illness: Implications for Social Practice." *Endocrinologie* 29 (1–2): 85–90.

LiPuma, Edward, and Benjamin Lee. 2004. *Financial Derivatives and the Globalization of Risk*. Durham, N.C.: Duke University Press.

Lock, Margaret. 2005. "Eclipse of the Gene and the Return of Divination." *Current Anthropology* 46: S47–70.

The Lost Tribes of Israel. 2000. *NOVA*, PBS. Originally broadcast February 22, 2000. http://www.pbs.org/wgbh/nova/israel.

Maity, Bhaswar, T. Sitalaximi, R. Trivedi, and V. K. Kashyap. 2004. "Tracking the Genetic Imprints of Lost Jewish Tribes among the Gene Pool of Kuki-Chin-Mizo Population of India." *Genome Biology* 6 (1): P1, doi:10.1186/gb-2004-6-1-p1.

Malkki, Lisa. 1992. "National Geographic—the Rooting of Peoples and the Territorialization of National Identity among Scholars and Refugees." *Cultural Anthropology* 7 (1): 24–44.

Margolis, E., J. Gurevitch, and D. Hermoni. 1960a. "Blood Groups in Ashkenazi Jews." *American Journal of Physical Anthropology* 18: 201–3.

———. 1960b. "Blood Groups in Sephardic Jews." *American Journal of Physical Anthropology* 18:197–99.

Margolis, E., J. Gurevitch, and E. Hasson. 1957. "Blood Groups in Jews from Morocco and Tunisia." *Annals of Human Genetics* 22:65–68.

Markell, Patchen. 2003. *Bound by Recognition*. Princeton: Princeton University Press.

Marks, Jonathan. 2002. *What It Means to Be 98% Chimpanzee: Apes, People, and Their Genes*. Berkeley and Los Angeles: University of California Press.

———. 2011. *The Alternative Introduction to Biological Anthropology*. Oxford: Oxford University Press.

Martin, Emily. 1994. *Flexible Bodies: Tracking Immunity in American Culture from the Days of Polio to the Age of Aids*. Boston: Beacon Press.

Masuzawa, Tomoko. 2005. *The Invention of World Religions, or, How European Universalism Was Preserved in the Language of Pluralism*. Chicago: University of Chicago Press.

Mausenbaum, Rufina Bernardetti Silva. 1999. "A Story of the Lemba and Me." *Kulanu* 6 (4): 7.

———. 2000. "Betar-Lemba Update." *Kulanu* 7 (3): 4.

Mayr, Ernst. 1982. *The Growth of Biological Thought: Diversity, Evolution, and Inheritance*. Cambridge, Mass.: The Belknap Press of Harvard University Press.

McDonald, Kim. 2005. "UCSD Study Shows 'Junk' DNA Has Evolutionary Importance." *ScienceDaily.com*, October 20.

M'charek, Amade. 2005. *The Human Genome Diversity Project: An Ethnography of Scientific Practice*. Cambridge: Cambridge University Press.

Michaels, Walter Benn. 1992. "Race into Culture: A Critical Genealogy of Cultural Identity." *Critical Inquiry* 18 (4): 655–85.

Mills, Charles W. 1998. *Blackness Visible: Essays on Philosophy and Race*. Ithaca: Cornell University Press.

Mordechai, Nissim. 1998. "An Update of the Bnei Menashe in Israel." *Kulanu* 5 (3): 9.

Morris, Benny. 1989. *The Birth of the Palestinian Refugee Problem, 1947–1949*. Cambridge: Cambridge University Press.

Morris-Reich, Amos. 2006. "Arthur Ruppin's Concept of Race." *Israel Studies* 11 (3): 1–30.

Mosse, George L. 1978. *Toward the Final Solution: A History of European Racism*. New York: H. Fertig.

Mourant, A. E. 1954. *The Distribution of Human Blood Groups*. Springfield: C. C. Thomas.

———. 1961. "Evolution, Genetics and Anthropology." *The Journal of the Royal Anthropological Institute of Great Britain and Ireland* 91 (2): 151–65.

Mourant, A. E., Ada C. Kopec, and Kazimiera Domaiewska-Sobczak. 1958. *The ABO Blood Groups: Comprehensive Tables and Maps of World Distribution*. Oxford: Blackwell Scientific Publications.

———. 1978. *The Genetics of the Jews*. New York: Clarendon Press.

Muhsam, Helmut. 1964. "The Genetic Origin of the Jews." *Genus* 20:36–63.

Murray, Charles. 2007. "Jewish Genius." *Commentary* 123 (4): 29–35.

Nash, Catherine. 2006. "Irish Origins, Celtic Origins." *Irish Studies Review* 14 (1): 11–37.

Nebel, A., D. Filon, D. A. Weiss, M. Weale, M. Faerman, A. Oppenheim, and M. G. Thomas. 2000. "High-Resolution Y Chromosome Haplotypes of Israeli and Palestinian Arabs Reveal Geographic Substructure and Substantial Overlap with Haplotypes of Jews." *Human Genetics* 107 (6): 630–41.

Need, Anna C., Dalia Kasperavičiūtė, Elizabeth T. Cirulli and David B. Goldstein. 2009. "A Genome-Wide Genetic Signature of Jewish Ancestry Perfectly Separates Individuals with and without Full Jewish Ancestry in a Large Random Sample of European Americans." *Genome Biology* 10 (1), doi:10.1186/gb-2009-10-1-r7.

Nelkin, D. 1963. "Blood Groups in Jewish Communities." In *The Genetics of Migrant and Isolate Populations: Proceedings of a Conference on Human Population Genetics in Israel, Held at the Hebrew University, Jerusalem*, edited by Elisabeth Goldschmidt, 18–20. Baltimore: Published for the Association for the Aid of Crippled Children by the Williams and Wilkins Co.

Nelkin, D., and M. Susan Lindee. 1995. *The DNA Mystique: The Gene as a Cultural Icon*. New York: Freeman.

Nelson, Alondra. 2008a. "Bio Science: Genetic Genealogy Testing and the Pursuit of African Ancestry." *Social Studies of Science* 38 (5): 759–83.

———. 2008b. "The Factness of Diaspora: The Social Sources of Genetic Genealogy." In *Revisiting Race in a Genomic Age*, edited by Barbara Konig, Sandra Soo-Jin Lee, and Sarah Richardson, 253–68. New Brunswick: Rutgers University Press.

———. 2012. "Reconciliation Projects: From Kinship to Justice." In *Genetics and the Unsettled Past: The Collision between Race, DNA and History*, edited by Keith Wailoo, Alondra Nelson, and Catherine Lee. New Brunswick: Rutgers University Press.

Novas, Carlos, and Nikolas Rose. 2000. "Genetic Risk and the Birth of the Somatic Individual." *Economy and Society* 29 (4): 485–513.

Novick, Peter. 1988. *That Noble Dream: The "Objectivity Question" and the American Historical Profession*. Cambridge: Cambridge University Press.

———. 1999. *The Holocaust in American Life*. Boston: Houghton Mifflin.

O'Brien, Tim. 1998. *The Things They Carried: A Work of Fiction*. New York: Broadway Books.

Olender, Maurice. 1992. *The Languages of Paradise: Race, Religion, and Philology in the Nineteenth Century*. Cambridge, Mass.: Harvard University Press.

Olson, Steve. 2002. *Mapping Human History: Discovering the Past through Our Genes*. Boston: Houghton Mifflin.

Ong, Aihwa, and Stephen J. Collier. 2005. *Global Assemblages: Technology, Politics, and Ethics as Anthropological Problems*. Malden, Mass.: Blackwell Publishing.

Ostrer, Harry. 2001. "A Genetic Profile of Contemporary Jewish Populations." *Nature Reviews Genetics* 2 (11): 891–98.

Palmié, Stephan. 2007. "Genomics, Divination, 'Racecraft.'" *American Ethnologist* 34: 205–22.

Pálsson, Gísli. 2007. *Anthropology and the New Genetics*. Cambridge: Cambridge University Press.

Pálsson, Gísli, and Paul Rabinow. 1999. "Iceland: The Case of a National Human Genome Project." *Anthropology Today* 15 (5): 14–18.

Parfitt, Tudor. 1985. *Operation Moses: The Untold Story of the Secret Exodus of the Falasha Jews from Ethiopia*. London: Weidenfeld and Nicolson.

———. 1992. *Journey to the Vanished City: The Search for a Lost Tribe of Israel*. London: Weidenfeld and Nicholson.

———. 2002. *The Lost Tribes of Israel: The History of a Myth*. London: Weidenfeld and Nicolson.

———. 2003. "Constructing Black Jews: Genetic Tests and the Lemba—the 'Black Jews' of South Africa." *Developing World Bioethics* 3 (2): 112–18, doi: 10.1046/j.1471-8731.2003.00066.x.

Parfitt, Tudor, and Emanuela Trevisan Semi. 2002. *Judaising Movements: Studies in the Margins of Judaism in Modern Times*. New York: RoutledgeCurzon.

Parfitt, Tudor and Yulia Egorova. 2006. *Genetics, Mass Media, and Identity: A Case Study of the Genetic Research on the Lemba and Bene Israel*. New York: Routledge.

Petryna, Adriana. 2002. *Life Exposed: Biological Citizens after Chernobyl*. Princeton: Princeton University Press.

———. 2009. *When Experiments Travel: Clinical Trials and the Global Search for Human Subjects*. Princeton: Princeton University Press.

Petryna, Adriana, Andrew Lakoff, and Arthur Kleinman. 2006. *Global Pharmaceuticals: Ethics, Markets, Practices*. Durham: Duke University Press.

Phimister, E. G. 2003. "Medicine and the Racial Divide." *New England Journal of Medicine* 348 (12): 1081–82.

Pickens, D. K. 1968. *Eugenics and the Progressives*. Nashville: Vanderbilt University Press.

Pinker, Stephen. 2005. *Jews, Genes and Intelligence*. New York: YIVO Institute for Jewish Research.

Pottage, Alain. 1998. "The Inscription of Life in Law: Genes, Patents, and Bio-Politics." *The Modern Law Review* 61 (5): 740–65.

———. 2002. "Unitas Personae: On Legal and Biological Self-Narration." *Law and Literature* 14 (2): 275–308.

Povinelli, Elizabeth A. 2002. *The Cunning of Recognition: Indigenous Alterities and the Making of Australian Multiculturalism*. Durham: Duke University Press.

———. 2006. *The Empire of Love: Toward a Theory of Intimacy, Genealogy, and Carnality*. Durham: Duke University Press.

Primack, Karen, ed. 1998. *Jews in Places You Never Thought Of*. Hoboken: Ktav.

———. 2000. "Betar and Lemba Youth Meet." *Kulanu* 7 (2): 1, 8.

Proctor, Robert. 1988. *Racial Hygiene: Medicine under the Nazis*. Cambridge, Mass.: Harvard University Press.

Rabinow, Paul. 1992. "Artificiality and Enlightenment: From Sociobiology to Biosociality." In *Zone 6: Incorporations*, edited by Jonathan Crary and Sanford Kwinter, 234–52. New York: Zone Books.

———. 1996a. *Essays on the Anthropology of Reason*. Princeton: Princeton University Press.

———. 1996b. *Making PCR: A Story of Biotechnology*. Chicago: University of Chicago Press.

———. 1999. *French DNA: Trouble in Purgatory*. Chicago: University of Chicago Press.

Rabinow, Paul, and Talia Dan-Cohen. 2005. *A Machine to Make a Future: Biotech Chronicles*. Princeton: Princeton University Press.

Rabinowitz, S. 1997. "Who Is a Jew: Prime Minister Begin and the Jewish Question." *Judaism* 46 (3): 293–301.

Rapp, Rayna. 1999. *Testing Women, Testing the Fetus: The Social Impact of Amniocentesis in America*. New York: Routledge.

Reardon, Jenny. 2001. "The Human Genome Diversity Project: A Case Study in Coproduction." *Social Studies of Science* 31 (3): 357–88.

———. 2004. "Decoding Race and Human Difference in Genomic Age." *Differences: A Journal of Feminist Cultural Studies* 15 (3): 38–65.

———. 2005. *Race to the Finish: Identity and Governance in an Age of Genomics*. Princeton: Princeton University Press.

Relethford, John. 2001. *Genetics and the Search for Modern Human Origins*. New York: Wiley-Liss.

———. 2003. *Reflections of Our Past: How Human History Is Revealed in Our Genes*. Boulder: Westview.

Renfrew, Colin, and Martin Jones. 2004. *Traces of Ancestry: Studies in Honour of Colin Renfrew*. Cambridge: McDonald Institute for Archaeological Research.

Rheinberger, Hans-Jörg. 1997. *Toward a History of Epistemic Things: Synthesizing Proteins in the Test Tube*. Stanford: Stanford University Press.

Risch, Neil, Esteban Burchard, Elad Ziv, and Hua Tang. 2002. "Categorization of Humans in Biomedical Research: Genes, Race and Disease." *Genome Biology* 3 (7), doi:10.1186/gb-2002-3-7-comment2007.

Rogin, Michael Paul. 1996. *Black Face White Noise: Jewish Immigrants in the Hollywood Melting Pot*. Berkeley and Los Angeles: University of California Press.

Rose, Nikolas. 1999. *Powers of Freedom: Reframing Political Thought*. Cambridge: Cambridge University Press.

———. 2001. "The Politics of Life Itself." *Theory, Culture & Society* 18 (6): 1–19.

———. 2007. *Politics of Life Itself: Biomedicine, Power and Subjectivity in the Twenty-First Century*. Princeton: Princeton University Press.

Rose, Nikolas, and Carlos Novas. 2005. "Biological Citizenship." In *Global Assemblages: Technology, Politics, and Ethics as Anthropological Problems*, edited by A. Ong and S. J. Collier , 439–463. Oxford: Blackwell.

Said, Edward W. 1975. *Beginnings: Intention and Method*. New York: Basic Books.

Salaman, Redcliffe Nathan. 1911. "Heredity and the Jew." *Journal of Genetics* 1 (3): 273–92.

Saletan, William. 2007. "Jewgenics: Jewish Intelligence, Jewish Genes, and Jewish Values." *Slate Magazine*, http://www.slate.com/id/2177228.

Samra, Myer. 2005. "Praise for Freund and Avichail." *Kulanu* 12 (1): 7.

Sand, Shlomo. 2009. *The Invention of the Jewish People*. Translated by Yael Lotan. New York: Verso.

Sankar, Pamela, and Jonathan D. Kahn. 2005. "Bidil: Race Medicine or Race Marketing?" *Health Affairs*, October 11.

Schmelz, Uziel. 1993. "Terminological Problems in Research on Jewish Communities" [Hebrew]. *Pe'amim* 56:125–39.

Schorsch, Ismar. 1989. "The Myth of Sephardic Supremacy." *Leo Baeck Institute Yearbook* 34 (1): 47–66.

Schorske, Carl E. 1998. *Thinking with History: Explorations in the Passage to Modernism*. Princeton: Princeton University Press.

Schwartz, Bryan. 2001. "Bryan Schwartz in Manipur." *Kulanu* 8 (3): 2.

Schwartz, R. S. 2001. "Racial Profiling in Medical Research." *New England Journal of Medicine* 344 (18): 1392–93.

Scott, David. 2004. *Conscripts of Modernity. The Tragedy of Colonial Enlightenment*. Durham: Duke University Press.

Sela, Maya. 2009. "Israeli Wins French Prize for Book Questioning the Origins of Jewish People." *HaAretz*, December 3.

Senior, Jennifer. 2005. "Are Jews Smarter?" *New York Magazine*, October 16, http://nymag.com/nymetro/news/culture/1478.

Shafir, Gershon. 1989. *Land, Labor, and the Origins of the Israeli-Palestinian Conflict, 1882–1914*. Cambridge: Cambridge University Press.

Shapin, Steven. 1994. *A Social History of Truth: Civility and Science in Seventeenth-Century England*. Chicago: University of Chicago Press.

———. 2008. *The Scientific Life: A Moral History of a Late Modern Vocation*. Chicago: University of Chicago Press.

Sheba, Chaim. 1960. "An Attempt to Re-Construct the Wanderings of the Children of Israel Using Biochemical Tests" [Hebrew]. *Mada* 4:34–39.

———. 1971. "Jewish Migration in Its Historical Perspective." *Israel Journal of Medical Sciences* 7 (12): 1333–41.

Sheleg, Yair. 2005a. "Amar: Bnei Menashe Are Descendants of Ancient Israelites." HaAretz.com, April 1; http://www.haaretz.com/news/amar-bnei-menashe-are-descendants-of-ancient-israelites-1.154715.

———. 2005b. "In Search of Jewish Chromosomes in India." *HaAretz English Edition*, April 1.

Shenhav, Yehouda A. 2006. *The Arab Jews: A Postcolonial Reading of Nationalism, Religion, and Ethnicity*. Stanford: Stanford University Press.

Sherman, Daniel J. 1999. *The Construction of Memory in Interwar France*. Chicago: University of Chicago Press.

Shohat, Ella. 1988. "Zionism from the Standpoint of Its Jewish Victims." *Social Text* 19/20:1–35.

———. 1989. *Israeli Cinema: East/West and the Politics of Representation*. Austin: University of Texas Press.

Shvarts, Shifra, Nadav Davidovitch, Rhona Seidelman, and Avishay Goldberg. 2005. "Medical Selection and the Debate over Mass Immigration in the New State of Israel (1948–1951)." *Canadian Bulletin of Medical History* 22 (1): 5–34.

Silverman, Lawrence. 2009. "Jews, Like All Other Peoples, Have a Right to Self-Determination." *Financial Times*, December 11.

Sklare, Marshall, and Joseph Greenblum. 1967. *Jewish Identity on the Suburban Frontier: A Study of Group Survival in the Open Society*. New York: Basic Books.

Skorecki, K., S. Selig, S. Blazer, R. Bradman, N. Bradman, P. J. Waburton, M. Ismajlowicz, and M. F. Hammer. 1997. "Y Chromosomes of Jewish Priests." *Nature* 385 (6611): 32.

Sommer, Marianne. 2008. "History in the Gene: Negotiations between Molecular and Organismal Anthropology." *Journal of the History of Biology* 41 (3): 473–528.

———. 2010. "DNA and Cultures of Remembrance: Anthropological Genetics, Biohistories and Biosocialities." *Biosocieties*, 5 (3): 366–90. Special Issue, Biohistories, edited by Soraya de Chadavarian.

Soo-Jin, Sandra, Deborah A. Bolnick, Troy Duster, Pilar Ossorio, and Kimberly TallBear. 2009. "The Illusive Gold Standard in Genetic Ancestry Testing." *Science* 325: 38–39.

Sorkin, David. 1987. *The Transformation of German Jewry, 1780–1840*. Oxford: Oxford University Press.

Spurdle, A. B., and T. Jenkins. 1996. "The Origins of the Lemba 'Black Jews' of Southern Africa: Evidence from P12f2 and Other Y-Chromosome Markers." *American Journal of Human Genetics* 59 (5): 1126–33.

Steedman, Carolyn. 1995. *Strange Dislocations: Childhood and the Idea of Human Interiority, 1780–1930*. Cambridge: Harvard University Press.

———. 2001. *Dust*. Manchester: Manchester University Press.

Steichen, Edward. 1955. *The Family of Man*. New York: The Museum of Modern Art.

Stepan, Nancy. 1982. *The Idea of Race in Science: Great Britain, 1800–1960*. Hamden, Conn.: Archon Books.

———. 1998. "Race, Gender, Science and Citizenship." *Gender & History* 10 (1): 26–52.

Stevens, Jacqueline. 2002. "Symbolic Matter: DNA and Other Linguistic Stuff." *Social Text* 70 (1): 105–36.

Stewart, Caro-Beth. 2000. "The Powers and Pitfalls of Parsimony." In *Shaking the Tree: Readings from Nature in the History of Life*, edited by Henry Gee, 48–62. Chicago: University of Chicago Press.

Stocking, George W. 1968. *Race, Culture, and Evolution: Essays in the History of Anthropology*. New York: Free Press.

———. 1987. *Victorian Anthropology*. New York: Free Press.

Stoler, Ann Laura. 1995. *Race and the Education of Desire: Foucault's History of Sexuality and the Colonial Order of Things*. Durham: Duke University Press.

———. 2009. *Along the Archival Grain: Epistemic Anxieties and Colonial Common Sense*. Princeton: Princeton University Press.

Stone, Linda, and Paul F. Lurquin. 2005. *A Genetic and Cultural Odyssey: The Life and Work of L. Luca Cavalli-Sforza*. New York: Columbia University Press.

Stotz, Karola, Paul E. Griffiths, and Rob Knight. 2004. "How Biologists Conceptualize Genes: An Empirical Study." *Studies in History and Philosophy of Science, Part C: Studies in History and Philosophy of Biological and Biomedical Sciences* 35 (4):647–73.

Sunder Rajan, Kaushik. 2005. "Subjects of Speculation: Emergent Life Sciences and Market Logics in the United States and India." *American Anthropologist* 107 (1): 19–30.

———. 2006. *Biocapital: The Constitution of Postgenomic Life*. Durham: Duke University Press.

Swirski, Shlomo. 1989. *Israel: The Oriental Majority*. Translated by Barbara Swirski. Atlantic Highlands, N.J.: Zed Books.

Sykes, Bryan. 2001. *The Seven Daughters of Eve*. London: Bantam Press.

———. 2006. *Saxons, Vikings, and Celts: The Genetic Roots of Britain and Ireland*. 1st American ed. New York: W. W. Norton and Company.

TallBear, Kimberly M. 2007. "Narratives of Race and Indigeneity in the Genographic Project." *Journal of Law Medicine & Ethics* 35 (3): 412–24.

———. 2008. "Native-American DNA.coms: In Search of Native American Race and Tribe." In *Revisiting Race in a Genomic Age*, ed. Barbara Koenig, Sandra Soo-Jin Lee, and Sarah Richardson. 232–52. Piscataway, N.J.: Rutgers University Press.

Tapper, Michael. 1995. "Interrogating Bodies: Medico-Racial Knowledge, Politics, and the Study of a Disease." *Comparative Studies in Society and History* 37:76–93.

Taussig, Karen-Sue, Rayna Rapp, and Deborah Heath. 2003. "Flexible Eugenics: Technologies of the Self in the Age of Genetics." In *Genetic Nature/Culture: Anthropology and Science Beyond the Two-Culture Divide*, edited by Alan H. Goodman, Deborah Heath, and M. Susan Lindee, 58–76. Berkeley and Los Angeles: University of California Press.

Taussig, Karen-Sue. 2009. *Ordinary Genomes: Science, Citizenship, and Genetic Identities, Experimental Futures*. Durham: Duke University Press.

Taylor, Charles. 1989. *Sources of the Self: The Making of the Modern Identity*. Cambridge, Mass.: Harvard University Press.

———. 1992. *The Ethics of Authenticity*. Cambridge, Mass.: Harvard University Press.

Taylor, Charles, and Amy Gutmann. 1992. *Multiculturalism and "the Politics of Recognition": An Essay*. Princeton: Princeton University Press.

———. 1994. *Multiculturalism: Examining the Politics of Recognition*. Princeton: Princeton University Press.

Tekiner, Roselle. "Race and the Issue of National Identity in Israel." *International Journal of Middle East Studies* 23 (1991): 39–55.

Thomas, M. G., T. Parfitt, D. A. Weiss, K. Skorecki, J. F. Wilson, M. le Roux, N. Bradman, and D. B. Goldstein. 2000. "Y Chromosomes Traveling South: The Cohen Modal Haplotype and the Origins of the Lemba—the 'Black Jews of Southern Africa.'" *American Journal of Human Genetics* 66 (2): 674–86.

Thomas, M. G., K. Skorecki, H. Ben-Ami, T. Parfitt, N. Bradman, and D. B. Goldstein. 1998. "Origins of Old Testament Priests." *Nature* 394 (6689): 138–40.

Thomas, Mark G., Michael E. Weale, Abigail L. Jones, Martin Richards, Alice Smith, Nicola Redhead, and Antonio Torroni. 2002. "Founding Mothers of Jewish Communities: Geographically Separated Jewish Groups Were Independently Founded by Very Few Female Ancestors." *American Journal of Human Genetics* 70 (6): 1411–20.

Thompson, Thomas L. 1992. *Early History of the Israelite People*. New York: E. J. Brill.

———. 1999. *The Mythic Past: Biblical Archaeology and the Myth of Israel*. New York: Basic Books.

Tigay, Alan. 2003. "In Praise of Amishav." *Kulanu* 10 (2): 8.

Traubmann, Tamara, and Ruti Suni. 2000. "And Who Is the Most Similar Genetically to the Jews? Palestinians" [Hebrew]. *HaAretz*, May 9.

UNESCO. 1952a. *The Race Concept: Results of an Inquiry*.

———. 1952b. *What Is Race? Evidence from Scientists*.

Van Warmelo, N. J. 1937. "The Classification of Cultural Groups," In *The Bantu-Speaking Peoples of Southern Africa*, edited by W. D. Hammond-Tooke 68–81. London: Routledge.

Van Warmelo, N. J., and W.M.D. Phophi. 1948. *Venda Law. Part 1: Betrothal, Thakha, Wedding*. Ethnological publications 23. Pretoria, South Africa: Department of Native Affairs.

Vickers, Steve. 2010a. "The Lemba." February 25. BBC World Service. http://www.bbc.co.uk/worldservice/programmes/2010/02/100225_outlook_lemba.shtml (accessed April 1, 2011).

———. 2010b. "Lost Jewish Tribe 'Found in Zimbabwe.'" BBC News. March 8. http://news.bbc.co.uk/2/hi/africa/8550614.stm (accessed April 1, 2011).

Wade, Nicholas. 1999. "DNA Backs a Tribe's Tradition of Early Descent from the Jews." *The New York Times*, May 9.

———. 2002. "In DNA, New Clues to Jewish Roots." *The New York Times*, May 14.

———. 2006. *Before the Dawn: Recovering the Lost History of Our Ancestors*. New York: Penguin Press.

———. 2008. "A Dissenting Voice as the Genome Is Sifted to Fight Disease." *The New York Times*, September 15.

———. 2010. "Studies Show Jews' Genetic Similarity." *The New York Times*, June 9.

Wailoo, Keith. 2003. "Inventing the Heterozygote: Molecular Biology, Racial Identity and the Narratives of Sickle Cell Disease, Tay Sachs and Cystic Fibrosis." In *Race, Nature and the Politics of Difference*, edited by Donald S Moore, Jake Kosek, and Anand Pandian, 235–53. Durham N.C.: Duke University Press.

Wailoo, Keith, and Stephen Gregory Pemberton. 2006. *The Troubled Dream of Genetic Medicine: Ethnicity and Innovation in Tay-Sachs, Cystic Fibrosis, and Sickle Cell Disease*. Baltimore: Johns Hopkins University Press.

Waldby, Catherine. 2002. "Stem Cells, Tissue Cultures and the Production of Biovalue." *Health* 6 (3): 305–23.

Washburn, S. L. 1963. *Classification and Human Evolution*. Viking Fund Publications in Anthropology. Chicago: Aldine Pub. Co.

Weber, Max. 1958. "Science as a Vocation." In *From Max Weber: Essays in Sociology*, edited by H. H. Gerth and C. Wright Mills, 129–56. Oxford: Oxford University Press.

Weiss, Meira. 2001. "The Immigrating Body and the Body Politic: The 'Yemenite Children Affair' and Body Commodification in Israel." *Body and Society* 7 (2–3): 93–109.

Wells, Spencer. 2002. *The Journey of Man: A Genetic Odyssey*. Princeton: Princeton University Press.

———. 2006. *Deep Ancestry: Inside the Genographic Project*. Washington, D.C.: National Geographic.

Wertheimer, Jack 1994. "Family Values and the Jews." *Commentary* 97 (1): 30–34.

———. 2009. "Time for Straight-Talk about Assimilation." *The Forward*, October 2.

Weston, Kath. 1997. *Families We Choose: Lesbians, Gays, Kinship*. New York: Columbia University Press.

White, Hayden. 1973. *Metahistory: The Historical Imagination in Nineteenth Century Europe*. Baltimore: Johns Hopkins University Press.

Wiendling, Paul. 2006. "The Evolution of Jewish Identity: Ignaz Zollschan between Jewish and Aryan Race Theories, 1910–45." In *Jewish Tradition and the Challenge of Darwinism*, edited by Geoffrey Cantor and Mark Swetlitz, 116–36. Chicago: University of Chicago Press.

Wiener, Julie. 2011. "Plight of the Patrilineals." *The Jewish Week*. January 20.

Winstein, Keith J. 2007. "Harvard's Gates Refines Genetic-Ancestry Searches for Blacks." *The Wall Street Journal*, November 15.

Woolfson, Adrian. 2004. *An Intelligent Person's Guide to Genetics*. London: Duckworth Overlook.

Wright, Susan. 1986. "Recombinant-DNA Technology and Its Social Transformation, 1972–1982." *Osiris* 2:303–60.

Zalloua, Pierre, Daniel E. Platt, Mirvat El Sibai, Jade Khalife, Nadine Makhoul, Marc Haber, and Yali Xue. 2008. "Identifying Genetic Traces of Historical Expansions: Phoenician Footprints in the Mediterranean." *American Journal of Human Genetics* 83 (5): 633–42.

Zalloua, Pierre A., Yali Xue, Jade Khalife, Nadine Makhoul, Labib Debiane, Daniel E Platt, Ajay K. Royyuru. 2008. "Y-Chromosomal Diversity in Lebanon Is Structured by Recent Historical Events." *American Journal of Human Genetics* 82 (4): 873–82.

Zeller, Jack. 2000. "From the President: Proving One's Jewishness." *Kulanu* 7 (1): 2.

———. 2001. "Overly Ambitious?" *Kulanu* 8 (2): 6.

Zerjal, Tatiana, Yali Xue, Giorgio Bertorelle, R. Spencer Wells, Weidong Bao, Suling Zhu, and Raheel Qamar. 2003. "The Genetic Legacy of the Mongols." *American Journal of Human Genetics* 72 (3): 717–21.

Zerubavel, Yael. 1995. *Recovered Roots: Collective Memory and the Making of Israeli National Tradition*. Chicago: University of Chicago Press.

Ziv, E., and E. G. Burchard. 2003. "Human Population Structure and Genetic Association Studies." *Pharmacogenomics* 4 (4): 431–41.

Zoossmann-Diskin, Avshalom. 2000. "Are Today's Jewish Priests Descended from the Old Ones?" *Homo: Journal of Comparative Human Biology* 51:156–62.

Zuckerkandl, Emile. 1963. "Perspectives in Molecular Anthropology." In *Classification and Human Evolution*, ed. Sherwood L. Washburn, 243–72. Chicago: Aldine.

Zuroff, Avraham, 2011. "Bring Bnei Menashe Home, Knesset Committee Urges. Former Minister's Comments Surprise Shavei Israel," *The Jewish Tribune* (Canada), January 12. www.jewishtribune.ca

NOTES

INTRODUCTION

1. For scientific studies, see Capelli et al. 2003; Bamshad 2001; Zalloua, Platt et al. 2008; Zalloua, Xue et al. 2008; Sommer 2010.
2. www.pbs.org/wnet/facesofamerica; accessed October 11, 2010.
3. The four chemical components of DNA are adenine, cytosine, guanine and thymine, otherwise known as A, C, G, T.
4. The field I am working on goes by many names, including anthropological genetics, genetic anthropology, genetic history, and population genetics. Since I want to distinguish contemporary work from earlier versions of population genetics, I reserve the term "population genetics" for work on group-based biological differences in the 1950s and 1960s, which I discuss in chapters 1 and 2. Moreover, insofar as none of the main scientists involved in the research projects I discuss in this book is trained as an anthropologist, and insofar as anthropologists ask, in addition to questions about population-specific genealogies, much broader questions about the relationship between culture and biology, I chose not to name this field "genetic anthropology." I have chosen "genetic history," a term that has emerged over the last couple of years as one way of referring to the kind of research I explore in this book.
5. Rose 2001; Ong and Collier 2005; Rabinow 1996a, 1996b, 1999; Fortun 2008; Sunder Rajan 2006; Haraway 1997; Taussig 2009; Rabinow and Dan-Cohen 2005; Petryna, Lakoff, and Kleinman 2006; Petryna 2009; Wailoo and Pemberton 2006.
6. For similar efforts, see Nelson 2008a; Brodwin 2002; Pálsson 2007; TallBear 2007, 2008; Parfitt and Egorova 2006; Nash 2006. See also the Special Issue of *Developing World Bioethics* 2, no. 3 (2003).
7. See for example, http://www.webdubois.org/dbAtlantaConfs.html; paragraph 5 summarizes funding issues and the scarcity of intellectual resources for efforts such as that of Du Bois to study the "Negro" racial self. Moreover, and quite importantly, Du Bois developed a critical understanding of race in his emphasis on shared culture and history, one quite distinct from the racial self-understanding promoted by Jewish scientists and political figures in this same period.
8. The mitochondrion is not found in the nucleus of the cell but in the surrounding material that is the cell's energy source.
9. I discuss the specific attributes of the Y-chromosome at length in chapter 1, and of mtDNA in chapters 1 and 3.

10. Recombinant DNA is not a naturally occurring strand of DNA. It is a molecular sequence produced in a laboratory and used as a tool to manipulate DNA molecules. For example, the polymerase chain reaction, a recombinant DNA technology, allows scientists to rapidly replicate DNA sequences, making them available for analysis and manipulation.
11. Genes, by definition, "code for" proteins. Noncoding regions are not, technically speaking, genes. Within anthropological genetics, there is an increasing recognition of shifting understandings of noncoding regions of DNA. Nevertheless, in broad terms focusing on noncoding regions has not been abandoned; see the postscript to chapter 1.
12. Marianne Sommer likewise argues that one cannot characterize the anthropological gene and genome as a retrograde notion, but she does so for a different reason than the one I propose here. She argues that anthropological geneticists are not alone in continuing to understand the gene as the most significant source of information—and component part—of the genome. Those assumptions still characterize the work of practitioners in a variety of disciplines, including some claims made by criminologists and psychiatrists, for example (see Stotz, Griffiths, and Knight 2004; Sommer 2008).
13. See "Remarks by the President, Prime Minister Tony Blair of England (Via Satellite), Dr. Francis Collins, Director of the National Human Genome Research Institute, and Craig Venter, President and Chief Scientific Officer, Celera Genomics Corporation, on the Completion of the First Survey of the Entire Human Genome Project," The White House, Office of the Press Secretary, June 26, 2000.
14. In 2001–2002 I conducted extensive interviews with several African American scholars involved in this work, and I did some fieldwork at genealogy conferences and in a laboratory doing work on African American genetics. That preliminary research has been important in informing my analysis of the broad implications of genetic history as a proliferating scientific and social practice, even though I do not incorporate it as a case study here. For a sustained analysis of African American ancestry projects, see Nelson 2008a.
15. The prevalence of self-studies is contingent upon the group and its socio-economic and political standing in U.S. society. Studies of Native American origins, for example, are not generally self-studies, and they have generated a very different set of conversations and ethical and political concerns (Koenig, Lee, and Richardson 2008; Tallbear 2006 and 2007).
16. Although far less widely known, disease and pathology were not the only paradigms for the study of the Jewish race. Jewish and non-Jewish scientists also produced an alternative account of the Jewish body: "the Healthy Jew" who was rooted in religious traditions and practices and who stood as a lesson to the Gentile world (Hart 2007).
17. For a different perspective on the making of a distinction between Jews and Arabs and on the emergence of Mizrahi identity, see Gil Eyal, *The Disenchantment of the Orient* (2006). Eyal examines the history of Israeli Orientalists, providing another perspective on the cardinal categories of distinction in the Israeli state.
18. There are newer techniques, such as "admixture maps" and "genome wide surveys," that rely on technologies that work differently from mapping the Y-chromosome and mtDNA, which I address at certain points to give an account of what kinds of differences—or not—they might make. Nevertheless, this book focuses on scientific research projects published in the late 1990s and early 2000s that relied on

the two genetic systems, mtDNA and the Y-chromosome. And both the findings and techniques of those studies continue to circulate as public knowledge and commodity forms today (see Fullwiley 2008a for an extended discussion of admixture mapping).

19. See Palmié 2007, 206–7. As he correctly points out, the U.S. racial system has never been based on visible difference alone. The one-drop rule grounded a system that sought to police the boundaries between white and black. Nevertheless, until the birth of molecular biology, the *science* of race relied on what were presumably *visible* biological signs of racial difference. By the 1920s, that began to shift with the discovery of the "first molecular disease," sickle-cell anemia, which was understood to be a "Negro disease" (see Tapper 1995; Wailoo 2003). Early work in molecular biology laid the groundwork for the scientific turn to internal signs of race, which emerged as the privileged form of biological evidence by the 1950s and 1960s in the work of population genetics. As I discuss at length in chapter 1, however, the meaning of inward signs of human biological difference was recast in significant ways by the mid-twentieth century.
20. On the phenomenon of genetic evidence of ancestry helping to sustain or undermine identity claims more generally, see Hauskeller 2004.
21. In using the term "racism," Foucault is referring to a broader phenomenon: to citizen-subjects understood to be "biological threats" more generally, not to specific historico-biological *races* per se. Thus, madness, criminality, and "other abnormalities" can be—and were in Nazi Germany and in "Soviet style states"—"conceptualized in racist terms" (Foucault 2003, 258).
22. For a sustained discussion of the question of genomics' potential "world-historical importance," the public "hype" surrounding the Human Genome Project, and the subsequent "anti-climax," see Barnes and Dupré 2008.

CHAPTER ONE

1. Like Cohanim, Levites are a paternally defined group of male Jews who had, according to biblical accounts, responsibilities in religious practice. In Second Temple times, they were understood to be secondary (in status) to the Cohanim.
2. The two genetic loci were defined by (1) the Y Alu Polymorphic (YAP) insert (an addition or insertion on the Y chromosome), and (2) DYS19.
3. The Cohen modal haplotype is composed of six binary polymorphisms and six microsatellites (see Thomas et. al. 1998). Binary polymorphisms are a kind of "mutation"—or allelic variation—found in the genome. An "allele" is the technical term for a polymorphism, the different "variants" in nucleotide sequences present at a given locus on a given chromosome. A "microsatellite" is a DNA sequence that is composed of blocks of two to six bases (or nucleotides) that repeat. An "allele frequency" of a microsatellite refers to the number of such repeats, (say, of the sequence A-C-G).
4. The authors do not specify whether they are referring to the First or Second Temple. However, later in the paper, they date the coalescence time for the CMH "to between the Exodus and the destruction of the First Temple in 586 B.C." Presumably then the "dispersion" to which they refer is that of the Jewish priesthood following the destruction of the First Temple in 586 B.C.E.
5. The complications associated with dating in genetic history go well beyond the factors listed here. As one researcher explained in an interview, "When people estimate coalescence dates in genetic data they will usually publish a point estimate and a

confidence estimate. I've never seen an estimate of a date that takes into account all the sources of error. There are multiple sources of error—not just mutation rate and generation time. It's a stochastic process, basically—[it presents] a whole range of different problems." It is worth noting that there are newer and more complex models for calculating the mutation rates that do allow it to vary over time. Those models, however, were not used in this study.

6. I do not deal with the question of the uncertainty in the mutation rate in this chapter but will revisit it in chapter 6. It is important to keep in mind that the mutation rate is crucial to understanding the problem of dating in genetic history: confidence intervals can end up stretching over many a thousand year period. Such time spans may not matter much to studies of evolution. In contrast, however, when studying the histories of specific populations—which often involve, as is true in the case of the Cohen study, dating to within rather narrow intervals—it may make quite a difference.
7. Gene flow, also called admixture, refers to the genes introduced into one population from a second population (exogamy, in anthropological terms).
8. The precise numbers are: 0.132 and 0.098 respectively for Ashkenazi and Sephardic Cohanim; n = 9 and 5 respectively (Thomas et al. 1998: Supplementary Information).
9. I put prevalent in quotation marks because one of the striking things in this research is the wide statistical range of "modal haplotypes" and other Y-chromosome markers considered "shared." Modal haplotypes can be identified in 10 percent of a population sample or 50 percent or more, although the latter is rare; moreover, Y-chromosome types found at 5 percent can be seen as significant, a sign of shared ancestry as easily as ones found at 70 percent, even if the one found at 5 percent is not actually "modal." There is, apparently, no debate about what constitutes a level above which a marker can be considered a "modal haplotype." As one researcher explained to me, "modal just means most common." In a similar way, "shared" and "prevalent" remain mathematically underspecified terms.
10. Geneticists who contest the privileging of mtDNA and the Y-chromosome prefer autosomal markers because they reflect a greater portion of a person's ancestry. Having said that, reconstructing specific "histories"—as I will illustrate below, constructing what are considered "clean phylogenetic trees"—is far more possible with mtDNA and the Y-chromosome, and that seems to be a key reason why so many anthropological geneticists interested in reconstructing the past have for the past two decades or so favored those two loci.
11. Meiosis refers to the cell divisions that produce a gamete (a mature reproductive cell).
12. There is a "pseudo-autosomal" region at each end of the Y-chromosome and those regions do recombine. The focus of phylogenetic Y-chromosome analysis, however, is the largest region of the Y-chromosome, which is its non-recombining section (the NRY) (see Jobling and Tyler-Smith 1995).
13. The nucleotides are the structural components of DNA, containing four different chemical bases—adenine, cytosine, guanine, and thymine. Deciphering the nucleotide sequences involves determining the ordering of the chemical components at any stretch of a given chromosome—A, C, G, T or A, G, C, T, and so forth.
14. There are other possible explanations of what is being measured by "genetic distance," which I discuss in chapter 6.
15. This is a retrospective narrative because, as recent scholarship has shown, Darwin's

theorization of evolution did not immediately undermine racial thinking, neither for Darwin nor among his contemporaries.

16. By way of contrast, if one studied ancient DNA one might well find genetic diversity that has not survived.
17. This has also emerged as a problem with the analysis of microsatellites (also called short-tandem repeats or STRs) on the Y-chromosome, as I discuss below.
18. There are attempts to resolve this problem by looking at a combination of the control region (hypervariable regions) of the mitochondrion in tandem with the coding regions in order to control for the possibility of convergence. I revisit this issue in chapter 3.
19. This phrase comes from the title of Bradman et al. 2004.
20. For a critical perspective on this account of genomics and the organism, see Barnes and Dupré 2008, chap. 2.
21. There are some geneticists, and probably an increasing number, who would disagree with Cavalli-Sforza's assessment. Researchers who develop and use Ancestry Informative Markers (AIMs) often target skin pigmentation genes when searching for AIMs to differentiate "European" ancestry from "African" ancestry. They do so because there are SNPs (single-nucleotide polymorphisms) in those genes that occur at very different frequencies in (northern) European and (sub-Saharan) African populations, and they take those SNPs as evidence of one's proportions of "European" and "African" ancestry. (I discuss this technology in chapter 4.) Nevertheless, as AIMs, what researchers are looking for are signs of ancestry rather than biological markers that "cause" cognitive and cultural differences. In other words, even phenotypic differences, and even those that were so cardinal to racial distinctions, perform a different evidentiary function in anthropological genetics than they did for race science. In addition, it is worth noting that one criticism of AIMs technology is that it is not entirely clear if what one is finding is "ancestry" (recent common ancestry, that is) or the effects of natural selection. (In other words, if your ancestors lived close to the equator, they were more likely to evolve darker skin and pass that trait on to you.) (I would like to thank Deborah Bolnick for elaborating this contemporary critique of Cavalli-Sforza's analysis.)
22. It is worth noting that this language of environment versus biological significance parallels the logic of what was until recently mainstream genetics quite well: that DNA is the site of "life"—of "code" and "command structure" in one (Keller 2000). As Richard Lewontin has argued in *The Triple Helix* (2000), such an understanding of the genome fails to integrate an organism with its environment in its understanding of evolution. It renders the environment subsidiary by framing the evolutionary question in the following terms: how does an organism, already formed, adapt to its environment?
23. If there is a severe bottleneck—i.e., if at some point in time the population went through a severe contraction and descendants of only a few men or women survived—then the ancient signal could be eliminated. Nevertheless, the assumption that the signal does survive, at least in most instances, remains fundamental to the work of genetic historical reconstruction.
24. Genetic distance values generally correlate with geographic distance (see Stone and Lurquin 2005, 67).
25. Jewish groups, in fact, "had the lowest ratio of genetic-to-geographic distance of all groups in this study" (Hammer et al. 2000, 6772).
26. There have been subsequent studies that likewise seek to infer Jewish origins via

comparisons with Middle Eastern populations in general, and the Palestinian population(s) in particular. I discuss those additional studies in chapter 3.

27. One of the strongest objections the drafters of the second Statement had with regard to the first was its assertion that humanity is essentially a cooperative species.
28. The specter of race was raised most centrally for genetic anthropologists with respect to the Human Genome Diversity Project, which I discuss at more length in chapter 3. See Reardon 2005 for an extended discussion.
29. The transformative nature of anthropological genetic knowledge is not generally the focus of existing scholarship. For example, Paul Brodwin writes: "The changes could unfold in two ways: (1) such knowledge may undermine received wisdom about family, ethnic and racial identity or (2) it may shore up conventional understandings of identity. Of course, knowledge itself doesn't change anything. Particular people use such knowledge either to undermine or buttress conventional understandings" (2002, 326). See also Palmié 2007.

    Alondra Nelson's work is a subtle engagement with genetic genealogical self-testing and the potentially transformative practices of self-fashioning that such tests generate. Her focus, however, is different than mine. She focuses on categories, reception, and self-fashioning rather than exploring in depth the variety of epistemological commitments with which genetic history operates (Nelson 2008a).
30. Genetic historical work can generate new groupings or identifications (ones that may supplement rather than replace longer-standing ones). For example, some African Americans are self-testing to find their "ethnic" roots in specific African communities. See, for example, *African American Lives 1 and 2* (Gates 2006); see also the website of AfricanAncestry.com, and the BBC 2 production, *Motherland: A Genetic Journey*, broadcast February 14, 2003; available at http://www.rootsforreal.com/motherland_en.php; accessed March 16, 2011.
31. For an extended discussion of the different ways in which laboratory practices produce different definitions of what a population is, see M'Charek 2005, chap. 2.

CHAPTER TWO

1. In Israeli social classification, the Druze (in the 1950s and 1960s, referred to as "Druses") are not considered Arabs. Rather, both their religion and their nationality—two separate criteria of identification within the Israeli state—are Druze. In making the distinction here, I invoke that system of classification, which is used by the geneticists.
2. On the centrality of the Rockefeller Foundation in establishing molecular biology as a discipline, see *The Molecular Vision of Life* (Kay 1993).
3. Population genetics continues in Israel today. I have chosen to focus on the first two decades of that work as it represents a particular scientific-political problem space—as does the earlier European-Jewish race science project—in modern Jewish politics.
4. In the context of this book, I do not provide a comprehensive account of Jewish race science. I intend merely to outline the parameters of a discussion about the nature of the Jews insofar as it is relevant to my broader interest in continuities and reconfigurations between race science and genetic history. In so doing, I also lay the foundations for comparisons between moments in modern Jewish culture and politics as they have simultaneously manifested in and been fashioned out of the work of the biological sciences, historically and today. There is a growing literature on the topic of Jewish race science. See especially, John Efron, *Defenders of the Race* (1994)

and the excellent historical studies by Mitchell Hart, *Social Science and the Politics of Modern Jewish Identity* (2000) and *The Healthy Jew* (2007). See also the recent interesting work by Etan Bloom (2007) and Dafna Hirsch (2009, 2008)

5. As Mitchell Hart points out, in contrast to Fishberg's earlier writings, in which Fishberg emphasized the environmental influence on supposedly permanent racial traits, by the time he published his magnum opus in 1911, he put far more emphasis on the hereditary nature of particular physical characteristics. If Jews displayed diversity even in those characteristics that are "wholly the product of heredity," there were no grounds to argue that Jews are a race. In so arguing, Fishberg accepted the terms of race science in order to argue against reigning anthropological assessments of the Jews (Hart 2000, 163).
6. Fishberg served as the chief medical examiner for the United Hebrew Charities of New York. In that capacity, he conducted anthropometric studies of Eastern European Jewish immigrants and he examined them for physical and mental illnesses. In addition, Fishberg traveled as a consultant for the U.S. Bureau of Immigration to Eastern Europe in 1905 and 1907 to conduct studies there. He was allied with Franz Boas in the debates over immigration to the U.S. Boas used Fishberg's data from Eastern Europe and on Eastern European Jewish immigrants to the U.S. in his famous work on headforms (Hart 2000, 158).
7. Over the past several decades, Fishberg argued, evidence from "Western" countries and even from some Eastern European countries ("Hungary, and of late, Russia and Poland, since the recent revolution when intermarriage between Christian and Jews was finally permitted"), Jews—generally, counted as Jewish men—were increasingly marrying Christians. For example, the rate in Berlin was 44.05 percent (1923, 127).
8. According to Eric L. Goldstein, following 1897 and the rise of political Zionism, new Eastern European Jewish immigrants to the U.S. embraced the Zionist cause. That made "establishment Jewry even more wary of the political implications of Jewish racial identity" (1997, 54). After all, as Goldstein has argued with respect to the rise of race-talk in the Jewish community in the 1870s, race talk emerged as a way of deflecting the specter of "dual loyalty" that was raised in the aftermath of the American Revolution when Jews referred to themselves as a nation. (The language of religion had been adopted as an alternative to "nation" until the 1870s when the Jewish community began to identify itself in racial terms.) See Goldstein 1997, 2006.
9. Fishberg's argument with Zionist Jews was far more encompassing than this. Invoking Renan's definition of a nation as involving "'actual consent, the desire to live together, the will to preserve worthily the undivided inheritance which has been handed down'" (1911, 484), Fishberg argued that the vast majority of Jews exhibit no such inclinations, even if Zionist institutions are working hard to foster them (486). He also argued that the Zionist insistence on Jewish nationhood—and, in particular, on religion as the basis for nationhood—was against the interests of the Jews: it is akin to asserting that "the Jews are aliens in the Occident," thus justifying "the claims of the Anti-Semites"(481).
10. It is worth noting that Fishberg did harbor anxiety about the future of *Judaism*: while it may have been better for Jews *as individuals* to assimilate into U.S. and European societies—better from the perspective of health and wealth, so to speak—it was not better for *Judaism*: "From the standpoint of the preservation of Judaism, intermarriage has been considered a disaster, robbing as it does the Jews of its best adherents. . . . Without the separative tenets of its religious practices, Judaism is

inconceivable and in danger of extinction through absorption by the surrounding majority of other faiths" (1923, 129–30). Fishberg recognized: "The recent nationalist movement among the Jews is mainly the result of these conditions. They face extinction among the European and American nations" (131). Nevertheless, that was a price that Fishberg was willing—however ambivalently—to pay.

11. The term "merely"—or some equivalent—is widely used in this debate over the Jewish Question, and not just by scholars of the time. See, for example, Goldstein (1997, 29) for a similar framing of the issue.
12. The "Edict Concerning the Civic Conditions of the Jews in the Prussian State," for example, refers to "'persons of the Jewish faith." Jews began to engage in similar linguistic shifts, thus marking a transformation in the ways in which Jews self-identified. For example, the journal *Sulamith* changed its subtitle (in 1810) to "A Journal for the Promotion of Culture and Humanity among the Israelites," from its original version of "within the Jewish Nation" (see Markell 2003, 234; Sorkin 1987). For insightful reading of the Emancipation, see Markell, *Bound by Recognition*, chapter 5.
13. Joseph Jacobs' analysis of Jewish facial expression was different from Salaman's, although it was structurally homologous. Jacobs argued that there is little "intermixture" in these mixed marriages ("only in five families out of forty-nine" [1891, xxiiii]). He concluded, however, that the "prepotency" is on the Jewish side. In other words, he maintained that *most* offspring of mixed Jewish-Christian marriages *look Jewish* (ibid.), in contrast to Salaman's conclusion. On the basis of that argument, Jacobs concluded that there is little "Christian" intermixture in "Jewish blood." Given the prepotency of Jewish blood—i.e., its likelihood to survive in "mixed marriages," in which the offspring is of one type or another and not an intermixture between the two types—Christian blood did not survive.
14. According to Endelman, Salaman collected photographs and genealogies of Jewish families as well as of families into which Jews had married. Initially, his collection came from family and friends, and their networks. When serving in Palestine as the medical officer to the British Army's Jewish Regiment in 1918, he added to his collection by photographing Jewish recruits from elsewhere: the Middle East, Eastern Europe, and the United States (2004, 59–60).

    According to Mitchell Hart, Jewish scientists, in order to supplement arguments "adduced from statistics," turned to pictures: "Images preserved in ancient Egyptian and Mesopotamian monuments . . . were seen to offer the best empirical evidence of the racial purity or at least, the racial distinctiveness of the Jews." That iconographic evidence demonstrated that "Jews were not part of Germany or Europe. . . . These images provided evidence that the Jews did in fact *belong*, in a fundamental way, in Palestine" (Hart 2000, 182). Jewish racial distinctiveness was read off of the surface of Jewish bodies, and it was best evidenced—most easily "seen"—in pictorial representations, be they photographs of living or recently deceased Jewish persons or depictions of Jews etched on archaeological remains and presented to readers of these scientific texts.
15. On the centrality of being in the right "environment"—in one's homeland, that is—for racial health in scientific theories of race in the early twentieth century, see Warwick Anderson, *The Cultivation of Whiteness* (2006).
16. Who it was that occupied the position of "degenerate" Jew is more complex than this statement suggests. In the late nineteenth and early twentieth centuries, it was

Eastern European Jews who were the exemplars of Jewish degeneracy—for Jewish scientists and political figures. But by the early twentieth century, for Zionist social scientists and political leaders, the "problem" of Western Jewry emerged. With the rapid tide of assimilation and mixed marriages among Western Jews, the future of the Jewish race was endangered. In reality, however, for Zionists neither Eastern nor Western Jews were solutions to the problem of Jewish survival. As the Austrian Zionist Ignaz Zollschan maintained, there were only two solutions to the Jewish problem: "either the preservation of the ghetto or the abolition of the Diaspora. And 'the first alternative can only mean a continued morbid existence'" (Hart 2000, 92; Aschheim 1982).

17. The institutionalization of Jewish social science also occurred in other locales. The American Jewish Committee established a statistical bureau in 1914. Joseph Jacobs, who had moved from Britain to the U.S. in 1900 to become the editor of the *Jewish Encyclopaedia,* traveled to Berlin and met with Bruno Blau (Berlin's Statistical Bureau's second director) as part of planning for the establishing of a parallel institution in the U.S. Jacobs was its first director, and he modeled the American Bureau on its Berlin counterpart (Hart 2000, 69). In Britain, institutionalization took a somewhat different form and had a very specific focus. In 1922, Salaman, together with fellow Jewish scientists (including the anthropologist Charles Seligman, who was his brother-in-law) founded the Jewish Health Organization of Great Britain, which aimed to "improve the health of Anglo-Jewry." In practice, the Organization's work focused on the East European Jewish immigrant communities of London's East End (Endelman 2004).
18. My discussion of Israeli population genetics in the early years of statehood is brief, and like my account of Jewish race science in Europe and the U.S. in the late nineteenth and early twentieth centuries, it could be far more extensive. There is far less of an existing scholarly literature on this topic than on Jewish race science, and as such I have done a lot of research with primary sources for this part of the chapter. Having said that, an entire book could be written on the topic. The most comprehensive account, told from the perspective of an Israeli geneticist, is Raphael Falk's very interesting book, *Tsiyonut vehabiologia shel hayehudim* (Zionism and the Biology of the Jews; 2006b). See also Nurit Kirsh's essays on the topic, from which I have learned a great deal (Kirsh 2003, 2007).
19. Palestine's Arab population had dropped dramatically after the war of 1948, with approximately 750,000 Arabs expelled from or fleeing the newly founded Jewish state (Morris 1989).
20. The largest years of immigration were 1949–51, prior to the enactment of the first "Law of Return" and of official medical selection rules, which I will discuss below (Shvarts et al. 2005, 10). And over those years, one witnesses the shift from a primarily Euro-American Jewish immigration to a primarily "Asian" and "African" one (11). To give a sense of both the scale of immigration and the shift in demography: in 1949, there were 239,076 immigrants, 51.5 percent of which were Euro-American; in 1950, 169,405, with 50 percent Euro-American; in 1951, 173,901, 30 percent of whom were Euro-American. The percentage of Jewish immigrants from "Asia" rose from 30 percent in 1949 to 59 percent in 1951 (11).
21. For a discussion of nomenclature in the classification of Jewish groups that deals with this issue of East and West, see Schmelz 1993.
22. Built into the Prussian edict was a distinction between Prussian and foreign Jews.

The latter were banned from employment in the Prussian state, thus ending a longstanding practice whereby the wealthier communities of German Jews had employed Eastern immigrants as domestics and laborers. Moreover, as Markell points out, insofar as the Edict required Prussian Jews to use German and/or some other "living language" in all communication, it helped to undermine "communication and identification between Prussian Jews and *Ostjuden*" (Markell 2003, 148).

23. It is worth emphasizing here that in Mandate Palestine the term "Ashkenazi Jews" did not generally refer to longstanding Orthodox Ashkenazi communities in Palestine. They were part of the "old Yishuv" in contrast to which members of the "new Yishuv" defined themselves (see Hirsch 2009, 606).
24. Those favoring medical selection made a variety of arguments, including eugenic ones. One such argument emphasized the need for able-bodied men and women who could be both laborers and soldiers. In addition, there was the ongoing fear that too many ill or weak immigrants would be more than the fledgling state's health system could bear. It would be better, the argument went, if they were treated in their countries of origin before arriving on Israel's shores (see Shvarts et al. 2005; Davidovitch and Shvarts 2004).
25. Blood group systems identify specific antigens, substances which cause reactions when exposed to different antigens. They are important for establishing compatibility between donors and recipients of blood transfusions, and therefore their discovery was key for enabling organ transplants. The first blood antigens to be discovered were "A" and "B," although the subsequent discovery of "O" completed the initial ABO system. Other substances were subsequently discovered: M, N, and then P, that came to be known as the MN system; and then the Rhesus blood groups system (Rh, which can be either positive or negative) (see Mourant 1961).
26. Explaining a significant departure from previous anthropological studies of group-based biological difference, Mourant wrote: "since there was no necessary distinction between individuals of one population and of another, the populations themselves became units of study, and statistical methods, which could still perhaps be regarded as an extra embellishment in classical anthropometric work, became an essential feature of the new type of investigation" (Mourant 1961, 155).
27. For example, Mourant asked whether or not Jews who migrated to Israel were more "traditional" than their brethren who migrated, say, to Canada, and if so, if they really were a representative sample (Mourant 1954).
28. According to Schmelz (1993), the term "Sephardim" has a history of being used in two ways: first, to refer to a specific Jewish population, the descendants of Jews exiled from Portugal and Spain in the fifteenth century, and second, to refer to any Jew who is not Ashkenazi (a residual category, so to speak). In the latter usage, those Jews of Portuguese and Spanish descent can end up falling under the Ashkenazi category.
29. There is no specific evidence for how blood samples were collected for these studies. I presume, from the work on medical management of Israel's new immigrants, blood samples were collected as part of medical examinations, as various immigrants entered the country. What I have been able to find out is that the data on Ashkenazi Jewry was based on Jews who had arrived in Israel during the years 1949–57 (Kirsh 2003, 642). In addition, we know from Bondi's biography of Chaim Sheba (whose work on population genetics I discuss later in the chapter), that he collected blood in a variety of ways: for example, at the airport when new immigrants landed,

using his connections with the Jewish agency (Bondi 1981, 302); he also gathered together youth from Yemeni settlements in the hills of Jerusalem, asked them about their customs and way of life, and "took a bit of their blood" (307).

Tel Hashomer Hospital was the institutional center of much of this work of blood data collection and analysis, along with the Hadassah Hospital. According to Raphael Falk, the reason why Tel Hashomer was so central to work on genetics was because of Chaim Sheba's role as director of the hospital (personal communication). Chaim Sheba was a Zionist activist who became Chief of Medicine for the Haganah and later for the Israel Defense Forces, and subsequently the head of the Ministry of Health. He was also quite a eugenicist.

It is worth noting that these studies were conducted before the practice of informed consent was standardized in biomedical research, and not just in Israel.

30. There were two main Israeli journals in the 1950s and 1960s in which work on population genetics was published: *The Israel Medical Journal* (in English), and *Harefuah* (in Hebrew).
31. I assume that what they knew of what Baghdad or Kurdish Jews believed about themselves came from what the researchers had been told by their research subjects. But there is no source cited for that information in the paper.
32. The source of comparative data on European populations is not provided.
33. The mathematical language of these papers is very unspecific: "similar to" and "more of a difference" are typical phrases. According to Nurit Kirsh, there was a general absence of statistical analysis in the Israeli field before the late 1950s (2003, 642–43).
34. There is no explanation in this paper or any of the others for why certain comparative data and not others are presented.
35. There is one other comparison made in the paper, between Kurdish and Persian Jewish women, in which they note that Kurdish women have "an even lower percentage" of Rh negatives than do Persians, the "lowest so far in a Jewish community."
36. As with various other oriental Jewish communities, the Cochin Jews are actually made up of two distinct populations: "Black" Jews and "White" Jews. The latter are descendants of Jews from Spain and Portugal who came to India much later. There is some presumption of "'crossing'" (i.e., intermarriage) between "these two groups," as well as with mixing with various European colonizers over the centuries (Gurevitch, Hasson, Margolis, and Poliakoff 1955a, 254).
37. The two communities, that is, "Black" Jews and "White" Jews, are not separated in this study. The blood group data is collected as data of "Cochin Jews" (Gurevitch, Hasson, Margolis, and Poliakoff 1955a).
38. "The population of these villages lived an isolated life for generations in a segregated community, kept its traditions and its Hebrew-Arabic dialect" (Gurevitch, Hasson, Margolis, and Poliakoff 1955b, 260).
39. By way of contrast, for example, Northern Europeans have characteristically high frequencies of the cDE chromosome (the "north European" chromosome), and Africans have high cde chromosomes (Margolis 1957).
40. Muhsam points out that all "recent authors" who have grappled with this question agree on two points. First, "In any society where a Jewish minority has lived for several generations, the Jewish group . . . resembles to a certain degree the non-Jewish majority." Second, "eidoth differ considerably from one another" (1964, 36). It is worth noting that, in contrast to most other papers, he explicitly uses the term "race."

41. On the basis of skeletal evidence from Lachish, Muhsam compared the cephalic index of presumably ancient Israelites with those of Jews of Prague and their genetic environment (i.e., non-Jewish residents of Prague). The results he got were exactly what he expected: contemporary Jews of Prague have a cephalic index that is intermediate between that of the "original Jewish race" and that of "their genetic environment" (1964, 38).
42. There is no expectation that these "vectors" will actually join. Muhsam was looking for "concurrent" vectors that show some overlap in the *direction* they exhibit, going from current genetic environment and eidoth to a point of origin.
43. Contradicting his earlier argument that the only possible assumption one can make is that the appropriate genetic environment for each eidah is its present one, Muhsam suggests that such an assumption might have distorted the results. "For example, many Jugoslav Jews are descendants from refugees from Spain" (1964, 50).
44. While a random sample was taken of each genetic environment, according to Muhsam it might have been that a very particular subset of that population married or raped Jews.
45. When setting up the model, Muhsam argued that the ABO locus provides good evidence for such a study precisely because "the selection which is known to act on the ABO locus [cannot] be expected to disturb the gene frequencies very appreciably" (1964, 46). Once the conclusions did not match his hypothesis, he revised that initial hypothesis: "any selectivity at the ABO locus would also render our model inapplicable. Now it is well known that some selectivity actually exists at the ABO locus but it is generally assumed to be small. If this is not so—our model would again be invalidated" (1964, 52).
46. For example, Sheba was an advocate of limiting immigration from Morocco, since he believed the Jewish community of Morocco to be particularly diseased. He cast his argument in pragmatic terms: Israel was not in a position to deal with them (Bondi 1981, 154). Pragmatics alone cannot explain his attitude towards Moroccan Jewish immigration, however. Eugenic commitments were always shot through with class and racial hierarchies and clearly not only in Israel (see Kevles 1985; Pickens 1968).
47. This address is a compilation of various studies that Sheba published.
48. See Dafna Hirsch (2008) for a similar argument about the complexity of both the shifting boundaries of membership—and of East and West—in the context of Hadassah-funded nursing practices during the Yishuv.
49. Nurit Kirsh also makes the point that Israel's Arab citizens were, by and large, excluded from genetic studies: "In the 1950s the population of the State of Israel was multifarious and, for the genetics researcher, truly fascinating. However, some of Israel's population groups were completely ignored by these scientists. In addition to the Jewish population, Israel had about 140,000 Muslim Arabs, about 44,000 Christian Arabs, about 20,000 Druze, and a similar number of Muslim Bedouins. While the genetics of the Jewish 'communities' was researched very thoroughly and comprehensively, other groups, no less fascinating from a genetic standpoint, were hardly studied at all" (2003, 645).

CHAPTER THREE

1. For the persistence of this "problem" in Israeli population genetics well beyond the 1950s and 1960s, see Batsheva Bonné-Tamir 1980.

2. Of course, this is not universally true. As I discuss in chapter 5, there are groups for whom genetic facts can make all the difference. On the potential consequences of genetic genealogical testing for membership in Native American communities, for example, see TallBear 2006; on efforts to use genetic evidence in legal suits for reparations for slavery, see Nelson 2012.
3. The comparison did not involve a direct comparison of data from Y-chromosomes and mtDNA to one another: the genetic systems are too different to make that kind of a comparison viable. Instead, the two sex-specific genetic histories were "compared" by reading the Y-chromosome data of Jewish and non-Jewish male populations, on the one hand, and the mtDNA data of Jewish women in comparison with non-Jewish female populations, on the other. The results of each of those comparisons were then analyzed in relation to each other in order to assess male versus female Jewish genetic histories and to try and construct a plausible historical narrative out of the data.
4. The communities studied were: Ashkenazi Jews, Bene Israel (Indian Jews), Beta Israel (Ethiopian Jews), Bukharin Jews, Georgian Jews, Iranian Jews, Iraqi Jews, Moroccan Jews, and Yemenite Jews (Thomas et al. 2002, 1412).
5. Two Jewish communities shared the same modal haplotype, the Cambridge Reference Sequence [CRS]. The CRS is used as the standard mitochondrial sequence, first produced by researchers at Cambridge in England. It is also too ubiquitous a marker to be helpful in determining the precise geographic origins of these two communities.
6. Each Jewish community was compared to a "local" host population, "self-identifying members of the various populations in the countries and regions indicated by" the labels of the Jewish communities. The authors report far more mtDNA diversity in those non-Jewish populations: "No host population in our sample has a mtDNA frequency > 12% (mean 7.7%) whereas seven of the Jewish communities have frequencies > 12% (with a mean of 22.6%)" (Thomas et al. 2002, 1414).
7. In general, the researchers were not able to identify the source populations for the maternal line. In the case of two communities, Ethiopian Jews and the Bene Israel (a Jewish community in India) the data indicate a probable provenance of the mtDNA gene pool in the local population (Thomas et al. 2002, 1417).
8. Thomas and colleagues report: "Apart from the CRS, none of the other [mtDNA] Jewish modal haplotypes are represented in the Israeli Arab/Palestinian data set, in contrast to the similarities between Ashkenazic Jews, Sephardic Jews, Israeli Arabs/Palestinians, and Lebanese populations reported for the Y chromosome" (Thomas et al. 2002, 1415).
9. The study recognizes the practice of converting to Judaism: "How Jewish identity was determined in antiquity is also unclear. Conversion to Judaism was not uncommon in the pre-Christian Roman Empire and, in the 1st millennium C.E., the ruling classes of more than one polity adopted Judaism as the state religion (e.g., Himyar and Khazaria)" (Thomas et al. 2002, 1411).
10. According to interpretations of the genetic data, once these women chose Judaism it was not just to their faith but also to their men that they remained loyal. "'People say that Jews look like Europeans because there's been hanky-panky through the generations,' [Michael] Hammer says, 'But not that much, according to our study. You might expect that, if there were a lot of Jewish women taking non-Jewish men as husbands, then their children would still be Jewish but their Y-chromosomes

would be non-Jewish. But we didn't see that, which indicates that there's been isolation'" (quoted in Garifo 2000; see also Entine 2007).

In his book *Jacob's Legacy*, David B. Goldstein tells the following story: while giving a public talk on his work at a synagogue in California, "one of the women in the audience pointed out that the story was really about the wives, not the priests: if there had been even a small amount of funny business going on over the years, there could be no continuity of the priestly Y chromosome line. I had simply never thought about that before" (2008, xiv–xv). It is worth emphasizing that "hanky panky" and "funny business" here refer to sex—and/or marriage—with non-Jewish men, i.e., exogamy.

11. For a different perspective on genomics and eugenics, see Barnes and Dupré 2008.
12. For a broader range of positions and debates about genomics and its implications, see Goodman et al. 2003 and Lock 2005.
13. For an extended discussion of this divergence, see Sunder Rajan 2006; see also Rose 2007.
14. According to Nikolas Rose and Carlos Novas, post-genomic medicine "operates in a political and ethical field in which individuals are increasingly obligated to formulate life strategies to seek to maximize their life chances, to take actions or refrain from actions to increase the quality of their lives, and to act prudently in relation to themselves and to others" (2005, 487). In arguing for the question of ethical responsibility, I am not denying that there are forms of hierarchy and discrimination inherent in this epistemological logic of risk. Knowledge of genetic risks may well lead to discrimination, social stigma, and medical, psychiatric, and legal surveillance (Phelan 2005). It may also, however, generate novel rationales of exclusion. As I argued elsewhere, if for neoliberalism "rational action" is not just an ontology but also a norm to be adhered to in all spheres of human life (Brown 2003), then those citizens who fail to perform that norm—who fail to act prudently to avoid or at least try to deter the "diseased" outcome indicated by their genetic profiles—could well be produced as citizen-subjects and patients worthy of moral reprobation and social abandonment (Abu El-Haj 2007b).
15. Ostrer's paper was published prior to the Thomas et al. 2002 study; nevertheless, the older studies were inconclusive on the matter of a maternal origin in ancient Palestine. The second half of the paper relies on genetic mutations of medical significance as the basis for positing a unified origin.
16. This conclusion is derived from a reading of mitochondrial evidence from Thomas et al. 2002 as well as subsequent studies that I discuss below. Nevertheless, even after those subsequent mtDNA studies one needs to privilege the Y-chromosome in order to tell the story that way. I return to this below.
17. Palestinians served as the control group for grounding Jewish origins in ancient Palestine. See chapter 1 for a discussion of the Hammer et al. 2000 study that was reported on in this article.
18. The question of a strong founder effect has important implications for medical debates regarding the cause of the high incidence of "more than 20 known recessive disease alleles in Ashkenazi populations" (Behar et al. 2004, 356). The debate is over whether this fact is a consequence of "heterozygote advantage" or "both recent and more ancient founder effects" (356). In other words, did these recessive disorders survive and spread because they conferred some selective advantage on the population or because, due to a strong founder effect, certain deleterious alleles became established in the population and, via endogamy, spread?

19. "While the genetic ancestry of the Ashkenazim has been investigated recently in some depth in terms of both male and female lineages . . . the comparative data currently available on the non-Ashkenazi Jews is scant" (Behar et al. 2008, 1).
20. Goldstein then explains how this "aspect" of genetic history is quite different from other genetic fields: "Medical genetics is now using many of the same population-genetic tools employed to study genetic history. The difference is that in medical genetics you often know what you are after. In my own work on the genetics of epilepsy or infectious diseases, for example, I study patients who do and do not respond well to a particular treatment or I study individuals who do and do not naturally control the virus responsible for HIV/AIDS. There is little ambiguity about what I am looking for in that context" (2008, 90).
21. There is more substance to the story he tells: the first Gulf War, Goldstein's frustration over the U.S. telling Israelis "to sit tight," his desire to join the IDF, and then his guilt over "having chosen my career over Israel" (ibid.).
22. Daston and Galison distinguish between "mechanical objectivity" and "judgment" as two different and successive genres of scientific epistemology. As they point out, however, the former is not simply displaced by the latter. It persists even as it is refigured in light of subsequent scientific epistemologies, including the emergence of trained judgment as a mode of being and knowing in the latter half of the twentieth century. In the rhetoric surrounding genetic history, "objectivity"—as detachment and a willingness to submit to evidence—is merged with the question of expertise or trained judgment. It is an objectivity no longer understood to be "mechanical," that is, no longer governed by epistemological convictions, scientific practices, and "moral comportment that aimed to quiet the observer so nature could be heard" (2007, 120). If genetic history's moral comportment is a different one, so too is the centrality of expertise—of the scientific "priesthood," in the words of one informant—in being able to produce and interpret the genetic evidence.
23. So too, for example, is a research project that sought to specify the origins of African Americans buried in the African Burial Ground in Manhattan. That original project—directed, after some initial conflict, by African American scholars in cooperation with what they called the "descendant community"—generated work on African American origins more generally, including the initial idea to found an African American genetic ancestry-testing company (African Ancestry), which I discuss below. (Interviews conducted by the author; see also Alondra Nelson 2008a, 2008b, 2012).
24. One other crucial difference for the Jewish genetic history studies is that, as studies to figure out whether or not contemporary Jews descend from an ancient population in Palestine, they begin with *self-identifications* as the basis for their classificatory logics, thus sidestepping the rather fraught debate about what a population is that confronted the Diversity Project and continues to be debated in diversity-based projects such as the HapMap project and race-based medical studies (see M'Charek 2002; Reardon 2005).
25. Whether or not research is conducted as a self-study, of course, is tied into the particular political-economic status of any given "group." Jewish self-studies, whether in biomedicine or in genetic history, are the most pervasive. That fact testifies both to the elite standing of Jewish communities in the U.S. and in England—specifically, to Jewish overrepresentation, statistically speaking, in biomedical and other academic fields—as well as to the centrality of Israeli institutions to this research. Moreover, identity politics does not involve the same sorts of *politics* for groups in different

political or socio-economic positions. I address that issue in chapter 5. Even if there is an overriding grammar, there is nevertheless no single political meaning or effect of work in genetic history.

CHAPTER FOUR

1. I borrow the title for this chapter from Jean and John Comaroff's recent book, *Ethnicity, Inc.*
2. Jacobson's larger argument is about the ways in which refashioning the U.S. as an immigrant nation effectively rearticulated "white supremacy." The immigrant-nation narrative renders invisible, yet again, "the gradual and violent history of this settler democracy in the making long before the first immigrants of the Cattle Garden-Ellis Island variety ever came ashore" (2006, 9).
3. Family Tree DNA offers a variety of Y-chromosome and mitochondrial tests for paternal and maternal ancestry and a combination of the two. In addition, they offer two X-chromosome tests in order to establish "siblingship." In offering these specific products, Family Tree DNA advertises its ability to provide information on particular ancestries: Native American and African in addition to Jewish. Likewise, you can test yourself to see if you match Thomas Jefferson's lineage, and to see if you carry the Cohen modal haplotype. More broadly, there is a service offered for individuals who are adopted and want to locate their ethnic ancestry and to connect with "genetic cousins."
4. To quote Greenspan on the cooperation: "They have a commercial division which we've convinced to run tests for profit so they make a profit on the thousands of thousands of thousands of samples we send to them every year. It's all transparent. We have a contract that's signed by the Board of Regents of U of A [University of Arizona]. They know they're making money and that's not a bad thing. And I'm glad because now that means that a university, that actually has the capabilities, actually giving something back in the way of testing services that were otherwise not available before the U of A had the foresight to say yeah, we don't mind providing accurate, honest answers and making a buck on it, so that's actually worked out pretty well."
5. The full mitochondrial sequence is a test of both hypervariable and coding regions of mitochondrial DNA. The hypervariable regions are highly polymorphic regions of the mitochondrial genome and as such are the regions classically used for genealogical testing. Including the coding region as well in this test provides customers with a sequence of their entire mitochondrial genome: "This test is for all three regions of the mitochondrial DNA: HVR1(16001–16569), HVR2(00001–00574), and the coding region (00575–16000). The entire mitochondrial genome is tested, and this is the last mtDNA test that a person would need to take." Because the mtDNA has turned out to be so highly variable, it is possible that two individuals who share similar mutations in the hypervariable regions may not actually be biologically related. Including the coding region—which mutates far more rarely—can help scientists disentangle genealogical relationships from seemingly genealogical ones.
6. A haplogroup is "a genetic population group associated with early human migrations and which can today be associated with a geographic region" (familytreedna.com, glossary). In other words, haplogroups are mathematical constructions created by looking for similar results of Y-chromosome and mtDNA tests and lumping

them together. They are used to decipher the various early migrations of male and female lineages out of Africa and across the globe.

7. As Lily Kay demonstrates in a remarkable study of the development of molecular biology (1993), the blurring of the boundary between basic and applied research characterized the field from the start. Nevertheless, the 1980s saw a qualitative shift with its commercialization. In the early decades of the twentieth century, funding came primarily from foundations and, after the Second World War, from state agencies such as the NSF. For a broader discussion and critical reassessment of the relationship between commerce and basic science in the U.S. during the twentieth century, see Shapin 2008.
8. http://venturebeatprofiles.com/company/profile/dnatraits, accessed 1/20/2011. DNATraits' "philosophy".

   Your Health.
   As scientists continue to unlock the mysteries of the human genome, we understand more about inherited disorders caused by mutations in our genes. By testing for mutations that are known to cause genetic disorders, we can make informed decisions about ourselves and the ones we love. Whether you are sick, concerned about your risk for a hereditary disease, interested in the risks and benefits of certain medications, or plan to start a family, a simple at-home saliva test can help inform and prepare you for a lifetime of better health and wellness.

   Your Right to Know.
   DNATraits was founded on the principle that DNA testing can tell us important information about ourselves and our family, and we have a right to that information. You have the right to answers about your personal genetic makeup free from the oversight of insurance companies, employers, and even your own doctor. The results of your DNATraits test will be available only to you in accordance with the spirit of the Genetic Information Nondiscrimination Act (GINA) currently under consideration by the US Congress. In addition, DNATraits does not accept reimbursement from insurance companies as this might limit our ability to ensure your privacy.

9. Max Blankfeld, interview in *The Genetic Genealogist*, July 23, 2008, http://www.thegeneticgenealogist.com/2008/07/23/tgg-interview-series-viii-max-blankfeld/. Blankfeld added: "We want to change the paradigm in this field, allowing people to get tested for substantially less than what the current market price is. We want for DNATraits to make a difference in this area, and by being very affordable, allow the widest number of people to get tested so that the quantity of people with inherited diseases can be reduced."
10. In addition to mitochondrial and Y-chromosome tests, the other most commonly purchased genealogical tests are "admixture" ones. For example, DNA Print Genomics offers a series of Ancestry Informative Marker tests that are used to estimate one's percentage of parentage from four major "bio-geographical groups." The "AncestryByDna" test is described as follows: "The test surveys 176 Ancestry Informative Markers (AIMs) to provide an inference of genetic ancestry or heritage. The AIMs were carefully selected from large-scale screens of the human genome;

and are characterized by sequences of DNA that are more prevalent in people from one continent than another. Using complex statistical algorithms, the test can determine with confidence to which of the major bio-geographical ancestry groups, Sub-Saharan African, European, East Asian or Native American, a person belongs, as well as the relative percentages in cases of admixed peoples" (www.dnaprint.com, accessed February 22, 2010). As this test and tests of its kind rely on autosomal DNA and not just the Y-chromosome and the mtDNA, they cannot provide information about specific lineages. Instead, one presumably learns about the "mix" of one's ancestry. For an extended discussion of this technology, which was developed for medical purposes, see Fullwiley 2008a.

11. www.familytreedna.com/landing/discover-your-past.aspx, accessed June 20, 2001.
12. "Geographic" and "ethnic" origins do not function as distinct or alternative explanations. For example, the following is a mock question-and-answer exchange in Family Tree's mtDNA tutorial: "What type of ancestry do I have (ex. Native American, African, etc.)? The historic geographic origin of your direct maternal line is described under haplogroup in your results page." Mine was described as: "The mitochondrial haplogroup T is best characterized as a European lineage. With an origin in the Near East greater than 45,000 years ago, the major sub-lineages of haplogroup T entered Europe around the time of the Neolithic 10,000 years ago. Once in Europe, these sub-lineages underwent a dramatic expansion associated with the arrival of agriculture in Europe. Haplogroup T2 is one of the older sub-lineages and may have been present in Europe as early as the Late Upper Paleolithic."
13. After receiving two different test results from genetic ancestry companies—the first claiming his maternal line traced to Egypt and the second to Europe—Henry Louis Gates Jr, as he tells the story, decided to "start his own DNA-tracing company, one that he says will be able to take a more refined look at African-American ancestry" (Winstein 2007) to assess the genetic evidence and its plausibility, as well as to help provide historical narratives for its customers (Gates 2007). African DNA works in partnership with Family Tree DNA. According to Greenspan, Family Tree DNA manages the website, and, like Family Tree DNA, the samples are run by Michael Hammer's commercial laboratory at the University of Arizona.
14. This information is based upon email interviews I conducted with people who had contacted me about a match. I told them about my research, and several agreed to be interviewed.
15. If one adds the Hypervariable 2 (HRV2) region to the test, the odds increase to a 50 percent probability of sharing a common maternal ancestor within the past 28 generations or 700 years. The full mitochondrial sequence is designed to give customers information that might be more informative with regard to more recent generations, although in the mtDNA tutorial offered by Family Tree DNA they do not specify a time frame for "recent."

    The following is the explanation of a match sent along with one's mtDNA results: "We compare your specific sequence to other people in our database of mtDNA results. If we find someone who has exactly the same changes within their sequence as you do and you have signed the Family Tree DNA release form, we will inform both of you by email. In this way we are able to link you up with your 'genetic cousins.' However, despite the fact that the HVR-1 region experiences mutations more frequently, the mutations do not occur often enough for us to determine whether two individuals are more closely or more distantly related. As a result, we cannot tell you whether you share a common ancestor with these individuals in a genealogical time frame or further back in time. Testing the HVR2 region helps to pick out the individuals with whom you most likely share a common ancestor in a genealogical time frame. You may then employ conventional genealogical techniques to see whether you can find where that person might actually fit on your family tree!"

16. For an ethnographic account of producing matches in forensic genetics, see M'Charek 2005.
17. Marianne Sommer has published an article that deals with the European branch of Family Tree DNA (2010).
18. See, for example, Need et al. 2009.
19. I am using their online names.
20. Together with a few collaborators, Michael Hammer has launched the "DNA Shoah Project," which is dedicated to using genetic testing to find relatives of Holocaust survivors. According to a student working on this project, it is a "cousin" to Family Tree DNA. The idea is to collect as much material now from survivors or their descendants so that such information is available in the future. As of the summer of 2008, they were still in the phase of "pilot" testing the project and trying to convince survivors or their descendants to participate. This is a noncommercial venture to be funded through donations or grants (interview).
21. There is a small and growing group of customers on Family Tree DNA's web forum who are sophisticated enough in their understanding of genetics that they go back to scientific papers and evaluate their credibility. But as is evident from their online signatures, these are only four or five individuals. The rest are far more likely to defer—to the test results and, sometimes, to those four or five individuals.
22. And even that critical evaluation assumes that the databases are extensive enough to know that there are not higher frequencies of a given haplotype in an untested—or under-tested—community, in this instance, in West Africa.
23. Since her husband is and her own upbringing was Christian, they also celebrated the Christian holidays. As she told me, intermarriage is very common today among Jews. My family was just "a few generations ahead of the curve."
24. Critics of ancestry testing have pointed out that such results are derived from databases that falsely assume "that contemporary groups are reliable substitutes for ancestral populations" (Soo-Jin et al. 2009).
25. Such identifications are based not upon Y-chromosome or mitochondrial DNA tests offered by Family Tree DNA. They are based upon admixture maps offered by several other companies.
26. http://www.familytreedna.com/products.aspx, accessed 12/13/2010.

CHAPTER FIVE

1. I am not arguing that genetic historical evidence has any legal standing today. There are, however, a lot of web-based discussions about whether it does, whether it should, and at least one court case (if not a successful one) of someone trying to use genetic evidence as the basis for invoking the Law of Return in Israel.
2. In the larger field of anthropological genetics, "linguistic communities" are widely accepted as the primary criteria of classification. The practice of equating language and race was articulated in nineteenth-century philology, which built an entire corpus of knowledge on the basis of the distinction between Indo-European/Aryan versus Semitic language families and races (Olender 1992; Masuzawa 2005). For discussions of the arguments within anthropological genetics over how to define a population, see Reardon 2005; M'Charek 2005; Pottage 2002.

3. For an extensive discussion of the history of the discourse of lost tribes, see especially Ben-Dor Benite 2009; see also Parfitt 2002.
4. It is worth highlighting the racial terminology all over this paper, something quite typical of the larger field of contemporary anthropological genetics, even in works of scholars who explicitly deconstruct the biological reality of race. See for example, Cavalli-Sforza 2000.
5. The primary reference for their "cultural" data is an anthropologist who seems to have worked for the apartheid regime and certainly worked within its framework of cultural and tribal difference. See Van Warmelo and Phophi 1948. The other main source of information on Lemba culture is an article by the same author, published a decade earlier (Van Warmelo 1937).
6. For a detailed account of Lemba oral tradition about the community's origins, see Parfitt and Egorova 2006, chap. 3.
7. Genetic tests have repeatedly concluded that Ethiopian Jews are not phylogenetically related to other contemporary Jewish communities and that they do not descend from the ancient Levant. That evidence, however, does not intervene in assessing the "truth-value" of Lemba accounts of their origins.
8. The four markers are: p12F2/*TaqI* RFLP (restriction fragment length polymorphism), p49a/*TaqI*, pDP31, and the Y-*Alu* polymorphism (YAP). These are all polymorphisms that had been studied before in terms of their (relative) frequencies in various populations—generally, European, African, and Asian—relying on particular subgroups of each population as the basis for the evidence for the larger group. Generally speaking, the number of populations within each of these categories that had been tested for each marker—and especially with reference to non-European populations—is quite small. According to several African American scholars who work on African genetics, sampling of African populations has been strikingly slow in its development (interviews with author). It is worth noting that these are all polymorphisms based upon techniques quite different from the nucleotide sequencing that is the basis of Thomas et al. 2000.
9. The question of African gene flow or the "genetic distance" between Lemba and Africans is complex. Spurdle and Jenkins's conclusion, for example, that Lemba are closer to (some) Caucasoids than to African populations is based in large part on the absence of the p12F2 RFLP in "Africans" and its presence (26 percent) in Lemba. Spurdle and Jenkins explain that the Y-chromosome polymorphism is "generally" absent in African populations. But that claim is based upon the following data sets: 182 South African Bantu-speaking Negroids and 90 Khoisan, used in their study, plus two samples from previous studies, one which included 15 "African males" in its sample, and the other 65 Senegalese (Spurdle and Jenkins 1996, 1127). "Since this allele is absent in Africans and also was not observed in a sample of 60 Polynesians, it would appear to be specific to Caucasoids" (ibid.). Needless to say, although I will say it anyway, there is not nearly a large enough—and expansive or diverse enough—African data set to support such a sweeping conclusion.
10. Spurdle and Jenkins used restriction fragment length polymorphisms (RFLPs). Those polymorphisms are found by using specific enzymes that cut DNA at particular (known) points, thereby revealing patterns of difference between different samples. Different fragment lengths indicate base substitutions, additions, deletions, or sequence rearrangements (different kinds of polymorphisms) at specific points on the chromosome, indicating different genetic sequences between one individual and another. Thomas et al. 2000 use nucleotide sequencing to generate detailed

information regarding the polymorphisms: they read nucleotide sequences and thereby know precisely which allelic differences exist between any two samples.

11. For a critical reassessment of the 1998 identification of a *single* Cohen modal haplotype, see Hammer et al. 2009. In their later work, they identify multiple paternal lineages rather than a single Cohen lineage.
12. A more extensive explanation for the presence of the CMH among Lemba men is given by Parfitt in a later book:

> The date of entry of the haplotype into the Lemba is unknown. It could be that a carrier of the haplotype—perhaps a Cohen—was responsible for introducing this genetic element in relatively recent times. However, the haplotype was found in Buba communities geographically quite distant from each other. It is more likely then that the gene flow occurred over a hundred years ago and thereby predates contact between the Lemba and Jews, which started on a small scale from the time of the colonization of the northern Transvaal around the turn of the last century. Prior to that we know of no Jews who penetrated the inland area inhabited by the Lemba. It is altogether more likely that when some of the ancestors of the Lemba arrived in Africa they were carrying this haplotype with them (Parfitt and Trevisan Semi 2002, 49)

13. The evidence on the CMH among Lemba men is as follows. The CMH appears in the Lemba with a frequency similar to that in the general Jewish population (in just under one out of every ten men). In addition, the CMH was found at a 50 percent frequency in one clan, the Buba (n = 13 of whom 7 carried the CMH). (Remember that the CMH shows up in approximately 50 percent of Jewish priests in the studies by Skorecki et al. 1997 and Thomas et al. 1998.) "Of particular interest is the Buba clan, since membership of this clan and possession of the CMH are significantly associated." While no such person as Buba exists in Jewish sources, the Lemba believe that their ancestors left Judea "under the leadership of Bhuba" (Thomas et al. 1998, 682–83).
14. All quotations attributed to Vivian Moses in this chapter are taken from a speech he gave before the nineteenth annual meeting of the International Association of Jewish Genealogical Societies. The title of the talk was "Are We All Jacob's Children?" I transcribed the talk from a tape-recording of the session.
15. See Zoossmann-Diskin 2000 for a critique.
16. Tudor Parfitt has conducted research on how the results of genetic historical studies have impacted the Lemba (see Parfitt and Trevisan Semi 2002; Parfitt and Egorova 2006). It is worth mentioning that many Lemba are practicing Christians, and they see no contradiction between asserting Jewish origins (or even being Jews) and their Christian faith.
17. "Tudor Parfitt to Visit North America," *Kulanu* 6, no. 4 (2000): 1.
18. http://www.haaretz.com/news/report-dna-tests-support-zimbabwe-tribe-s-claim-of-jewish-roots-1.264278.
19. It is significant that Parfitt references the Chief Rabbinate in Israel and not just some rabbinical council elsewhere. I return to the issue of Israel's centrality to the question of recognizing "returning" Jews at the end of the chapter.
20. http://test.kulanu.org/lemba/lemba1.php, accessed 1/20/2011.
21. A major figure in Kulanu insisted in an interview that he did not care if communities converted. For example: "There was a great debate between Rabbi Bernard, the

most welcoming of the rabbis in South Africa, and a professor. . . . Bernard told his congregation to welcome the Lemba. . . . He should have stopped there. . . . [But] he took the traditional position. He said you'll be Jewish when you undergo Halakhic conversion. I was so disheartened by his position that I gave up [on the debates]." Nevertheless, the organization's activities—as well as the commitments of activists evidenced in the newsletter—contradict this statement. *As an organization*, facilitating conversion is a key part of what they do.

22. One could push the roots of this movement further back in time, to the advocacy by U.S. Jews on behalf of Soviet Jewry, and the ways in which, as Matthew Frye Jacobsen (2006) argues, that movement was part of resisting the rising conviction among Black Nationalist and African American Civil Rights leaders more generally that American Jews were, quite simply, white. In Soviet Jewry, American Jews found their own continued identity as part of an oppressed group.
23. For a more extensive discussion of the reactions of South African Jews to the genetic tests, see Parfitt and Egorova 2006, chap. 4.
24. For Bruno Latour (1987), translation involves the appropriation and partial reconfiguration of a particular scientific object or fact in order to further one's own interests, which is precisely what groups such as Kulanu have done.
25. Povinelli has argued that postcolonial (perhaps more accurately named anticolonial) struggles and multiculturalism need to be differentiated (they operate with very different "demand" structures, so to speak). But it is also true that "multiculturalism" as a form of liberal politics emerges in the wake of anticolonial movements and postcolonial critiques. It is in terms of that specific history that I frame Kulanu's work as articulating a postcolonial politics, which, as I demonstrate below, ends up furthering specifically colonial aims.
26. For one example of the tolerance lesson from a South African activist: "I want to share an experience with you of what happened recently. One of the many visitors who were interested in finding out more about the Lemba duly went on a fact-finding mission to the Northern Province. I received a telephone call from a distraught person, who was '*shocked that the Lemba were wearing beads and living in huts.*' !! <grin> I continue to be amazed at our ability to 'not hear' or comprehend the situation. The separation of over 2,000 years—the cultural difference of being African instead of our version of perceived 'Jewishness.' I am pleased to say that the experience was a positive one with firm friendships developing" (Mausenbaum 2000, 4).
27. Avichail's relationship with Rabbi Kook is testified to by the former Head of the Jewish Fund's director for Aliyah; see Committee of Immigration, Absorption, and Diaspora Affairs, Protocol 1490, 5 October 2000 (Hebrew; http://www.knesset.gov.il/protocols/data/html/alia/2000-01-11.html); see also Parfitt and Trevisan Semi 2002.
28. Committee of Immigration, Absorption, and Diaspora Affairs, Protocol 1490, 5 October 2000 (Hebrew; http://www.knesset.gov.il/protocols/data/html/alia/2000-01-11.html).
29. As indicated by the representative of the Interior Ministry at the Absorption Committee hearing, the Right of Return does not apply to the Bnei Menashe.
30. As reported in a subsequent *Kulanu* newsletter (Tigay 2003), there is also an *ulpan* (Hebrew immersion program) for Anusim in Efrat.
31. "Avichail Leads Seminar for Bnei Menashe," Kulanu Briefs, *Kulanu* 8, no. 2 (2001): 10.

32. ". . . And a Bnei Menashe Community Center in Israel," Kulanu Briefs, *Kulanu* 12, no. 2 (2005): 2.
33. Parfitt and Egorova have written that it is unclear whether or not the Indian genetic study influenced Rabbi Amar. Nevertheless, they note, "the general media and Internet discourse about the Bnei Menashe has altered significantly. . . . A good deal of the entirely justifiable skepticism about the origins of the group has withered on the vine" (2006, 124–25).
34. See Committee for Immigration, Absorption, and Diaspora Affairs, Protocol 117, December 21, 2010 (Hebrew; http://www.knesset.gov.il/protocols/data/html/alia/2010-12-21.html). This meeting to discuss the "Immigration and Absorption of the Bnei Menashe," before which Michael Freund was the first to testify, was held in Kiryat Arba (a settlement near Hebron) rather than in the Knesset. As explained by the committee chair, they decided to hold the hearings in Kiryat Arba for two reasons: first, to discuss how to bring more Jews to Israel, and second, to understand the absorption needs of the Bnei Menashe. The fact that the meeting was held at Kiryat Arba says a lot about the integration of the immigration of the Bnei Menashe to "Israel" into the settler movement.
35. "Are These Bnei Menashe Immigrants the Last?" *Kulanu* 10, no. 2 (2003): 1, 7.
36. Among Kulanu and other lost tribe hunters there is a discourse about colonialism, but a very different one: of Christian colonialism that, for example, forced the Bnei Menashe, among other communities, to convert to Christianity. It is, in part, to right that historical wrong that groups such as Kulanu understand their work.
37. Kulanu is deeply committed to helping the Falash Mura "return" to Israel.
38. Eliyahu Salpeter, "New Communities," *HaAretz*, 2005 (online).

CHAPTER SIX

1. I borrow this title from a short story by Tim O'Brien, "The Things They Carried" (1998).
2. Through the study of ancient DNA, it is now possible for anthropological geneticists to study extinct species, even though ancient DNA researchers are limited by the length of time that DNA can survive after death (about one million years). See Callaway 2010.
3. Etienne Geoffrey Saint-Hilaire is believed to have coined the term "phylogeny" in 1831. Its first English usage is attributed to Lyell in *Principles of Geology* (1832) (Steedman 2001, 11). "Ontogeny," by way of contrast, refers to the evolution or development of the individual.
4. According to Steedman, "associations between littleness and interiority and between history and childhood" were further developed in psychological theories of childhood and historical theories of origins in the late nineteenth and early twentieth centuries. More broadly, over time Victorian society embraced an understanding of events past (childhoods, histories) as "something [that], though lost and gone, has left behind memories and traces," developed and articulated most fully, perhaps, by Freud (1995, 88).
5. See deCODE (www.decode.com) and 23andme (www.23andme.com) for examples of the information being sold by genome-testing companies.
6. For a similar circularity of evidence, proof, and narrative in biblical archaeology, see Abu El-Haj 2001, chapter 5.
7. Stewart also points to the opposite possibility: "two closely related taxa may appear quite different from each other" (2000, 49).

8. There is also the question of whether or not the historical explanations the researchers provide are necessarily parsimonious. Are there events that go unrecognized in their own accounts of what counts as an event, for example? Who gets to decide what counts as an event?
9. I thank Rayna Rapp for drawing my attention to this distinction.
10. This sampling technique is typical of anthropological genetic work. For example, in conducting anthropological genetic studies of Native American communities, researchers look for individuals whose four grandparents are from the same community or tribe (interview). See also Behar et al., 2010. They are less explicit in their methods section that they took samples only from individuals who had four grandparents from the same Jewish community. But given two of the central questions they set out to answer, it seems logical that they did: "What are the genetic distances between contemporary Jewish communities, their Diaspora neighbours and Middle Eastern populations? Can the genetic origin of Jews be pinpointed within the Middle East?" (238). In addition, later on in the paper the authors write, "Our conclusion favouring common ancestry over recent admixture is further supported by the fact that our sample contains individuals that are known not to be admixed in the most recent one or two generations" (240, emphasis added).
11. As I finish this manuscript, the Jewish HapMap project has published results of its efforts to study the genetic diversity of the Jewish World (Ashkenazim, Sephardim, and Mizrahim, the latter represented by individuals from Syria, Iran, and Iraq). The recent studies use a different technique called genome-wide genotyping, which targets a much larger array of single-nucleotide polymorphisms across the genome. Those SNPs are both coding and noncoding, although, as I discussed at the end of chapter 1, the epistemological struggle remains the same: how to separate out ancestry from selection, even if as post-genomic understandings of the genome become more and more complex, some coding regions may turn out to be indicative of the former (given the time frame in which researchers are interested) and some noncoding regions indicative of the latter. As has been true of all earlier "novel" biological methods for studying Jewish origins, these new techniques have been declared, by scientists and journalists, as possibly being able to finally resolve—or finally provide ample evidence for—the (shared) origins of contemporary Jewish communities. As summed up in a newspaper report: "This is not the first study attempting to isolate a genetic thread among Jews. However, it is the first to show clear and unequivocal genetic significance." According to one of the study's main authors, Harry Ostrer of New York University's Medical School, "The study supports the idea of a Jewish people linked by a shared genetic history" (Dudi Goldman, "Researchers locate Jewish genetic linkage," *Israel Jewish Scene*, 6 Sept. 2010; accessed at http://www.ynetnews.com/articles/0,7340,L-3902234,00.html).

INDEX

CHICAGO STUDIES IN PRACTICES OF MEANING
Edited by Jean Comaroff, Andreas Glaeser, William Sewell, and Lisa Wedeen
Published in collaboration with the Chicago Center for Contemporary Theory
http://ccct.uchicago.edu

Series titles, continued from front matter

The Making of Romantic Love: Longing and Sexuality in Europe, South Asia, and Japan, 900–1200 CE
by William M. Reddy

Laughing at Leviathan: Sovereignty and Audience in West Papua
by Danilyn Rutherford

Bengal in Global Concept History: Culturalism in the Age of Capital
by Andrew Sartori

Parité!: Sexual Equality and the Crisis of French Universalism
by Joan Wallach Scott

Logics of History: Social Theory and Social Transformation
by William H. Sewell, Jr.

Bewitching Development: Witchcraft and the Reinvention of Development in Neoliberal Kenya
by James Howard Smith

The Devil's Handwriting: Precoloniality and the German Colonial State in Qingdao, Samoa, and Southwest Africa
by George Steinmetz

Peripheral Visions: Publics, Power, and Performance in Yemen
by Lisa Wedeen

Printed and bound by CPI Group (UK) Ltd, Croydon, CR0 4YY

09/06/2025

14685762-0001